Hydrological Modelling for the Sustainable Management of Water Resources in River Basins

Hydrological Modelling for the Sustainable Management of Water Resources in River Basins

Guest Editors

Ioannis Panagopoulos
Pantelis Sidiropoulos

MDPI

Basel • Beijing • Wuhan • Barcelona • Belgrade • Novi Sad • Cluj • Manchester

Guest Editors

Ioannis Panagopoulos
School of Agriculture
Aristotle University of Thessaloniki
Thessaloniki
Greece

Pantelis Sidiropoulos
School of Rural and Surveying Engineering
Aristotle University of Thessaloniki
Thessaloniki
Greece

Editorial Office
MDPI AG
Grosspeteranlage 5
4052 Basel, Switzerland

This is a reprint of the Special Issue, published open access by the journal *Hydrology* (ISSN 2306-5338), freely accessible at: https://www.mdpi.com/journal/hydrology/special_issues/9P93109E5N.

For citation purposes, cite each article independently as indicated on the article page online and as indicated below:

Lastname, A.A.; Lastname, B.B. Article Title. *Journal Name* **Year**, *Volume Number*, Page Range.

ISBN 978-3-7258-5369-4 (Hbk)
ISBN 978-3-7258-5370-0 (PDF)
https://doi.org/10.3390/books978-3-7258-5370-0

Contents

Darío Xavier Zhiña, Alex Avilés, Lorena González, Ana Astudillo, José Astudillo and Carlos Matovelle
Effects of Climate Change and Changes in Land Use and Cover on Water Yield in an Equatorial Andean Basin
Reprinted from: *Hydrology* **2024**, *11*, 157, https://doi.org/10.3390/hydrology11090157 . . . . . . **1**

Sharad Aryal, Mukand S. Babel, Anil Gupta, Babak Farjad, Dibesh Khadka and Quazi K. Hassan
Attribution of the Climate and Land Use Change Impact on the Hydrological Processes of Athabasca River Basin, Canada
Reprinted from: *Hydrology* **2025**, *12*, 7, https://doi.org/10.3390/hydrology12010007 . . . . **22**

Samir Hakimi, Mohamed Abdelbaset Hessane, Mohammed Bahir, Turki Kh. Faraj and Paula M. Carreira
Development of Transient Hydrodynamic and Hydrodispesive Models in Semi-Arid Environments
Reprinted from: *Hydrology* **2025**, *12*, 46, https://doi.org/10.3390/hydrology12030046 . . . . **44**

Maria Sismanidi, Lamprini Kokkinaki, Sofia Kavalieratou, Haralampos Georgoussis, Kyriakos D. Giannoulis, Elias Dimitriou and Yiannis Panagopoulos
Assessing the Effects of Bioenergy Cropping Scenarios on the Surface Water and Groundwater of an Intensively Agricultural Basin in Central Greece
Reprinted from: *Hydrology* **2025**, *12*, 66, https://doi.org/10.3390/hydrology12040066 . . . . **70**

Fatima Daide, Thomas Hasiotis, Soumaya Nabih, Soufiane Taia, Abderrahim Lahrach, Eleni-Ioanna Koutsovili and Ourania Tzoraki
Assessing Hydrological Alterations and Environmental Flow Components in the Beht River Basin, Morocco, Using Integrated SWAT and IHA Models
Reprinted from: *Hydrology* **2025**, *12*, 109, https://doi.org/10.3390/hydrology12050109 . . . . . . **94**

Malgorzata Borowiecka, Mar Alcaraz and Marisol Manzano
An Application of the Ecosystem Services Assessment Approach to the Provision of Groundwater for Human Supply and Aquifer Management Support
Reprinted from: *Hydrology* **2025**, *12*, 137, https://doi.org/10.3390/hydrology12060137 . . . . . . **117**

Daniel Shepard, Norman L. Jones and Gustavious P. Williams
Application of the Groundwater Data Mapper Tool to Assess Storage Changes in a Groundwater-Driven Basin in the Klamath Watershed, Oregon, USA
Reprinted from: *Hydrology* **2025**, *12*, 140, https://doi.org/10.3390/hydrology12060140 . . . . . . **141**

Enrique Soriano, Luis Mediero, Andrea Petroselli, Davide Luciano De Luca, Ciro Apollonio and Salvatore Grimaldi
Assessment of the Impact of Climate Change on Dam Hydrological Safety by Using a Stochastic Rainfall Generator
Reprinted from: *Hydrology* **2025**, *12*, 153, https://doi.org/10.3390/hydrology12060153 . . . . . . **164**

Toritseju Oyen and Duke Ophori
Regional Groundwater Flow and Advective Contaminant Transport Modeling in a Typical Hydrogeological Environment of Northern New Jersey
Reprinted from: *Hydrology* **2025**, *12*, 167, https://doi.org/10.3390/hydrology12070167 . . . . . . **184**

Stefanos Sevastas, Ilias Siarkos and Zisis Mallios
A Comparative Analysis of Remotely Sensed and High-Fidelity ArcSWAT Evapotranspiration Estimates Across Various Timescales in the Upper Anthemountas Basin, Greece
Reprinted from: *Hydrology* **2025**, *12*, 171, https://doi.org/10.3390/hydrology12070171 . . . . . . **206**

hydrology

MDPI

Article

Effects of Climate Change and Changes in Land Use and Cover on Water Yield in an Equatorial Andean Basin

Darío Xavier Zhiña [1,2], Alex Avilés [1,2,*], Lorena González [1,2], Ana Astudillo [1,2], José Astudillo [1,2] and Carlos Matovelle [3,4]

[1] Carrera de Ingeniería Ambiental, Facultad de Ciencias Químicas, Universidad de Cuenca, Cuenca 010207, Ecuador; dario.zhina0608@ucuenca.edu.ec (D.X.Z.); lorena.gonzalez@ucuenca.edu.ec (L.G.); ana.astudillo@ucuenca.edu.ec (A.A.); jose.astudillob@ucuenca.edu.ec (J.A.)

[2] Grupo de Evaluación de Riesgos Ambientales en Sistemas de Producción y Servicios (RISKEN), Departamento de Química Aplicada y Sistemas de Producción, Universidad de Cuenca, Cuenca 010207, Ecuador

[3] HYDROLAB, Centro de Investigación, Innovación y Transferencia de Tecnología (CIITT), Universidad Católica de Cuenca, Cuenca 010108, Ecuador; cmmatovelleb@ucacue.edu.ec

[4] Grupo de Investigación, Ambiente Ciencia y Energía, Universidad Católica de Cuenca, Cuenca 010108, Ecuador

* Correspondence: alex.aviles@ucuenca.edu.ec

Abstract: Ecosystem services contribute significantly to human development, with water production being a crucial component. Climate and land use changes can impact water availability within a basin. In this context, researching water-related areas is essential for formulating policies to protect and manage hydrological services. The objective of this study was to estimate water yield in the sub-basins of the Tabacay and Aguilán rivers under climate change scenarios in 2030, 2040, and 2050, combined with scenarios of changes in land cover and land use. The InVEST model was employed to analyze water yield. The results show that crop areas were identified as the lowest water yield in future scenarios, and forested areas, particularly the region where the Cubilán Protected Forest is located, contribute the most to water yield in the subbasin. Besides, water yield has increased in the historic period (2016–2018) due to the conservation and reforestation initiatives carried out by the Municipal Public Service Company for Drinking Water, Sewerage, and Environmental Sanitation of the city of Azogues in 2018, the so-called Reciprocal Agreements for Water. Additionally, an increase in water yield is projected for future scenarios. This study can serve as a basis for decision-makers to identify areas that should prioritize protection and conservation.

Keywords: ecosystem services; basin water yield; InVEST model; equatorial Andean basins

1. Introduction

Ecosystems are crucial in human development and provide various ecosystem services (ES) [1]. These services include, among others, some essential resources such as water, air, and wood, as well as processes such as water purification and quality, carbon sequestration, and nutrient recycling [2,3]. The provision of fresh water is one of the most essential global ecosystems. These are crucial in developing society, agriculture, industry, and other sectors [4–6]. However, ES can be affected by climate change and changes in land cover and land use.

Numerous studies worldwide have estimated the impacts of climate change and land use changes on water availability. Positive relationships between precipitation and water availability have been found in the literature [6–8]. Conversely, some research has shown negative impacts between water availability and changes in land use, such as changes in surface water balance, evapotranspiration rates, runoff dynamics, groundwater flow, and other factors [9–12]. For this reason, it is crucial to estimate its effects through hydrological modeling to understand the impacts of climate change and changes in land cover and use.

Modeling hydrological processes within river basins is fundamental to helping decision-makers and administrators improve water resource management [13]. Hydrological models can help estimate the amount of water under various conditions, and these estimates can be made at different spatial scales, from hillside areas to basins [14–16]. However, developing hydrological models can prove challenging in some areas of the world.

Although hydrological modeling often requires large amounts of data, its accessibility is challenging [17]. Over the last two decades, with the development of new technologies and geographic information systems (GIS), it has become possible to develop hydrological systems capable of incorporating complex parameters [18]. Within this category of models, we find the Soil and Water Assessment Tool (SWAT) [19], the Topography Based Hydrological Model (TOPMODEL) [20], Variable Infiltration Capacity (VIC) [21], Artificial Intelligence for Ecosystem Services (ARIES) [22], Integrated Valuation of Ecosystem Services and Tradeoffs (InVEST) [23], and various additional ones. Although developing these models has helped simplify data requirements, having the necessary data to implement a particular model is still challenging.

In this regard, the InVEST model (based on annual precipitation and landscape characteristics) can be run with a reduced data set compared with other hydrological models [24], achieving good results in different locations where it has been used. Applications of this model include the study by [18], which determined the water discharge in the Xitiaoxi River basin and found slight differences compared with the actual discharge of the basin. A study in the UK by [25] validated the model's performance in water production and found differences of less than 10% between actual and modeled performance. On the other hand, the studies by [26,27], which were carried out in Altai Prefecture (China) and the alpine region of the Qinghai–Tibet Plateau (QTP), show a decreased water yield in the first study area and an increase in water production in the second study area due to changes in land use and climatic trends. The study by [28] found that the water retention in the forests of North Korea is lower than that in the forests of South Korea.

In North America, the study by [29] developed in 749 basins found a relatively poor performance compared with real-world conditions due to the study area's cryospheric variables, which, when included, significantly improved predictions. In Latin America, the study developed by [30] evaluated the performance of water production in 224 basins in southern Chile, finding that the annual estimates showed good agreement between the observed and modeled values. Another study [31], developed in southern Ecuador, allowed a classification according to the importance of water production.

A notable advantage of the InVEST model is its minimal data requirements, which facilitates its use when extensive data are scarce. This adaptability is particularly advantageous for challenging terrain, such as the mountainous regions of the Andean páramos, where complex topography and difficult access present obstacles to establishing hydrometeorological stations to measure the numerous variables required for many hydrological models.

In the Andean region, the páramo provide water to large cities, such as Quito and Bogotá [32]. However, water availability is influenced by various factors such as precipitation, temperature, soil permeability, slopes, and vegetation, among others [18]. Although there is high water production in the Andean region due to the near-constant rainfall and low consumption by vegetation, urban growth and changing land use and land cover have contributed to changes in water production and availability in these regions [31].

The purpose of this study is to evaluate water production in the sub-basins of the Tabacay and Aguilán rivers (Cañar Province, Southern Ecuador) using the InVEST model, taking into account the land use change in the years 2014, 2016, 2018, 2030, 2040, and 2050 and climate change for two representative concentration pathways (RCP): 4.5 and 8.5 scenarios in 2030, 2040, and 2050. The results will significantly help decision-makers and strengthen policies to protect water sources in the study area.

2. Materials and Methods

2.1. Study Zone

The study zone includes the Tabacay and Aguilán river sub-basins, which are located in southern Ecuador in the Cañar Province (Figure 1). Its total area is 86.55 km^2, and its altitude range is between 2480 and 3760 m above sea level. The climate in the study zone is characterized by relatively low temperatures (average temperature between 9–11 °C), which is typical in the Andean region. The average annual precipitation varies between 1115 mm in the upper part (Llaucay) and 876 mm in the lower part (Guapán). The sub-basins of the Tabacay and Aguilán rivers supply water to the city of Azogues, so slight changes in the sub-basins may result in a reduction in water supply. In addition, the study area is characterized by unsustainable land use (agriculture on steep slopes) [33], which disrupts the sector's land and water resources.

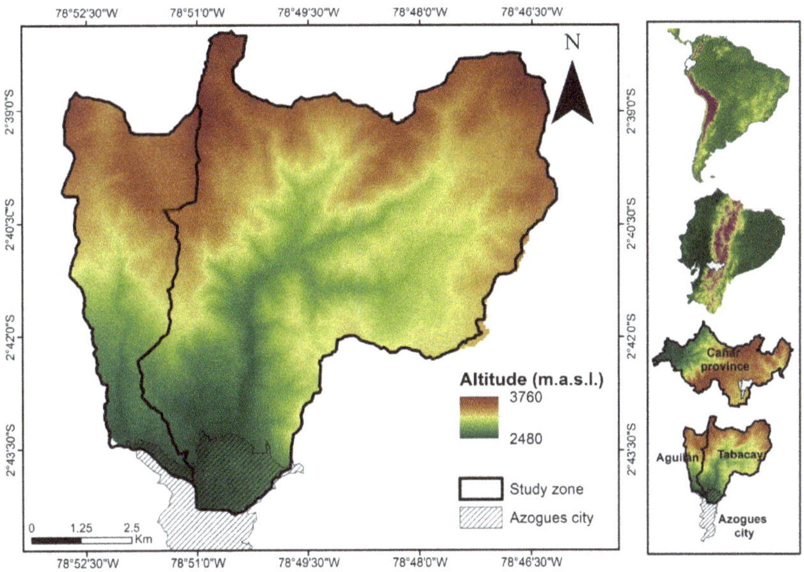

Figure 1. Study area localization: Tabacay and Aguilán river subbasins.

For this reason, and to protect water and other natural resources, the "Municipal Rules on the Conservation, Restoration of Water Sources, Water Recharge Zones, Sensitive Ecosystems, and Other Priority Protected Areas of Biodiversity, Environmental Services and Natural Heritage of the Canton of Azogues" apply to the study area. In addition, the Municipal Public Service Company for Drinking Water, Sewerage and Environmental Sanitation (EMAPAL-EP) uses the Reciprocal Water Agreements (ARAs, its Spanish acronym) as a protection tool. ARAs aim to provide agricultural technical assistance and financial compensation to landowners in the watersheds' upper areas, designated as water recharge areas. In exchange, the owners must allocate parts of their land as conservation areas. This initiative secures spaces dedicated to conservation and encourages good agricultural and livestock production practices in these areas, contributing to the region's sustainable development.

2.2. InVEST Model

The InVEST model V3.14.0 is a tool that can be used to examine how changes in ecosystems can lead to changing benefits for people. The InVEST model included various modules to quantify, map, and evaluate the benefits of marine, freshwater, and terrestrial ecosystems. The different modules of the model can be divided into four main categories:

support services, final ecosystem services, tools to facilitate the analysis of ecosystem services, and support tools [24]. This study used the "water yield production" module, which is located in the "ecosystem services" category.

InVEST—Water Yield Module

This module uses a grid map and the principle of water balance [annual precipitation (P) minus actual evapotranspiration (AET)] to determine the annual amount of water (Y) in each pixel (x). The model makes no distinction between surface and groundwater and assumes that at the end of the year, all water that falls into the basin, except what evaporates, leaves it (Equation (1)).

$$Y(x) = \left(1 - \frac{AET_{(X)}}{P_{(X)}}\right)P(x) \tag{1}$$

The relationship between $AET_{(X)}/P_{(X)}$ (Equation (2)) is based on the methodology developed by [34] and later adapted by [35,36].

$$\frac{AET(x)}{P(x)} = 1 + \frac{PET_{(X)}}{P_{(X)}} - \left[1 + \left(\frac{PET_{(X)}}{P_{(X)}}\right)^{\omega_x}\right]^{1/\omega_x} \tag{2}$$

where $PET_{(X)}$ represents potential evapotranspiration and ω_x is a nonphysical parameter that describes the inherent characteristics of both soil and climate.

The $PET_{(X)}$ can be calculated using Equation (3).

$$PET_{(X)} = Kc_{(X)} \cdot ETo_{(X)} \tag{3}$$

where $Kc_{(X)}$ represents the evapotranspiration coefficient for the vegetation linked to a particular land cover and land use (LULC).

The nonphysical parameter ω_x is calculated with Equation (4).

$$\omega_x = Z\frac{PAWC_{(X)}}{P_{(X)}} + 1.25 \tag{4}$$

where $PAWC_{(X)}$ is the water capacity available to the plant at pixel x and Z is a seasonality parameter that represents the depth and distribution of seasonal rainfall.

For more details describing the water production module of the InVEST model, see [24].

2.3. *Input Data for the Model*

The data required by the InVEST model to run the water production module are the average yearly precipitation (P, mm), average yearly reference evapotranspiration (ETo, mm), land cover and land use, available water fraction for plants (PAWC) and root restriction layer depth (mm), watershed profiles (which is explained later—Figure 2), a table with biophysical attributes, and the Z-parameter.

2.3.1. Precipitation (Current Data)

The current rainfall data (2014, 2016, and 2018) were obtained from the meteorological gauges of the Catholic University of Cuenca and EMAPAL-EP. The Kriging interpolation [37] was carried out to obtain spatial rainfall throughout the study area.

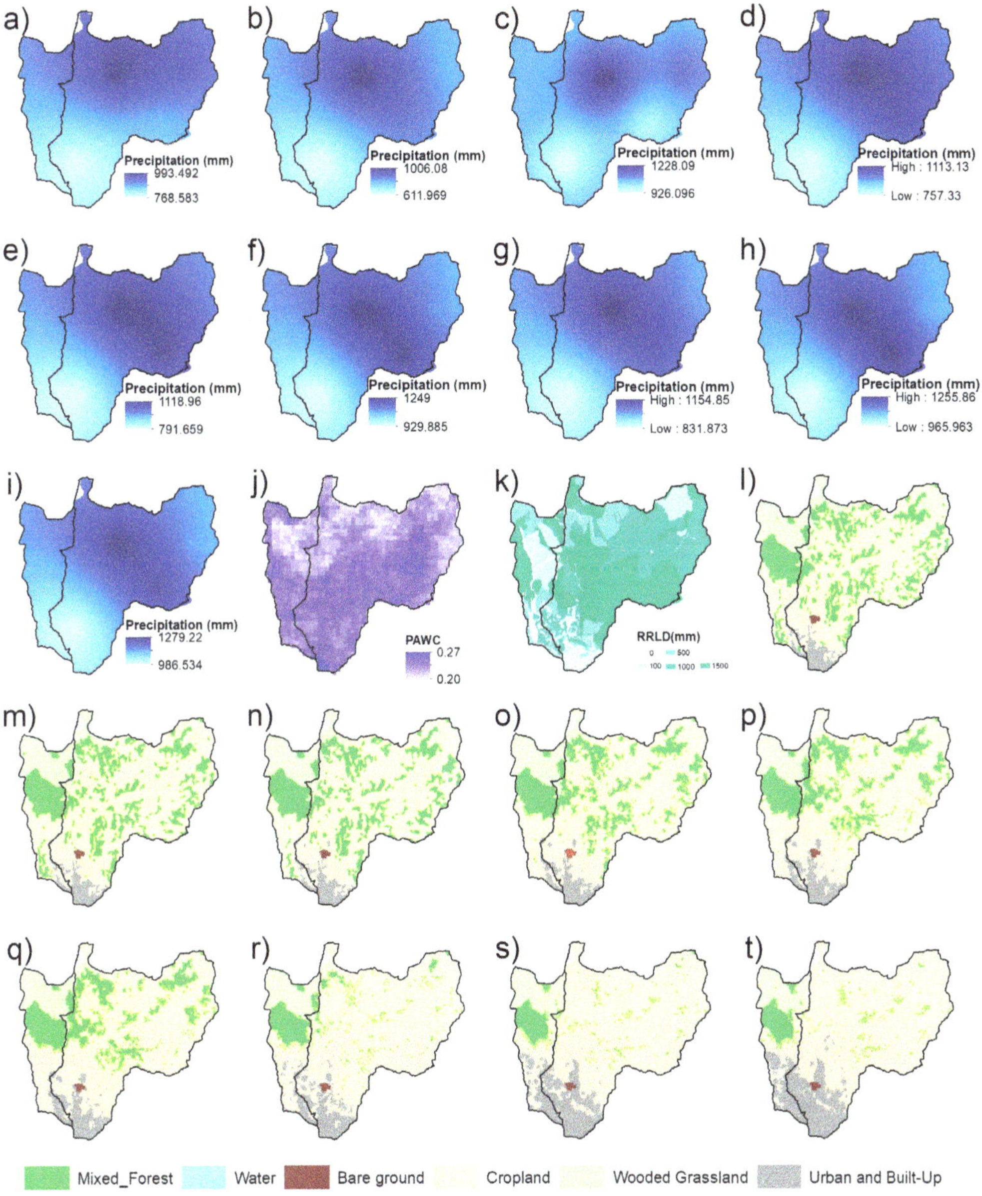

Figure 2. Variables required for the InVEST Water Yield Model: (**a**) precipitation (2014), (**b**) precipitation (2016), (**c**) precipitation (2018), (**d**) precipitation (RCP 4.5—2030), (**e**) precipitation (RCP 4.5—2040), (**f**) precipitation (RCP 4.5—2050), (**g**) precipitation (RCP 8.5—2030), (**h**) precipitation (RCP 8.5—2040), (**i**) precipitation (RCP 8.5—2050), (**j**) plant available water capacity, (**k**) soil depth, (**l**) land use and cover (2014), (**m**) land use and cover (2016), (**n**) land use and cover (2018), (**o**) land use and cover—trend scenario (2030), (**p**) land use and cover—trend scenario (2040), (**q**) land use and cover—trend scenario (2050), (**r**) land use and cover—pessimistic scenario (2030), (**s**) land use and cover—pessimistic scenario (2040), (**t**) land use and cover—pessimistic scenario (2050).

2.3.2. Precipitation (Future Data)

The Ecuador Regional Climate Model (RCM-EC) evaluated the future precipitation data. This model was developed by the Ministry of Environment, Water, and Ecological Transition (MAATE, its Spanish acronym) as part of the Third National Communication on Climate Change (TCN) [38]. The model was developed by downscaling an ensemble of global climate models (GCMs) using the Weather Research and Forecasting (WRF) model at 10 km resolution for Ecuador [39]. The base period for its development was 1981–2005, and the future is 2011–2100 for two RCP 4.5 and 8.5 scenarios. This study used the two scenarios to represent climate conditions in 2030, 2040, and 2050. The RCP 4.5 scenario, which reflects a stabilization path where CO_2 emissions peak around 2040 and then decline, suggests moderate mitigation efforts, while the RCP 8.5 scenario represents a high emissions scenario where no significant emissions reductions come [40]. Other studies at the national scale and in nearby catchments have also used the same scenarios to analyze the impacts of climate change [41–43]. This same approach has allowed for a shared analysis of the effects of climate change on water resources.

Since this data type may have significant systematic biases compared with the observed data, bias correction was performed using the Quantile Delta Mapping (QDM) method. This method preserves the relative changes in quantiles projected by the climate model and corrects for possible systematic biases in the quantiles of the modeled data concerning the observed values (Appendix A). According to the literature, the QDM correction is based on four steps [44]. Quantile Delta Mapping (QDM) is an advanced bias correction technique that adjusts the quantiles of modeled data based on the observed distribution over a historical period. The process consists of four steps: (i) The cumulative distribution function (CDF) is estimated for the modeled and observed data. (ii) The relative change in model quantiles between the historical and projected periods is calculated using the inverse of the CDF. (iii) This transformation is applied to the quantiles of the observed data in the historical period. (iv) The bias in the future projection is corrected at any time by applying the relative change multiplied by the observed data quantiles. As indicated by the literature [44], QDM is related to the equidistant CDF matching algorithm proposed by [45], and compared with Quantile Mapping (QM) algorithms, it is less prone to problems such as inflating trends relative to extreme data.

After the climate data were corrected, they were interpolated using the Kriging method [37] to obtain yearly spatial maps in different periods.

2.3.3. Evapotranspiration

Evapotranspiration was determined using the Holdridge method [46], in which a simple expression for its determination in tropical and subtropical regions was defined. This expression is based on the air temperature between 0 °C and 30 °C (biotemperature), which determines the rhythm and intensity of the physiological processes of plants (photosynthesis, respiration, and transpiration) as well as the rate of direct evaporation of water contained in the soil and vegetation. Holdridge's original model is shown in Equation (5).

$$ET_p = C_{HO} \cdot T \tag{5}$$

where ET_p is the potential evapotranspiration (mm·year^{-1}) and T is the air temperature (°C). The C_{HO} coefficient is used for annual estimates with a value of 58.93.

The current temperature of the study area was obtained from the meteorological gauges of the Catholic University of Cuenca and EMAPAL-EP. Future temperature data (Appendix A) were derived from the QDM downscaling process performed on RCM-EC. These data (historical and future) were interpolated using the Kriging methodology [37] to obtain the spatial temperature of the entire study area.

2.3.4. Plant Available Water Capacity

Plant available water content (Figure 2j) is defined as the difference between the volumetric field capacity fraction and the permanent wilting point. The data on available water content for plants come from the ISRIC database [47]. Since the database reports moisture content in mm and at seven depths, it is necessary to convert these values into a fraction using Equation (6).

$$AWC = \left(\frac{1}{b-a}\right)\left(\frac{1}{2}\right)\sum_{k=1}^{N-1}(x_{k+1}-x_k)(f(x_k)+f(x_{k+1})) \quad (6)$$

where N is the number of depths, x_k is the k-th depth, and $f(x_k)$ is the value of the water content at that depth. Since the result of the available water content is in a range of 0–100, which represents the percentage, it is necessary to divide by 100 to obtain the value in fraction and to be able to introduce these results into the InVEST model.

2.3.5. Root Restraint Layer Depth

It is the soil depth (mm) at which root penetration is inhibited due to chemical or physical characteristics of soil (Figure 2k). The data were obtained from Ecuador's National Information System of Rural Lands and Technological Infrastructure (SIGTIERRAS—http://geoportal.agricultura.gob.ec/index.php/visor-geo, accessed on 10 February 2024).

The available data had a qualitative classification and were transformed into quantitative data using the available scientific literature as a reference [48], achieving a more precise approximation to reality. Furthermore, to validate this new classification, field tests were carried out (Figure 3), providing an empirical validation of the categorization. The values obtained from these pits in the field for validation are presented in Table 1 for reference.

Figure 3. Fieldwork was carried out to validate the classification of root depth.

Table 1. Location of points used to validate root depth classification.

X	Y	Qualitative Classification	Literature Classification	Field Tests	Final Classification
738996.59	9705017.924	Superficial	<100	80	100
746905.33	9706492.673	Shallow	100–500	420	500
742884.14	9706173.413	Moderately deep	500–1000	650	1000
740168.82	9698183.620	Moderately deep	500–1000	700	1000
742064.25	9705251.796	Deep	1000–1500	>1000	1500

2.3.6. Land Use and Land Cover

The land use and land cover data used in the study were obtained from MAATE for 2014, 2016, and 2018 (Figure 2l–n). The methodology described for its classification can be found in [49].

In addition, land use change forecasts for two trending and other pessimistic scenarios were created for 2030, 2040, and 2050 (Figure 2o–t). The trend scenario assumes a population growth of 1.56% (population growth determined shortly before the analysis period), in addition to the temporary land use change trends observed between 2014 and 2018. On the other hand, the pessimistic scenario makes the following assumptions: with a population growth rate of 5.83% (corresponding to the highest growth rate in Ecuador during the study period), the Cubilán protective forest would no longer exist, and the existing paths in the study area become gravel roads.

Land use forecasts were created using the land change modeling software Dinamica EGO version 7.1.1. [50]. The Dinamica EGO methodology uses cellular automata to model changes in land use based on biological, physical, and socioeconomic variables obtained from land cover and use maps (for this study between 2016 and 2018). In this model, space is divided into a grid of cells, where each cell represents a unit of land and can be in different states that change according to specific rules. The simulation is based on the local interaction of cells, which enables the modeling of transformations and new spatial scenarios. In the Dinamica EGO model, windows of different sizes, from 1×1 to 11×11 pixels, are used to assess the quality of spatial simulations. Each window allows one to compare spatial patterns in specific areas of the map, from fine details with small windows (1×1) to general trends with large windows (11×11). This approach helps to identify how the model reproduces spatial patterns at different scales and ensures accurate and valid fitting of the simulations by comparing the obtained patterns with the observed patterns. This process measures the percentage of similarity, which means how closely the simulated map matches the observed map. The minimum and maximum similarity percentages represent the lowest and highest possible degree of agreement between the two, providing a range for model validation. A similarity percentage of 50% or more between the simulated and actual patterns indicates a good model fit [51]. For a more detailed explanation of the methodology, see [50]. The validation model was performed in 2018, and its results are presented in Appendix B.

For InVEST modeling, the trend and pessimistic scenarios were combined with the climate scenarios. The trend scenario was combined with the RCP 4.5 scenario, and the pessimistic scenario was used in conjunction with the RCP 8.5 scenario.

2.3.7. Biophysical Table

A biophysical table (Table 2) was created in text format (.csv) presenting the following attributes: Id, LULC classes, presence of vegetation (1 corresponds to the presence of vegetation and 0 without vegetation), evapotranspiration coefficient (Kc) and root depth. The LULC classes were developed by [24], the evapotranspiration coefficient was determined by [47], and the depth of the roots was determined from the previously presented cover and land use data.

Table 2. Biophysical table used in the InVEST water production model.

Id	Lucode	LULC_Classes	LULC_Veg	Kc	Root_Depth
1	7	Wooded grassland	1	0.15	500
2	11	Cropland (row crops)	1	0.90	1500
3	12	Bare ground	0	1	10
4	13	Urban and Built-Up	0	1	10
5	14	Water	0	1.05	10
6	16	Mixed forest	1	0.25	1500

2.3.8. Z-Parameter

The Z-parameter, which captures the local precipitation pattern, whose values range between 1 and 30, was taken from the literature [52]. This value was 7.33. This value within the model is used in Equation (4), as described previously.

2.4. Sensitivity Analysis

The model's sensitivity was assessed by varying different parameters according to the methods proposed in the literature [23,25,53], involving adjustments to various model parameters. For our study, each of the rasters used as inputs to the model was varied by ±10% and ±20%, allowing the comparison of five scenarios: scenario 1 (Variation −20%), scenario 2 (Variation −10%), scenario 3 (no variation), scenario 4 (variation +10%), and scenario 5 (variation +20%). The sensitivity analysis was carried out in 2018.

3. Results and Discussion

3.1. Variation in the Rainfall

As shown in Figure 2a–i, higher rainfall amounts are observed in the upper part of the study zone. In addition, in the base period, it is observed that the amount of rainfall is lower in 2016 compared with 2014 and then increases in 2018, which may be due to the large amounts of rainfall in these periods [54] and the humidity arises from transpiration the Amazon forests, which are transported by air masses from the Amazon basin and carried to the paramos in southern Ecuador [55]. Furthermore, it can be seen that in the two scenarios (RCP 4.5 and 8.5), the RCP 8.5 scenario is the one with the highest precipitation values.

3.2. Evolution of Land Use and Land Cover

As shown in Tables 3 and 4 and Figures 2 and 4, land cover and land use in the study area changed from the base period (2014) to 2018. As for forests, the area they covered decreased by 4.4% from 2014 to 2018. Urban regions recorded an increase of 7% compared with the base period. The areas earmarked for cultivation recorded a rise of 2.3%. In contrast, the water area experienced a decline of 9.4% and 15.6% in 2016 and 2018, respectively.

Table 3. Area covered by each cover and land use within the study area for the trend scenario.

Land Cover and Land Use	Base Period			Trending Scenario			Rate Change		
	2014	2016	2018	2030	2040	2050	2030	2040	2050
Mixed Forest	21.25	20.56	20.32	18.34	17.02	16.38	−9.7%	−16.3%	−19.4%
Wooded Grassland	11.34	11.20	10.88	9.77	8.95	8.53	−10.2%	−17.8%	−21.6%
Cropland	50.33	51.06	51.50	53.83	55.10	55.56	4.5%	7.0%	7.9%
Urban and Built-Up	3.24	3.34	3.47	4.28	5.15	5.74	23.4%	48.5%	65.5%
Wetland	0.03	0.03	0.02	0.02	0.02	0.02	−9.6%	−9.6%	−9.6%
Bare Ground	0.24	0.24	0.23	0.24	0.24	0.24	2.0%	2.0%	2.0%

Table 4. Area covered by each cover and land use within the study area for the pessimistic scenario.

Land Cover and Land Use	Base Period			Pessimistic Scenario			Rate Change		
	2014	2016	2018	2030	2040	2050	2030	2040	2050
Mixed Forest	21.25	20.56	20.32	7.60	5.48	5.40	−62.6%	−73.0%	−73.4%
Wooded Grassland	11.34	11.20	10.88	6.29	5.48	4.77	−42.1%	−49.6%	−56.2%
Cropland	50.33	51.06	51.50	65.39	64.57	62.89	27.0%	25.4%	22.1%
Urban and Built-Up	3.24	3.34	3.47	6.79	10.56	13.03	96.1%	204.7%	276.1%
Wetland	0.03	0.03	0.02	0.02	0.02	0.02	−9.6%	−9.6%	−9.6%
Bare Ground	0.24	0.24	0.23	0.23	0.23	0.23	−0.1%	−0.1%	−0.1%

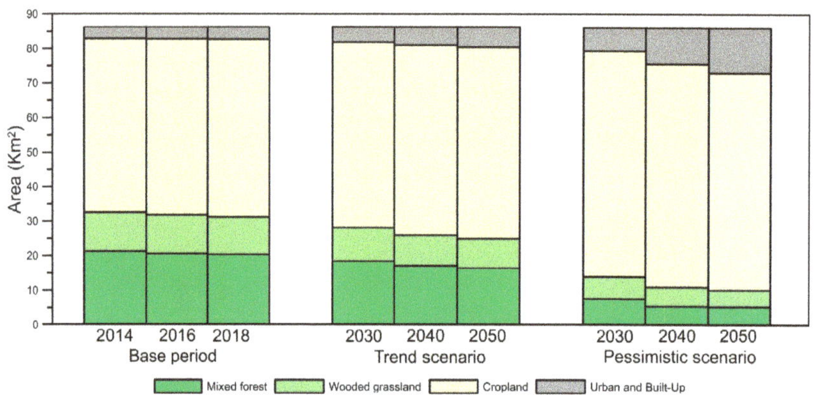

Figure 4. Area covered by each cover and land use within the study area in the different study periods and scenarios.

Let us compare the trend and pessimistic scenarios. We can find that the pessimistic scenario has the most considerable changes in land cover and use, with the increase in area of urban areas being the highest. By 2030, the urban area will almost double compared with 2018, and by 2050, the area will nearly triple. For the years 2030, 2040, and 2050, there was a significant decrease in forested grassland areas of 42.1%, 49.6%, and 56.2%. The areas earmarked for cultivation have increased compared with 2018. However, since 2030 is the year with the highest growth rate (27%), the growth rate starts to decrease after this year. Finally, the forests are experiencing the most significant decline in their area, namely by 62.6% by 2030 and 73.4% by 2050 compared with 2018.

Although this increase in the area dedicated to cultivation suggests that it could help ensure the food sovereignty of the area's residents in the future, several studies indicate that the conversion of forests to agricultural land will lead to modify their productivity [56,57] and changes in their physical and hydraulic properties that make them drier [32,58].

3.3. Annual Water Yield in the Study Zone

The InVEST model with the water yield module was used to determine water production in three current periods (2014, 2016, and 2018), future climate change scenarios (RCP 4.5 and 8.5), and future coverage and land use scenarios (trend and pessimistic). The model's only limitation is that it is impossible to perform seasonal analysis because it uses annual values for estimation.

As shown in Figure 5, the study area exhibits fluctuations in water yield in different periods. Although the region's rainfall determines the water available, the land's use and coverage determine the water yield. Significant changes in land cover and land use can affect, among other things, the hydrological systems of evaporation, transpiration, and water retention [53,59] and thus affect the water yield of a catchment. For this reason, the water yield module of the InVEST model estimates its water availability in each pixel as precipitation − actual evapotranspiration, which depends on cover and land use.

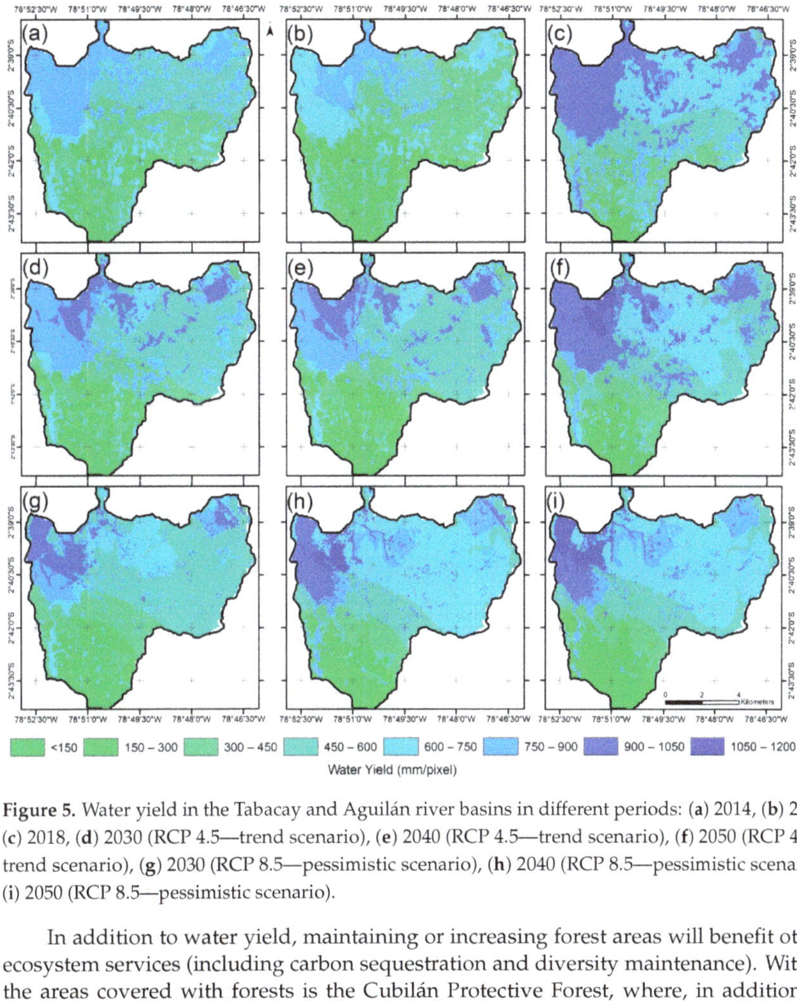

Figure 5. Water yield in the Tabacay and Aguilán river basins in different periods: (**a**) 2014, (**b**) 2016, (**c**) 2018, (**d**) 2030 (RCP 4.5—trend scenario), (**e**) 2040 (RCP 4.5—trend scenario), (**f**) 2050 (RCP 4.5—trend scenario), (**g**) 2030 (RCP 8.5—pessimistic scenario), (**h**) 2040 (RCP 8.5—pessimistic scenario), (**i**) 2050 (RCP 8.5—pessimistic scenario).

In addition to water yield, maintaining or increasing forest areas will benefit other ecosystem services (including carbon sequestration and diversity maintenance). Within the areas covered with forests is the Cubilán Protective Forest, where, in addition to providing various ecosystem services [60], sites with more outstanding water production are emerging, which is why it is necessary to carry out efforts between residents and government bodies for the protection of this area, which is responsible, among other things, for the production of water that supplies the city of Azogues (lower part of the Aguilán and Tabacay sub-basins).

On the other hand, in the current period (2014–2018), the coverage and land use that produces a lower amount of water are those destined for agriculture (Figure 7a). The lowest values of water yield in this area are less than 300 mm per year. These results are consistent with studies conducted in China by [61,62], where changes in land use, particularly in agricultural land, impact water yield due to increased evapotranspiration. Similarly, the study by [63] developed in the USA indicates that agricultural lands have lower water yields than forests, such as in our study area.

Changes or removal of vegetation cover to develop these activities can affect the water cycle by altering evapotranspiration and soil properties, resulting in drier soils [32]. Approximately 60% of the total area of the catchment is dedicated to agricultural and urban areas, which causes the catchments to lose their ability to regulate water, as stated in the literature [64]. In the same way, this can be demonstrated by analyzing the historical records of some streamflows in the area where the flow duration curve (FDC—Figure 6) is more pronounced, which is consistent with studies that analyzed the effects of land use changes [64,65]. These changes in land use and cover would impact water availability. However, they would also allow catchments to respond more quickly to rainfall events with higher and faster runoff, resulting in flooding in the lower-lying parts of the study area.

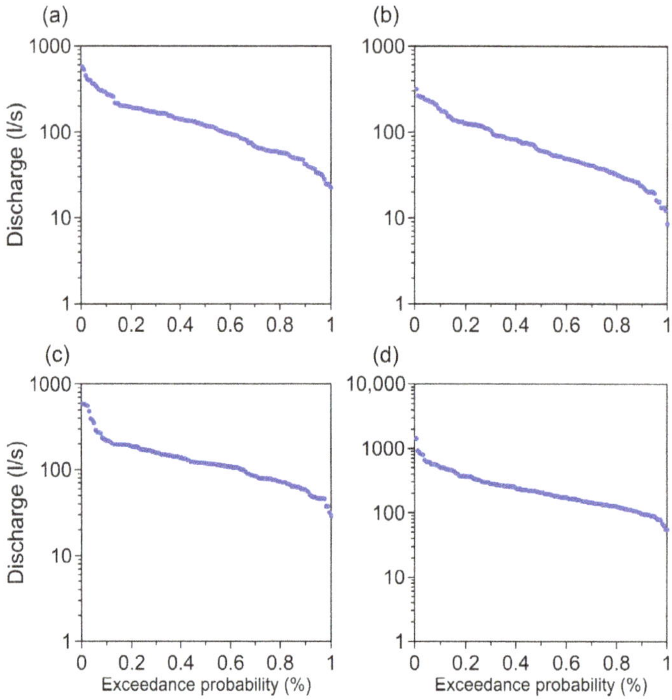

Figure 6. Flow duration curves of the streamflows that are within the study area: (**a**) Condoryacu streamflow, (**b**) Mapayacu streamflow, (**c**) Nudpud streamflow, and (**d**) Llaucay streamflow.

The areas with the highest water yield values were those covered with mixed forest and wooded grassland. Our results are consistent with the study by [66] which stated that forest cover is inextricably linked to rainfall, which increases the probability of rainfall events and increases water yield, and the study by [67] developed in New Zeeland, where tussock grasslands can maximize water yield compared with other vegetation cover types. Similarly, a study in the United States created by [63] found that forest covers also have higher water yields. When plotting these covers, the maximum values were 898 and 908 mm in specific locations for 2014, the maximum values were close to 891 and 857 mm for 2016, and there was more than 1000 mm of water yield in certain areas (Figure 7b,c) for 2018. During this period, there was a significant increase in water yield in various regions.

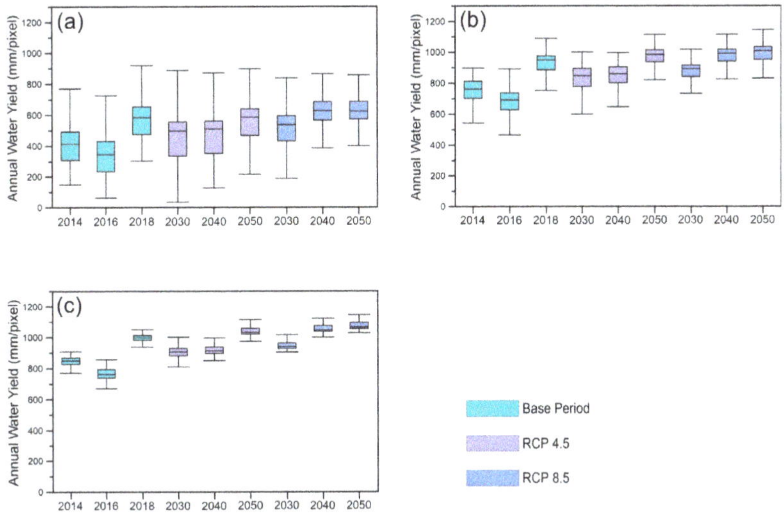

Figure 7. Boxplot of water yield for each land use and land cover in the study zone: (**a**) Cropland, (**b**) Mixed Forest, (**c**) Wooded Grassland.

Although it can be seen that in 2018 (Figure 7a), water yield increased in some agricultural areas, this increase in water yield may lead to a deterioration in water quality [68]. This fact leads to more significant salt movement and leaching of fertilizers used in these areas. For this reason, future studies are necessary to analyze the water quantity and quality.

If we compare the current water yield with the future in the study area (Figure 8), it can be seen that water yield in the Tabacay and Aguilán sub-basins will increase in the future. This behavior may be because climate extremes will increase in the future, as indicated in the literature [69]. Still, it must be considered that an increase in the amount of water can lead to a deterioration in water quality [68]. The annual water yield in the RCP 4.5 and RCP 8.5 scenarios shows fluctuations compared with the base period. Both scenarios indicate an overall increase in water yield over time; however, the RCP 8.5 scenario generally results in a higher average water yield than the RCP 4.5 scenario. An exception occurs in 2050, where the RCP 4.5 scenario is higher.

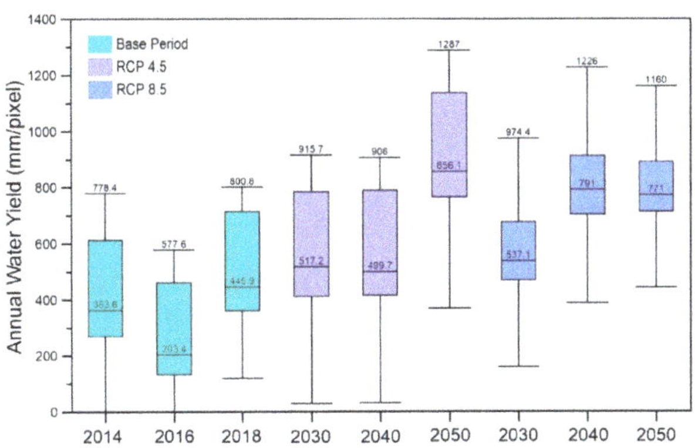

Figure 8. Boxplot of water yield for each period within the study zone.

3.4. Sensitivity Analysis

As shown in Figure 9, according to the sensitivity analysis, precipitation (Figure 9a) shows an apparent increase and decrease in the maximum and minimum water yield values with a variation of ±20%. In evapotranspiration (Figure 9b), there is no fluctuation in water yield maximum values, but there are fluctuations in water yield minimum values, which are not as noticeable as fluctuating precipitation. Since the evapotranspiration is obtained from temperature, it could indicate the model is not sensitive to temperature changes. Some studies have attempted to assess the response of water yield to precipitation and temperature, finding greater sensitivity to precipitation [6,70,71], consistent with that found in our study. No significant changes were observed in the other parameters (Figure 9c–e), which is consistent with previous findings in the literature [25,72].

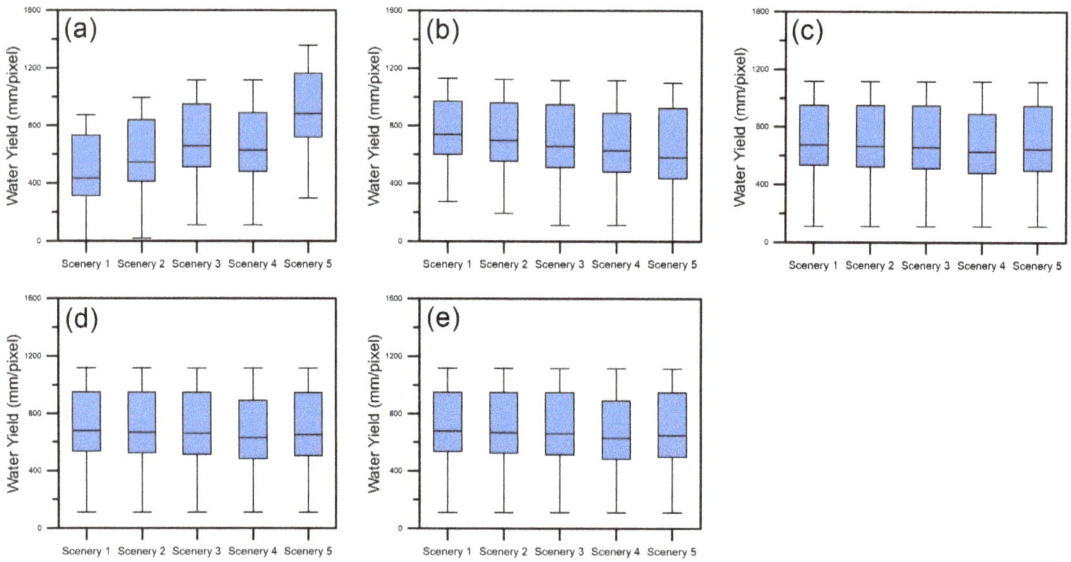

Figure 9. Sensitivity analysis of the different input parameters of the InVEST model with a variation of −10% (scenario 1), −20% (scenario 2), 0% (scenario 3), +10% (scenario 4), and +20% (scenario 5): (**a**) Precipitation variation, (**b**) ETo variation, (**c**) Root Restraint Layer Depth, (**d**) Plant available water capacity, and (**e**) Z-parameter.

3.5. Validation of the Obtained Annual Water Yield

The Catholic University of Cuenca and EMAPAL-EP carry out specific monthly flow measurements in different streams within the study area. These observations were used to compare the results of the modeling. However, as can generally be seen, using the InVEST software V3.14.0 (Table 5), the modeling tends to overestimate the observed discharge. There is only a slight coincidence in the Mapayacu discharge.

These differences can be attributed to the absence of some data in the modeling process that were not considered due to lack of access to them, such as water extractions and losses within the basin, as well as the presence of water-dependent activities such as fish farming and livestock watering, which influence the variation in river flow. These activities may impact water availability in the study area by diverting water flows for purposes not accounted for in the InVEST model. Something similar happened in the study by [29] in which, after evaluating 749 basins in five bioclimatic regions of North America, it was found that the correlation between the observed and modeled rivers was highly dispersed and relatively poor due to missing modeling considerations.

Table 5. Comparison between estimated and observed water yield within the study site.

Year 2014	**Discharge (m^3/s)**	
	Estimated	**Observed**
Condoryacu	0.138	0.07
Llaucay	0.230	0.06
Mapayacu	0.062	0.07
Nudpud	0.177	0.07
Year 2016		
Condoryacu	0.095	0.08
Llaucay	0.140	0.07
Mapayacu	0.042	0.07
Nudpud	0.115	0.07
Year 2018		
Condoryacu	0.166	0.06
Llaucay	0.258	0.13
Mapayacu	0.076	0.08
Nudpud	0.221	0.08

It is crucial to emphasize that the InVEST model plays a prominent role in environments with limitations in data availability [73], serving as a preliminary guiding tool. The initial use of the model allows for subsequent integration of more detailed data and additional considerations, essential for achieving a more accurate assessment of water flows in the catchment. Similar conclusions were reached in the studies by [74,75], where the InVEST model contributes to a general understanding of the hydrological patterns from the watershed due to the limited data availability and leaves the field of study open for complex modeling if more detailed data sets are available.

3.6. Public Policies and Decision Making

The water yield in 2018 increased compared with 2016 and 2018. This behavior may be due to the measures to protect water sources carried out by the EMAPAL-EP in conjunction with the Municipal Ordinance [76]. For instance, the Reciprocal Water Agreements (ARAs) serve as strategies to conserve and restore forests and provide technical and financial assistance to upland landowners. These initiatives have reduced agricultural land expansion, preserved natural areas, and improved productivity through targeted training.

The Tabacay and Aguilán sub-basin's water yield estimations would serve so decision-makers can set guidelines for caring for water-relevant areas. However, this is a crucial and complex challenge [77]. Conservation measures and policies focusing on a particular environmental service may endanger and degrade other environmental services [78]. This fact represents a challenge to improve hydro-ecological measurement systems to improve the accuracy of the models and make better decisions. However, for places where data availability is limited, the InVEST model has proven to be sufficient [79].

4. Conclusions

Our study aimed to determine the changes in water yield of two Andean sub-basins in southern Ecuador under climate change scenarios (RCP 4.5 and 8.5) and consider two future scenarios of land use and land cover. The results show water yield fluctuations in various years. The year 2016 had the lowest water yield in the base period. This behavior may be due to changes in land use and land cover in the study area. The highest water yield occurs in the upper part of the basins and, to a greater extent, in the Cubilán Protected Forest, which, in addition to providing various ecosystem services, also produces large amounts of water. It was observed that, in terms of coverage, cropland presents the lowest water yield compared with mixed forest coverage in both the trend and pessimistic scenarios. However, there has been an increase in water yield from 2016 to 2018. This increment may be due to the protection and conservation measures carried out in the study area by the EMAPAL-EP with its ARAs initiative.

Given the effects of climate change, the study area will experience higher water yield rates. However, when comparing both climate scenarios, the RCP 8.5 scenario will have higher water yield rates. These results suggest that future adaptation measures will be required to protect critical areas, ensure water security, and improve the resilience of watersheds in climate-extreme scenarios.

Although not enough data are available to validate the model's results, this study provides knowledge for delineating water-relevant areas and continuing or creating new research for their protection and conservation.

Author Contributions: Conceptualization, D.X.Z., A.A. (Alex Avilés), L.G. and A.A. (Ana Astudillo); methodology, D.X.Z.; software, D.X.Z. and L.G.; validation, D.X.Z., L.G., A.A. (Ana Astudillo) and J.A.; formal analysis, D.X.Z.; investigation, D.X.Z., A.A. (Alex Avilés), L.G., A.A. (Ana Astudillo), J.A. and C.M.; resources, D.X.Z., A.A. (Alex Avilés), L.G., A.A. (Ana Astudillo), J.A. and C.M.; data curation, D.X.Z. and L.G.; writing—original draft preparation, D.X.Z. and A.A. (Alex Avilés); writing—review and editing, D.X.Z., A.A. (Alex Avilés) and L.G.; visualization, D.X.Z. and A.A. (Alex Avilés); supervision, A.A. (Alex Avilés); project administration, A.A. (Alex Avilés), A.A. (Ana Astudillo) and J.A.; funding acquisition, A.A. (Alex Avilés). All authors have read and agreed to the published version of the manuscript.

Funding: This research received no external funding.

Data Availability Statement: The datasets used in the current study are not publicly available and must be requested from the Catholic University of Cuenca, EMAPAL-EP, and MAATE.

Acknowledgments: The authors thank the Dirección de Vinculación and Vicerrectorado de Investigación de la Universidad de Cuenca and the EMAPAL EP for the support provided to the project "Seguridad Hídrica de la subcuenca del río Tabacay en tiempos de cambios globales". We thank the Catholic University of Cuenca, EMAPAL-EP, and MAATE for providing this work's hydrometeorological and climate data.

Conflicts of Interest: The authors declare no conflicts of interest.

Appendix A

Data correction using the QDM method for precipitation (A1) and temperature (A2) for RCP 4.5 and 8.5 scenarios.

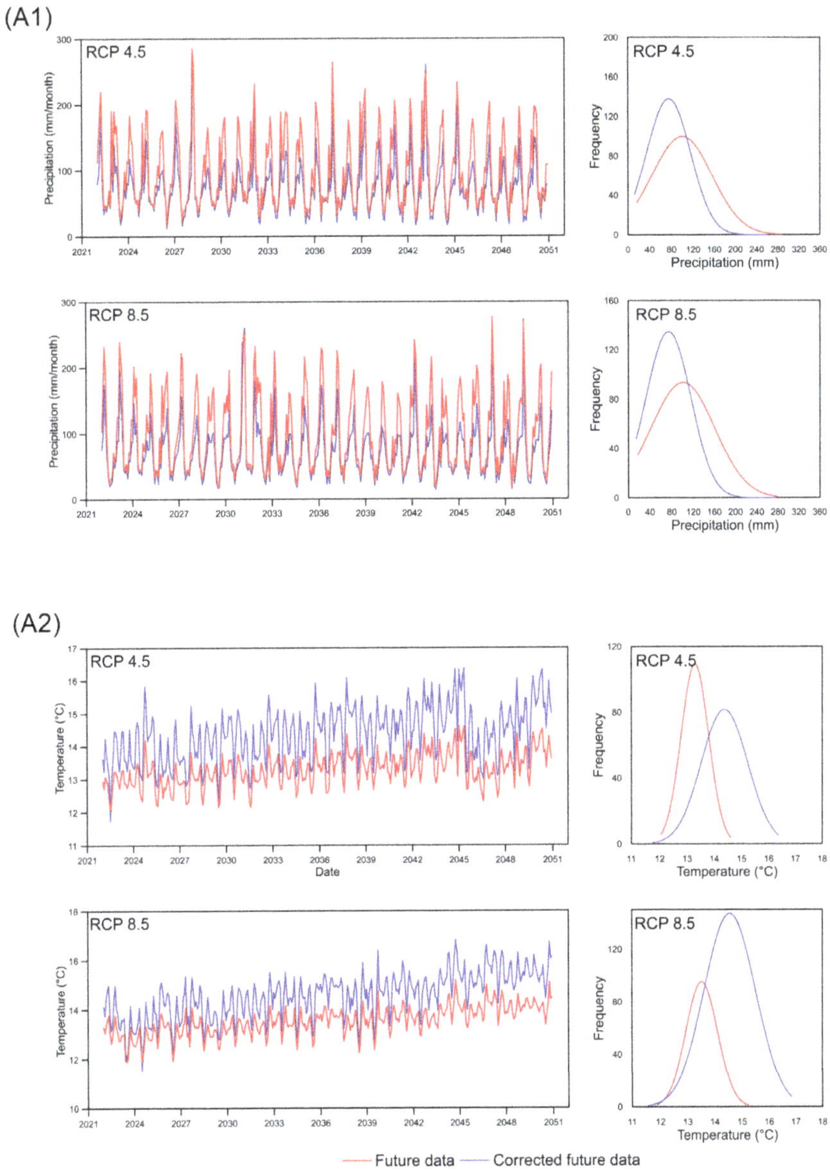

Appendix B

Validation of Dinamica Ego, (B1) land cover observed, (B2) land cover simulated, and (B3) similarity.

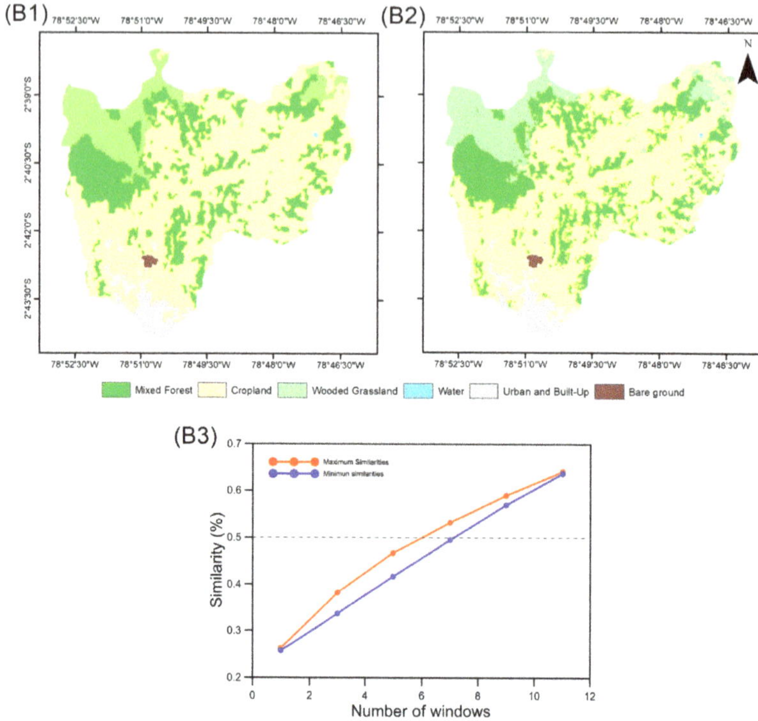

References

1. Maes, J.; Crossman, N.D.; Burkhard, B. Mapping ecosystem services. In *Routledge Handbook of Ecosystem Services*; Routledge: London, UK, 2017; p. 23, ISBN 978-954-642-829-5. [CrossRef]
2. Vaughn, C.C. Ecosystem services provided by freshwater mussels. *Hydrobiologia* **2017**, *810*, 15–27. [CrossRef]
3. Daily, G.C. Introduction: What are ecosystem services? In *Nature's Services: Societal Dependence on Natural Ecosystems*, 1st ed.; Island Press: Washington, DC, USA, 1997; pp. 1–10, ISBN 1-55963-475-8. Available online: https://islandpress.org/books/natures-services#desc (accessed on 31 August 2024).
4. Cudennec, C.; Leduc, C.; Koutsoyiannis, D. Dryland hydrology in Mediterranean regions—A review. *Hydrol. Sci. J.* **2007**, *52*, 1077–1087. [CrossRef]
5. Van Reeth, W. Ecosystem Service Indicators. In *Ecosystem Services*; Elsevier: Amsterdam, The Netherlands, 2013; pp. 41–61.
6. Pessacg, N.; Flaherty, S.; Brandizi, L.; Solman, S.; Pascual, M. Getting water right: A case study in water yield modelling based on precipitation data. *Sci. Total Environ.* **2015**, *537*, 225–234. [CrossRef] [PubMed]
7. Rohatyn, S. Differential Impacts of Land Use and Precipitation on "Ecosystem Water Yield". *Water Resour. Res.* **2018**, *54*, 5457–5470. [CrossRef]
8. Aber, J.D.; Ollinger, S.V.; Federer, C.A.; Reich, P.B.; Goulden, M.L.; Kicklighter, D.W.; Melillo, J.M.; Lathrop, R.G., Jr. Predicting the effects of climate change on water yield and forest production in the northeastern United States. *Clim. Res.* **1995**, *5*, 207–222. [CrossRef]
9. Zhang, L.; Cheng, L.; Chiew, F.; Fu, B. ScienceDirect Understanding the impacts of climate and landuse change on water yield. *Curr. Opin. Environ. Sustain.* **2018**, *33*, 167–174. [CrossRef]
10. Li, Y.; Piao, S.; Li, L.Z.X.; Chen, A.; Wang, X.; Ciais, P.; Huang, L.; Lian, X.; Peng, S.; Zeng, Z.; et al. Divergent hydrological response to large-scale afforestation and vegetation greening in China. *Sci. Adv.* **2018**, *4*, eaar4182. [CrossRef]
11. Sahin, V.; Hall, M.J. The effects of afforestation and deforestation on water yields. *J. Hydrol.* **1996**, *178*, 293–309. [CrossRef]
12. Foley, J.A.; DeFries, R.; Asner, G.P.; Barford, C.; Bonan, G.; Carpenter, S.R.; Chapin, F.S.; Coe, M.T.; Daily, G.C.; Gibbs, H.K.; et al. Global consequences of land use. *Science* **2005**, *309*, 570–574. [CrossRef]
13. Schröeter, D.; Cramer, W.; Leemans, R.; Prentice, I.C.; Araújo, M.B.; Arnell, N.W.; Bondeau, A.; Bugmann, H.; Carter, T.R.; Gracia, C.A.; et al. Europe Ecosystem Service Supply and Vulnerability to Global Change in Europe. *Science* **2005**, *310*, 1333–1337. [CrossRef]
14. Baker, T.J.; Miller, S.N. Using the Soil and Water Assessment Tool (SWAT) to assess land use impact on water resources in an East African watershed. *J. Hydrol.* **2013**, *486*, 100–111. [CrossRef]

15. McFarlane, D.; Stone, R.; Martens, S.; Thomas, J.; Silberstein, R.; Ali, R.; Hodgson, G. Climate change impacts on water yields and demands in south-western Australia. *J. Hydrol.* **2012**, *475*, 488–498. [CrossRef]
16. Blöschl, G.; Sivapalan, M. Scale issues in hydrological modelling: A review. *Hydrol. Process.* **1995**, *9*, 251–290. [CrossRef]
17. Bergström, S. Principles and Confidence in Hydrological Modelling. *Hydrol. Res.* **1991**, *22*, 123–136. [CrossRef]
18. Zhang, C.; Li, W.; Zhang, B.; Liu, M. Water Yield of Xitiaoxi River Basin Based on InVEST Modeling. *J. Resour. Ecol.* **2012**, *3*, 50–54.
19. Arnold, J.G.; Moriasi, D.N.; Gassman, P.W.; White, M.J. SWAT: Model use, calibration, and validation. *Trans. ASABE* **2012**, *55*, 1491–1508. [CrossRef]
20. Beven, K.J.; Kirkby, M.J.; Freer, J.E.; Lamb, R. A history of TOPMODEL. *Hydrol. Earth Syst. Sci.* **2021**, *25*, 527–549. [CrossRef]
21. Hamman, J.J.; Nijssen, B.; Bohn, T.J.; Gergel, D.R.; Mao, Y. The Variable Infiltration Capacity model version 5 (VIC-5): Infrastructure improvements for new applications and reproducibility. *Geosci. Model Dev.* **2018**, *11*, 3481–3496. [CrossRef]
22. Ceroni, M. ARIES (Artificial Intelligence for Ecosystem Services): A new tool for ecosystem services assessment, planning, and valuation. In Proceedings of the 11th Annual BIOECON Conference on Economic Instruments to Enhance the Conservation and Sustainable Use of Biodiversity, Venice, Italy, 21 September 2009.
23. Hamel, P.; Guswa, A.J. Uncertainty analysis of a spatially explicit annual water-balance model: Case study of the Cape Fear basin, North Carolina. *Hydrol. Earth Syst. Sci.* **2015**, *19*, 839–853. [CrossRef]
24. Sharp, J.; Tallis, R.; Ricketts, H.T.; Guerry, T.; Wood, A.D.; Chaplin Ennaanay, S.A.; Wolny, D.; Olwero, S.; Kramer, N.; Nelson, R.; et al. *InVEST 3.2.0 User's Guide*; Stanford University: Stanford, CA, USA; University of Minnesota: Minneapolis, MN, USA; The Nature Conservancy: Arlington, VA, USA; World Wildlife Fund: Gland, Switzerland, 2018.
25. Redhead, J.W.; Stratford, C.; Sharps, K.; Jones, L.; Ziv, G.; Clarke, D.; Oliver, T.; Bullock, J. Empirical validation of the InVEST water yield ecosystem service model at a national scale. *Sci. Total Environ.* **2016**, *569–570*, 1418–1426. [CrossRef]
26. Fu, Q.; Li, B.; Hou, Y.; Bi, X.; Zhang, X. Effects of land use and climate change on ecosystem services in Central Asia's arid regions: A case study in Altay Prefecture, China. *Sci. Total Environ.* **2017**, *607–608*, 633–646. [CrossRef] [PubMed]
27. Wei, P.; Chen, S.; Wu, M.; Deng, Y.; Xu, H.; Jia, Y.; Liu, F. Using the invest model to assess the impacts of climate and land use changes on water yield in the upstream regions of the shule river basin. *Water* **2021**, *13*, 1250. [CrossRef]
28. Kim, S.W.; Jung, Y.Y. Application of the InVEST model to quantify the water yield of North Korean forests. *Forests* **2020**, *11*, 804. [CrossRef]
29. Scordo, F.; Lavender, T.M.; Seitz, C.; Perillo, V.L.; Rusak, J.A.; Piccolo, M.C.; Perillo, G.M.E. Modeling Water Yield: Assessing the role of site and region-specific attributes in determining model performance of the InVEST Seasonal Water Yield Model. *Water* **2018**, *10*, 1496. [CrossRef]
30. Benra, F.; De Frutos, A.; Gaglio, M.; Álvarez-Garretón, C.; Felipe-Lucia, M.; Bonn, A. Mapping water ecosystem services: Evaluating InVEST model predictions in data scarce regions. *Environ. Model. Softw.* **2021**, *138*, 104982. [CrossRef]
31. Minga-León, S.; Gómez-Albores, M.A.; Bâ, K.M.; Balcázar, L.; Manzano-Solís, L.R.; Cuervo-Robayo, A.P.; Mastachi-Loza, C.A. Estimation of water yield in the hydrographic basins of southern Ecuador. *Hydrol. Earth Syst. Sci.* **2018**, *2018*, 1–18.
32. Buytaert, W.; Célleri, R.; De Bièvre, B.; Cisneros, F.; Wyseure, G.; Deckers, J.; Hofstede, R. Human impact on the hydrology of the Andean páramos. *Earth-Sci. Rev.* **2006**, *79*, 53–72. [CrossRef]
33. Estrella, R.; Cattrysse, D.; Van Orshoven, J. Comparison of Three Ideal Point-Based Multi-Criteria Decision Methods for Afforestation Planning. *Forests* **2014**, *5*, 3222–3240. [CrossRef]
34. Budyko, M.I. *Climate and Life*; Academic Press: Cambridge, MA, USA, 1974.
35. Fu, B.P. On the calculation of the evaporation from land surface. *Sci. Atmos. Sin.* **1981**, *5*, 23–31.
36. Zhang, L.; Dawes, W.R.; Walker, G.R. Response of mean annual evapotranspiration to vegetation changes at catchment scale. *Water Resour. Res.* **2001**, *37*, 701–708. [CrossRef]
37. Childs, C. Interpolating surfaces in ArcGIS spatial analyst. *ArcUser* **2004**, *3235*, 32–35.
38. MAE-PNUD. *Tercera Comunicación Nacional del Ecuador a la Convención Marco de las Naciones Unidas sobre Cambio Climático*; Ministerio de Ambiente y Agua: Quito, Ecuador, 2017.
39. Armenta, G.; Villa, J.; Jácome, P.S. Proyecciones Climáticas de Precipitación y Temperatura para Ecuador, Bajo Distintos Escenarios de Cambio Climático; [Internet] Quito, Ecuador. 2016. Available online: https://info.undp.org/docs/pdc/Documents/ECU/14%20Proyecciones%20de%20Clima%20Futuro%20para%20Ecuador%20en%20base%20a%20IPCC-AR5.pdf (accessed on 31 August 2024).
40. van Vuuren, D.P.; Edmonds, J.; Kainuma, M.; Riahi, K.; Thomson, A.; Hibbard, K.; Hurtt, G.C.; Kram, T.; Krey, V.; Lamarque, J.-F.; et al. The representative concentration pathways: An overview. *Clim. Change* **2011**, *109*, 5. [CrossRef]
41. Zhiña, D.; Avilés, A.; Pacheco, J.; Mendoza, D. Spatiotemporal Features Of Drought Conditions: A Case Study In An Equatorial Andean Basin Using The Spei Index (1982–2015). In *Estudios Teórico-Metodológicos en Ciencias Exactas, Tecnológicas y de la Tierra 4*; Atena Editora: Ponta Grossa, Brazil, 2024; pp. 50–63.
42. Campozano, L.; Ballari, D.; Montenegro, M.; Avilés, A. Future Meteorological Droughts in Ecuador: Decreasing Trends and Associated Spatio-Temporal Features Derived From CMIP5 Models. *Front. Earth Sci.* **2020**, *8*, 17. [CrossRef]
43. Senent-Aparicio, J.; Peñafiel, L.; Alcalá, F.J.; Jimeno-Sáez, P.; Pérez-Sánchez, J. Climate change impacts on renewable groundwater resources in the andosol-dominated Andean highlands, Ecuador. *Catena* **2024**, *236*, 107766. [CrossRef]
44. Cannon, A.J.; Sobie, S.R.; Murdock, T.Q. Bias correction of GCM precipitation by quantile mapping: How well do methods preserve changes in quantiles and extremes? *J. Clim.* **2015**, *28*, 6938–6959. [CrossRef]

45. Li, C.; Sinha, E.; Horton, D.E.; Diffenbaugh, N.S.; Michalak, A.M. Joint bias correction of temperature and precipitation in climate model simulations. *J. Geophys. Res. Atmos.* **2014**, *119*, 13153–13162. [CrossRef]
46. Holdridge, L.R. Simple Method for Determining Potential Evapotranspiration from Temperature Data. *Science* **1959**, *130*, 572. [CrossRef]
47. Hengl, T.; De Jesus, J.M.; Heuvelink, G.B.M.; Gonzalez, M.R.; Kilibarda, M.; Blagotić, A.; Shangguan, W.; Wright, M.N.; Geng, X.; Bauer-Marschallinger, B.; et al. SoilGrids250m: Global gridded soil information based on machine learning. *PLoS ONE* **2017**, *12*, e0169748. [CrossRef]
48. Martinez, A.; Matovelle-Bustos, C.M.; Astudillo, L. Geographic Information Systems in the protection of water resources in the micro-basin of Tabacay River, Ecuador. *Kill. Técnica* **2019**, *3*, 1–10.
49. MAE-MAGAP. *Protocolo Metodológico para la Elaboración del Mapa de Cobertura y uso de la Tierra del Ecuador Continental 2013–2014, Escala 1:100000*; Ministerio del Ambiente del Ecuador y Ministerio de Agricultura, Ganadería, Acuacultura y Pesca: Quito, Ecuador, 2015.
50. Espinoza-Mendoza, V. *Dinámica EGO: Una Herramienta Gratuita para Modelar y Brindar Soporte en el Análisis de CCUS*; Maimónides University: Buenos Aires, Argentina, 2017; pp. 1–20.
51. Piontekowski, V.J.; da Silva, S.S.; Mendoza, E.R.H.; Costa, W.L.d.S.; Ribeiro, F.C. Modelagem do desmatamento para o Estado do Acre utilizando o programa DINAMICA EGO. In Proceedings of the 4o Simpósio Geotecnologias no Pantanal, Bonito, MS, Brazil; 2012; pp. 20–24.
52. Xu, X.; Liu, W.; Scanlon, B.R.; Zhang, L.; Pan, M. Local and global factors controlling water-energy balances within the Budyko framework. *Geophys. Res. Lett.* **2013**, *40*, 6123–6129. [CrossRef]
53. Sánchez-Canales, M.; Benito, A.L.; Passuello, A.; Terrado, M.; Ziv, G.; Acuña, V.; Schuhmacher, M.; Elorza, F.J. Sensitivity analysis of ecosystem service valuation in a Mediterranean watershed. *Sci. Total Environ.* **2012**, *440*, 140–153. [CrossRef] [PubMed]
54. da Silva, R.M.; Lopes, A.G.; Santos, C.A.G. Deforestation and fires in the Brazilian Amazon from 2001 to 2020: Impacts on rainfall variability and land surface temperature. *J. Environ. Manag.* **2023**, *326*, 116664. [CrossRef] [PubMed]
55. Zhiña, D.X.; Mosquera, G.M.; Esquivel-Hernández, G.; Córdova, M.; Sánchez-Murillo, R.; Orellana-Alvear, J.; Crespo, P. Hydrometeorological factors controlling the stable isotopic composition of precipitation in the highlands of south Ecuador. *J. Hydrometeorol.* **2022**, *23*, 1059–1074. [CrossRef]
56. Zhang, Q.; Ban, Y.; Liu, J.; Hu, Y. Simulation and analysis of urban growth scenarios for the Greater Shanghai Area, China. *Comput. Environ. Urban Syst.* **2011**, *35*, 126–139. [CrossRef]
57. Zhang, Q.; Wallace, J.; Deng, X.; Seto, K.C. Central versus local states: Which matters more in affecting China's urban growth? *Land Use Policy* **2014**, *38*, 487–496. [CrossRef]
58. Li, H.; Yao, Y.; Zhang, X.; Zhu, H.; Wei, X. Changes in soil physical and hydraulic properties following the conversion of forest to cropland in the black soil region of Northeast China. *Catena* **2021**, *198*, 104986. [CrossRef]
59. Shirmohammadi, B.; Malekian, A.; Salajegheh, A.; Taheri, B.; Azarnivand, H.; Malek, Z.; Verburg, P.H. Impacts of future climate and land use change on water yield in a semiarid basin in Iran. *Land Degrad. Dev.* **2020**, *31*, 1252–1264. [CrossRef]
60. Pinos-Morocho, D.; Morales-Matute, O.; Durán-López, M.E. Suelos de páramo: Análisis de percepciones de los servicios ecosistémicos y valoración económica del contenido de carbono en la sierra sureste del Ecuador. *Rev. Cienc. Ambient.* **2021**, *55*, 157–179. [CrossRef]
61. Liu, M.; Tian, H.; Chen, G.; Ren, W.; Zhang, C.; Liu, J. Effects of land-use and land-cover change on evapotranspiration and water yield in China during 1900–2000. *J. Am. Water Resour. Assoc.* **2008**, *44*, 1193–1207. [CrossRef]
62. Li, S.; Yang, H.; Lacayo, M.; Liu, J.; Lei, G. Impacts of land-use and land-cover changes on water yield: A case study in Jing-Jin-Ji, China. *Sustainability* **2018**, *10*, 960. [CrossRef]
63. Ouyang, Y. New insights on evapotranspiration and water yield in crop and forest lands under changing climate. *J. Hydrol.* **2021**, *603*, 127192. [CrossRef]
64. Ochoa-Tocachi, B.F.; Buytaert, W.; De Bièvre, B.; Célleri, R.; Crespo, P.; Villacís, M.; Llerena, C.A.; Acosta, L.; Villazón, M.; Guallpa, M.; et al. Impacts of land use on the hydrological response of tropical Andean catchments. *Hydrol. Process.* **2016**, *30*, 4074–4089. [CrossRef]
65. Buytaert, W.; Iñiguez, V.; De Bièvre, B. The effects of afforestation and cultivation on water yield in the Andean páramo. *For. Ecol. Manag.* **2007**, *251*, 22–30. [CrossRef]
66. Ellison, D.; Futter, M.N.; Bishop, K. On the forest cover-water yield debate: From demand- to supply-side thinking. *Glob. Chang. Biol.* **2012**, *18*, 806–820. [CrossRef]
67. Mark, A.F.; Dickinson, K.J.M. Maximizing water yield with indigenous non-forest vegetation: A New Zealand perspective. *Front. Ecol. Environ.* **2008**, *6*, 25–34. [CrossRef]
68. Scanlon, B.R.; Jolly, I.; Sophocleous, M.; Zhang, L. Global impacts of conversions from natural to agricultural ecosystems on water resources: Quantity versus quality. *Water Resour. Res.* **2007**, *43*, 1–18. [CrossRef]
69. Zhiña, D.; Montenegro, M.; Montalván, L.; Mendoza, D.; Contreras, J.; Campozano, L.; Avilés, A. Climate Change Influences of Temporal and Spatial Drought Variation in the Andean High Mountain Basin. *Atmosphere* **2019**, *10*, 558. [CrossRef]
70. Lu, N.; Sun, G.; Feng, X.; Fu, B. Water yield responses to climate change and variability across the North-South Transect of Eastern China (NSTEC). *J. Hydrol.* **2013**, *481*, 96–105. [CrossRef]

71. Yin, G.; Wang, X.; Zhang, X.; Fu, Y.; Hao, F.; Hu, Q. InVEST model-based estimation of water yield in North China and its sensitivities to climate variables. *Water* **2020**, *12*, 1692. [CrossRef]
72. Yohannes, H.; Soromessa, T.; Argaw, M.; Dewan, A. Impact of landscape pattern changes on hydrological ecosystem services in the Beressa watershed of the Blue Nile Basin in Ethiopia. *Sci. Total Environ.* **2021**, *793*, 148559. [CrossRef]
73. Ningrum, A.; Setiawan, Y.; Tarigan, S.D. Annual Water Yield Analysis with InVEST Model in Tesso Nilo National Park, Riau Province. *IOP Conf. Ser. Earth Environ. Sci.* **2022**, *950*, 012098. [CrossRef]
74. Daneshi, A.; Brouwer, R.; Najafinejad, A.; Panahi, M.; Zarandian, A.; Maghsood, F.F. Modelling the impacts of climate and land use change on water security in a semi-arid forested watershed using InVEST. *J. Hydrol.* **2021**, *593*, 125621. [CrossRef]
75. Li, M.; Liang, D.; Xia, J.; Song, J.; Cheng, D.; Wu, J.; Cao, Y.; Sun, H.; Li, Q. Evaluation of water conservation function of Danjiang River Basin in Qinling Mountains, China based on InVEST model. *J. Environ. Manag.* **2021**, *286*, 112212. [CrossRef] [PubMed]
76. GAD del Cantón Azogues. Ordenanza para la Conservación, Restauración y Recuperación de las Fuentes de Agua, Zonas de Recarga Hídrica, Ecosistemas Frágiles y Otras Áreas Prioritarias para la Protección de la Biodiversidad, los Servicios Ambientales y el Patrimonio Natural del Can. *R.O. 171*, Azogues, Ecuador, 2014. p. 27. Available online: https://www.emapal.gob.ec/mdocuments-library/ (accessed on 31 August 2024).
77. Bagstad, K.J.; Semmens, D.J.; Winthrop, R. Comparing approaches to spatially explicit ecosystem service modeling: A case study from the San Pedro River, Arizona. *Ecosyst. Serv.* **2013**, *5*, 40–50. [CrossRef]
78. Power, A.G. Ecosystem services and agriculture: Tradeoffs and synergies. *Philos. Trans. R. Soc. B Biol. Sci.* **2010**, *365*, 2959–2971. [CrossRef]
79. Zheng, H.; Li, Y.; Robinson, B.E.; Liu, G.; Ma, D.; Wang, F.; Lu, F.; Ouyang, Z.; Daily, G.C. Using ecosystem service trade-offs to inform water conservation policies and management practices. *Front. Ecol. Environ.* **2016**, *14*, 527–532. [CrossRef]

hydrology

MDPI

Article

Attribution of the Climate and Land Use Change Impact on the Hydrological Processes of Athabasca River Basin, Canada

Sharad Aryal [1,2], Mukand S. Babel [2,3], Anil Gupta [4], Babak Farjad [4], Dibesh Khadka [2,3] and Quazi K. Hassan [1,*]

1 Department of Geomatics Engineering, Schulich School of Engineering, University of Calgary, 2500 University Dr. NW, Calgary, AB T2N 1N4, Canada; sharad.aryal@ucalgary.ca
2 Water Engineering and Management, Asian Institute of Technology, P.O. Box 4, Klong Luang, Pathumthani 12120, Thailand; msbabel@ait.ac.th (M.S.B.); dibesh@ait.ac.th (D.K.)
3 Centre for Water and Climate Adaptation, School of Engineering and Technology (SET), Asian Institute of Technology (AIT), P.O. Box 4, Klong Luang, Pathumthani 12120, Thailand
4 Alberta Environment and Protected Areas, Government of Alberta, Calgary, AB T2L 2K8, Canada; anil.gupta@gov.ab.ca (A.G.); babak.farjad@gov.ab.ca (B.F.)
* Correspondence: qhassan@ucalgary.ca; Tel.: +1-403-210-9494

Abstract: Climate change (CC) and land use/land cover change (LULCC) are significant drivers of hydrological change, and an effective watershed management requires a detailed understanding of their individual and the combined impact. This study focused on the Athabasca River Basin (ARB), Canada, and investigated how the basin responded to their changes using the MIKE SHE-MIKE Hydro River. Our findings revealed novel insights into ARB hydrological changes, including increment in non-vegetated lands (0.26%), savannas (1.28%), forests (0.53%), and urban areas (0.02%) while grasslands (2.07%) and shrublands (0.03%) decreased. Moreover, the basin experienced rising annual minimum (1.01 °C) and maximum (0.85 °C) temperatures but declining precipitation (6.2%). The findings suggested a significant impact of CC compared to LULCC as CC caused annual reduction in streamflow (7.9%), evapotranspiration (4.8%), and recharge (6.9%). Meanwhile, LULCC reduced streamflow (0.2%) and recharge (0.4%) but increased evapotranspiration (0.1%). The study revealed spatiotemporal variability across the ARB, with temperature impacts stronger in winter and precipitation influencing other seasons.

Keywords: hydrological modeling; cold region climate; MIKE SHE; water balance; climatic change

1. Introduction

The hydrological cycle is a naturally dynamic system comprising various hydrological processes driven by climatic factors, the watershed's physical properties, anthropogenic activities, and their mutual interactions varying spatially and temporally [1]. Consequently, changes in climatic conditions and land use/land cover have been identified as key drivers of shift in the hydrological regime [2]. Climate change (CC) alters the pattern of precipitation and temperature in terms of intensity and frequency, significantly impacting the hydrological cycle. Documented effects include flow regime alteration, changes in timing and frequency of hydrological extremes, seasonal shifts, increased snowmelt, earlier spring freshet, modified evapotranspiration, soil moisture content, groundwater resources, and shifts in groundwater–surface water interaction [3–7]. Meanwhile, land use dictates the response of a watershed to climatic input, and its changes are associated with alteration in hydrological processes like interception, infiltration, evapotranspiration, surface runoff, and groundwater recharge rates [8,9] which ultimately affects the spatiotemporal availability of water in the basin.

Climate and land use change affect different aspects of the hydrological cycle, and thus their combined effects are complex, often interacting to either amplify or counterbalance each other's impacts [10–12]. To develop effective, basin-specific water management strategies, it is essential to assess each stressor's individual influence on the hydrological cycle, as well as their integrated effects, within each basin [13]. While some studies have found that LULCC can exert a stronger influence on hydrological processes, most findings attribute CC as the primary driver of change [2,14,15]. However, the extent of each stressor's impact varies significantly by catchment [16]. For instance, urbanization was the dominant influence on runoff in six catchments in southeast Queensland, Australia [17] whereas in China's Weihe River Basin, CC had a considerably stronger impact on evapotranspiration, soil moisture, and streamflow compared to LULCC [2]. These variations underscore the need for individualized basin assessments to accurately capture localized hydrological responses, enabling more tailored water management approaches.

The impacts of these driving factors can also exhibit spatiotemporal variability. For instance, Zeng et al. [18] revealed that CC accounted for greater changes in runoff during the dry season, whereas changes induced by human activities were more dominant in the wet season in the Zhang River basin in north China. Furthermore, most studies exploring the drivers behind hydrological changes, however, have primarily focused on streamflow alone, which may not provide a comprehensive depiction of the entire hydrological cycle's dynamics [2]. Exploring other essential hydrological processes, such as evapotranspiration, overland flow, and groundwater recharge, can provide a more comprehensive understanding of the mechanisms governing the hydrological cycle and its responses to shifts in the surrounding environment. As water management increasingly emphasizes maintaining water availability across both spatial and temporal scales, it becomes essential to evaluate the impacts of these drivers on multiple hydrological components at diverse scales.

The ARB, home to Alberta's multi-billion-dollar oil sands industry, is vital to the province's economy, drawing considerable attention in hydroclimatic research due to its environmental and socio-economic significance. A consensus in past studies indicates declining river flow trend in the downstream [19–21] while noting increased flow at the headwaters. These trends have been linked with decline in precipitation [21–23], rising temperatures [22], and large-scale climatic oscillation cycles [24–26]. Predominantly, these findings are drawn from statistical analyses such as trend analysis [20], wavelet analysis [27], and multi-variate linear regression [22]. While hydrological models like SWAT [28], VIC [29], HydroGeosphere [30], and MISBA [31] have been applied in the ARB, they are primarily utilized to study future state of the basin's hydrology. This study presents a novel application of hydrological modeling to assess the historical impacts of CC and LULCC on the basin's hydrology, filling a critical gap in understanding past changes and their underlying mechanism in this vital region.

Hydrological modeling is widely used to distinguish individual impacts of stressors on hydrological systems by simulating baseline conditions and comparing them with scenarios and thereby isolating CC and LULCC effects [13]. Further, the modeling approach allows for evaluating effects at a higher temporal resolution as impact at daily, seasonal, and annual time scales can be assessed [18]. In this study, we have applied the coupled MIKE SHE-MIKE Hydro River model, which can comprehensively represent the primary hydrological process of the catchment. MIKE SHE is a deterministic, physics-based, fully distributed model that can demonstrate major hydrological cycle processes like evapotranspiration, overland flow, unsaturated flow, groundwater flow, and their interaction at varying spatial scales and complexity [32]. Meanwhile, MIKE Hydro River is a one-dimensional channel flow model for the simulation of rivers using fully dynamic Saint Venant equations. Numerous studies have used it to simulate the water balance of a watershed and predict its behavior in projected climate change, anthropogenic activities, and changes in land use [10,33,34]. In addition to this, the selection of a distributed hydrological model like MIKE SHE enables the evaluation to be concentrated and compared between smaller sub-watersheds within the basin.

Previous studies in the ARB have primarily focused on quantifying observed hydrological changes using statistical methods, with limited exploration of the underlying mechanisms [20,23,27]. Meanwhile, hydrological modeling efforts have largely been scenario-based, forecasting future changes rather than attributing past changes [3,28,29]. The objective of this study is to fill this gap by employing a physically based distributed hydrological model to ascertain the distinct and cumulative impacts of CC and LULCC on hydrological dynamics of the ARB. This study is the first to utilize the MIKE SHE in the ARB to specifically quantify the individual and combined effects of CC and LULCC on hydrological alterations, offering a novel perspective on hydrology of the ARB.

2. Materials and Methods

2.1. Study Area

The Athabasca River originates at the Columbia Icefield in Jasper National Park in Alberta, Canada, and flows northeast for more than 1500 km before finally draining into Lake Athabasca. The ARB encompasses an area of approximately 155,000 km^2 characterized by heterogenous hydroclimatic and physiographic properties. The elevation ranges from around 3600 m above mean sea level in the upper Rocky Mountains to around 200 m in the lower region. The river is the longest free-flowing river (with no regulating structures) in Alberta [35]. The flow regime is snow-dominated, with ice phenology playing a major role in the river flow [19]. Likewise, the region experiences four distinct seasons, winter (December to February), spring (March to May), summer (June to August), and fall (September to November). The major footprint of human activities includes agriculture, urbanization, livestock production, forestry, coal mining, oil and gas, and oil sands mining [36]. Although there has been a growing use of groundwater, the surface water from Athabasca and its tributaries serves as a primary source of water supply for anthropogenic activities [19,35,36]. The annual allocated water use is less than 5% of annual river flow, the bulk of which is withdrawal by the oil sands industry concentrated below Fort McMurray (Figure 1) [21,35,36].

2.2. Hydrological–Hydrodynamic Model

The availability of required data and computational efficiency were major factors that needed careful consideration for distributed models, particularly for a larger watershed like the ARB. These aspects influenced the choice of process and spatial resolution of process representation for the MIKE SHE-MIKE Hydro model. The relevant data and parameters used for the different modules, along with their details, are provided in the following sub-sections.

2.2.1. Model Domain and Grid

The topography of the model was defined by the digital elevation model obtained from the Canadian Digital Elevation Model (CDEM) at a resolution of 0.75 arc seconds. A cell size of 2000 m was selected as a computational grid representing the compromise between required information at the finest resolution and the computational requirement for the model.

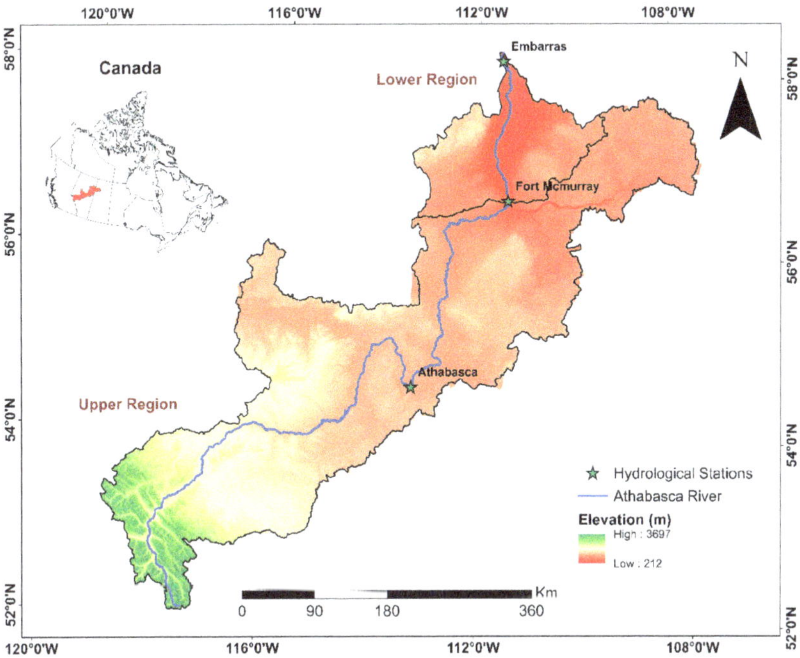

Figure 1. Location map and details of the Athabasca River Basin with the location of hydrological stations along the main river. The boundary line through Fort McMurray separates the lower region (Northern) and upper region (Southern), as adopted in the study.

2.2.2. Climate Input

Daily gridded observed precipitation, maximum temperatures, and minimum temperature from 1950 to 2019 available at a $0.1^{\circ} \times 0.1^{\circ}$ grid size [37] were obtained from Alberta Environment and Protected Areas (EPA). The Hargreaves–Samani (HS) method was adopted to calculate the daily potential evapotranspiration [38]. The applicability of the HS method was validated by comparing with potential evapotranspiration data at four climatic stations in the area (Appendix A). The modified degree day method was used for the snowmelt, which required information about the degree day melting coefficient. The initial values of the monthly degree day coefficient were selected based on [39] and were adjusted during the calibration.

2.2.3. Surface Water Component

We used Moderate Resolution Imaging Spectroradiometer (MODIS) Land Cover Type 3 (LULC) at a 500 m resolution to represent the LULC in 2001 and 2015. The dataset collected and processed for quality control by [40] was used in this study. Altogether, nine different land classes available in the ARB were grasslands, water bodies, shrublands, savannas, deciduous broadleaf forests, urban areas, deciduous needleleaf forests, evergreen needleleaf forests, and non-vegetated lands. Leaf area index (LAI), root depth (RD), Mannings roughness coefficient (M), and detention storage (D) were required to define the properties of each land use class. These values were initially obtained from the literature [39,41,42], and some were calibrated accordingly based on the sensitivity of the parameters.

2.2.4. River Component and Coupling with MIKE SHE

The MIKE Hydro River model was set up to represent the hydrodynamic component of the model. The details on the river network and cross-section were obtained from Alberta

Environment and Protected Areas. Altogether, 84 different branches with 514 cross-sections were extracted for the model and were used for the coupling of MIKE SHE and MIKE Hydro River. The interchange between the groundwater and river link was determined by the conductivity and head differential between the river and the grid cell.

2.2.5. Unsaturated Zone Component

The unsaturated zone was based on a two-layer water balance method. A soil map with a 1:5,000,000 spatial resolution was obtained from the Harmonized World Soil Database provided by the Food and Agriculture Organization (FAO), resulting in 22 different soil types for the study area. Saturated hydraulic conductivity, water content at field capacity, wilting point, and water content at saturation were input properties for each soil type adopted from the FAO database.

2.2.6. Saturated Zone Component

A simple linear reservoir model that represents the saturated zone by interflow and two parallel baseflow reservoirs was employed for the study. This technique, which lessens the amount of data required while boosting model execution time, has been proven especially helpful for large watersheds [43]. The time constants for the defined interflow and baseflow were calibrated during the model setup.

2.3. Model Calibration, Validation, and Sensitivity Analysis

The calibration and validation of the model were carried out against daily streamflow records for the period of thirty years from 1990 to 2019. The calibration of the model used LULC information from 2015 and a twenty-year period (2000–2019), while a ten-year period (1990–1999) was used to validate the model. The daily streamflow records from Environment Canada at three gauging stations (Athabasca, Fort McMurray, and Embarras Airport) along the Athabasca River (Figure 1) were used. However, flow data were only available for five years (2015 onwards) at Embarras Airport, so it was only used for model calibration. The performance of the model was evaluated using three statistical measures: Nash–Sutcliffe efficiency (NSE), percentage bias (PBIAS), and coefficient of determination (R^2).

As the distributed physical model includes numerous parameters, reducing the number of free parameters through sensitivity analysis was critical. However, applying automatic calibration for sensitivity analysis in such a model is not desirable [44]. Thus, calibration parameters were selected that have already been consistently identified as sensitive parameters from various studies [39,43,45,46]. The selected parameters were degree day snow melting coefficient, Manning's roughness coefficient for all land uses and riverbed, conductance between aquifer and river, and time constants for interflow and baseflow reservoirs. A local sensitivity analysis was conducted by individually varying each selected parameter within an appropriate range and evaluating the model's performance under each scenario using three statistical metrics. Those parameters that caused significant variations in model outputs were identified as the most sensitive and critical.

2.4. Scenario Settings and Hydrological Components

The scenario analysis technique was used in this study to attribute the individual impact of CC and LULCC. The climate data from 1960 to 2019 was divided into two 30-year periods: 1960–1989 (called hereafter baseline period) and 1990–2019 (called hereafter recent period). The selection of a thirty-year period for comparison also allowed the study to be focused on evaluating the impact of CC rather than climate variability. This study used the earliest available (2001) LULC record from MODIS to represent baseline LULC conditions. The changes observed in LULC primarily stemmed from human activities, particularly those associated with population growth. An initial examination of population trends in the basin between 1981 and 2021, using census data from Canada and municipal data from Alberta, revealed significant growth, particularly in the early 2000s (Appendix B). This suggests that the LULC pattern of 2001 can serve as a reasonable representation of the vegetation characteristics from 1960 to 1989, given that more substantial changes occurred after 2001. On the other hand, the LULC map obtained from MODIS for the year 2015 (called hereafter recent LULC map), also used for calibration, was used to represent the recent period. Furthermore, an analysis of the annual trends in LULC coverage over the years revealed gradual linear changes [47], suggesting that using land use maps from 2001 and 2015 is sufficient to represent the intermediate changes during this period. The year-to-year variations are minimal, and modeling the hydrological impacts of such small LULC changes may not accurately capture their effects within the model. Overall, four scenarios based on two climate periods and the LULC map were developed for the study, as shown in Table 1. As such, two scenarios reflecting real world situations (baseline and recent) and two reflecting hypothetical scenarios were developed and simulated in the model.

Table 1. Different scenarios with their climate period and LULC information used in the analysis.

Scenario	Climate Period	LULC	Remarks
S1	1960–1989	Baseline (2001)	Baseline
S2	1990–2019	Recent (2015)	Recent
S3	1960–1989	Recent (2015)	Hypothetical baseline
S4	1990–2019	Baseline (2001)	Hypothetical recent

The simulation results from individual scenarios were compared to identify the individual and combined impact of CC and LULCC for various selected hydrological indicators. The calculation was carried out by comparing the simulated value of hydrological indicators under different scenarios with respect to the baseline scenario's value. For instance, percentage change due to individual stressors and their combined impact is calculated using the following equation.

$$\text{Individual impact of LULCC} = \frac{S_{3i} - S_{1i}}{S_{1i}} \times 100$$

$$\text{Individual impact of CC} = \frac{S_{4i} - S_{1i}}{S_{1i}} \times 100$$

$$\text{Combined impact of CC and LULCC} = \frac{S_{2i} - S_{1i}}{S_{1i}} \times 100$$

where S_{1i}, S_{2i}, S_{3i}, and S_{4i} are the simulated values of the ith hydrological indicators under baseline, recent, hypothetical baseline, and hypothetical recent scenarios, respectively.

To depict the overall impact on the hydrological system, the analysis has been carried out for three different hydrological processes; streamflow, evapotranspiration, and groundwater recharge for annual and monthly time scales. Furthermore, as the streamflow is widely regarded as a master variable capable of reflecting different aspects of the hydrological cycle, we have analyzed not only the annual average magnitude of streamflow but also its annual extremes feature. Therefore, in addition to annual and monthly average values, we have selected four additional streamflow indicators from the indicators of hydrological alteration (IHAs), which include annual 7-day maximum, annual 7-day minimum, date of annual maximum, and date of annual minimum flow. The first two indicators show the maximum and minimum flow observed for seven consecutive days in a water year and reflect the extreme conditions in terms of high and low flow in the basin. The latter two indicators represent the timing of extreme flows as the date of occurrence of annual maximum flow and annual minimum flow for each water year. In order to address the spatial variability that may be present between areas with high human activities and with low human activities, this study has divided the basin into upper and lower regions, with Fort McMurray as the boundary (Figure 1). This allowed results to be evaluated for the upper region and lower region separately in addition to the entire watershed.

3. Results

3.1. Evolution of Land Use Pattern in the Basin

The distribution of LULC within the basin over two selected periods of 2001 and 2015 is presented in Figure 2. In general, it can be observed that savannas and forests of various types dominate over three-quarters of the total area in the basin. The areas of savannas were found have increased in 2015 compared to 2001. The distribution of urban areas demonstrates that human settlements are concentrated in a few urbanized areas, which have increased from 0.04% in 2001 to about 0.06% of the basin's area in 2015. This corresponds to an almost 50% increment in urban areas. Likewise, areas of evergreen forests and non-vegetated lands have also increased, while deciduous forest, grassland, and shrubland coverage decreased in these periods.

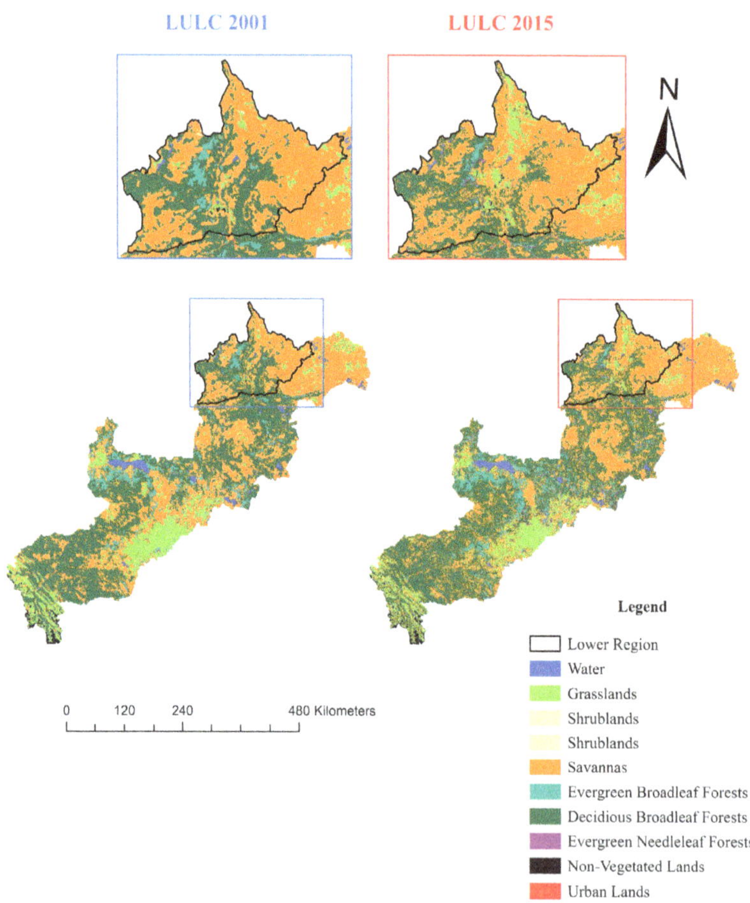

Figure 2. Spatial distribution of LULC classes in 2001 and 2015 in Athabasca River Basin, Canada.

The basin's overall landscape exhibits three distinct properties that align with the natural region division of Alberta [48]. The uppermost areas are characterized by the Rocky Mountains that remain mostly untouched by human activities, resulting in little to no change in land use patterns. The upper central part lies in the foothills region, with a gentle slope consisting of coniferous and mixed wood forests. The boreal forests cover the central and lower parts of the basin and play a pivotal environmental role in maintaining the natural biodiversity of the region [49,50]. The study found that while the upper region exhibited changes in LULC similar to those across the entire watershed, the lower region showed distinct patterns. From 2001 to 2015, the coverage of grasslands, savannas, and non-vegetated lands increased by about 5.5%, 4.0%, and 0.3%, respectively, in the lower region (Table 2). This is at the expense of broadleaf forests, whose coverage area was reduced by more than 10.0%. The oil and gas industries are mainly located in the lower region of the basin, leading to intense urbanization in these areas. Consequently, the lower region is very dynamic in terms of landscape changes. In addition, the basin is frequently subjected to wildfires, which have also contributed to alterations in the land use pattern of the area [51].

Table 2. Percentage coverage of LULC class in 2001 and 2015 for the entire watershed, upper region, and lower region.

LULC Class	Entire		Upper		Lower	
	2001	2015	2001	2015	2001	2015
Water Bodies	1.76	1.75	1.96	1.94	0.57	0.61
Grasslands	12.26	10.19	13.92	10.54	2.83	8.36
Shrublands	0.04	0.01	0.01	0.01	0.15	0.04
Savannas	44.88	46.16	42.13	42.92	60.41	64.27
Evergreen Broadleaf Forests	4.68	6.35	4.91	7.17	3.35	1.69
Deciduous Broadleaf Forests	35.60	33.85	36.21	35.68	32.51	23.87
Evergreen Needleleaf Forests	0.02	0.63	0.02	0.61	0.03	0.72
Non-Vegetated Lands	0.73	0.99	0.81	1.08	0.14	0.44
Urban Areas	0.04	0.06	0.04	0.07	0.01	0.01

3.2. Climate Change Observed in the Basin

CC in the basin exhibits a trend of reduced precipitation while increasing maximum and minimum temperatures (Figure 3). The average annual precipitation in the basin during the baseline period was 536 mm, and it reduced to 503 mm (−6.2%) during the recent period. Meanwhile, the average maximum temperature in the basin increased by 0.85 °C from its baseline value of 6.03 °C. The change in the average minimum temperature was even more pronounced, with a rise of almost 1.00 °C observed (increased from −5.35 °C to −4.34 °C) during the two periods.

The spatial distributions of observed climate variables in the baseline and recent periods, along with changes between these two periods, are shown in Figure 4. It is evident that the upper part of the basin (southwest) receives higher precipitation than the lower part. Although there is an overall decrease in precipitation, some scattered areas in the lower basin have experienced increased precipitation. The central basin is considerably warmer than other parts, with higher maximum and minimum temperatures. Surprisingly, the lower part of the basin, despite being at a lower elevation, is colder and experiences temperatures like those of the mountainous areas upstream. The comparison of temperature during the two periods reveals that the lower part of the basin is warming at a faster rate than other regions. Changes in the maximum and the minimum temperature in the lower part of the basin are up to 1.40 °C and 2.00 °C, respectively, in the considered period. This temperature increase aligns with findings by [23] who observed a consistent uptrend in maximum (0.23 to 0.28 °C per decade) and minimum (0.17 to 0.45 °C per decade) temperatures from 1950 to 2019 in the ARB.

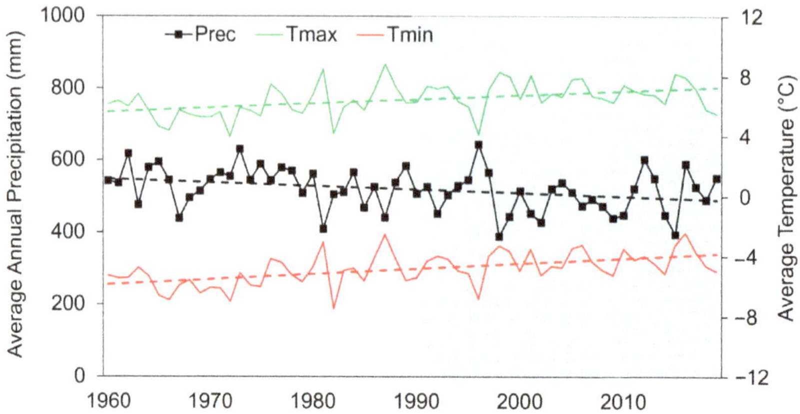

Figure 3. Average annual precipitation, maximum and minimum temperatures for the entire Athabasca River Basin from 1960 to 2019. Note: The shaded zone represents the recent 30-year period while unshaded zone represents baseline condition. The dashed lines represent best fit lines.

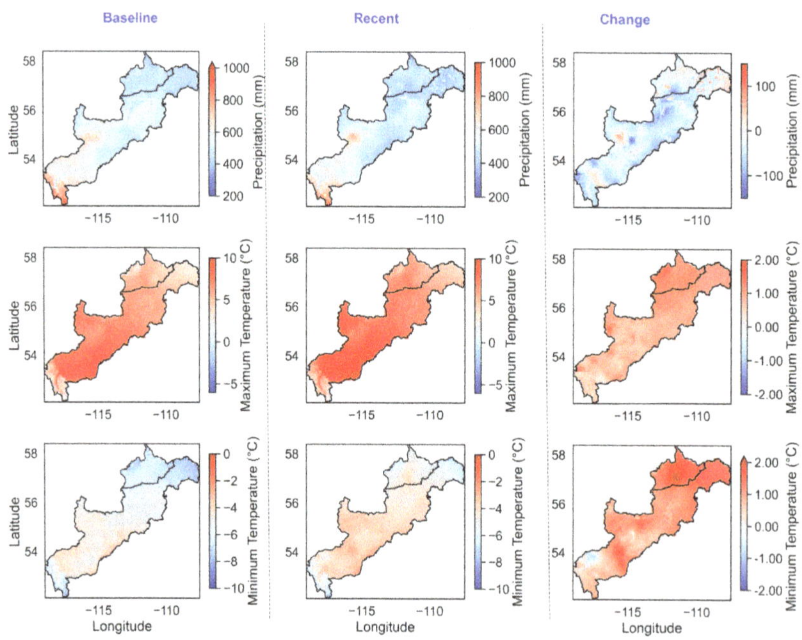

Figure 4. Spatial distribution of average annual precipitation, the maximum and minimum temperature in the baseline period (1960–1989), recent period (1990–2019), and changes in the recent period compared to the baseline period.

The seasonal distribution of the precipitation shows that the basin receives higher precipitation during the summer period, with almost 50% of annual precipitation falling between June and August (Figure 5). The average monthly precipitation in July (97 mm) was highest during the analysis period. Meanwhile, the precipitation during the late fall to early spring is lower and the least is generally observed in February (20 mm).

The comparison of the two periods reveals that precipitation in the basin decreases throughout the year except during April (13.68% increment) and June (5.09% increment). The change is particularly significant during the winter, with a reduction of up to 21% (4.75 mm) observed in February. Although the precipitation is still higher in summer and early fall months, they also experience a reduction ranging from almost 3.75 mm in July to 9.15 mm in September.

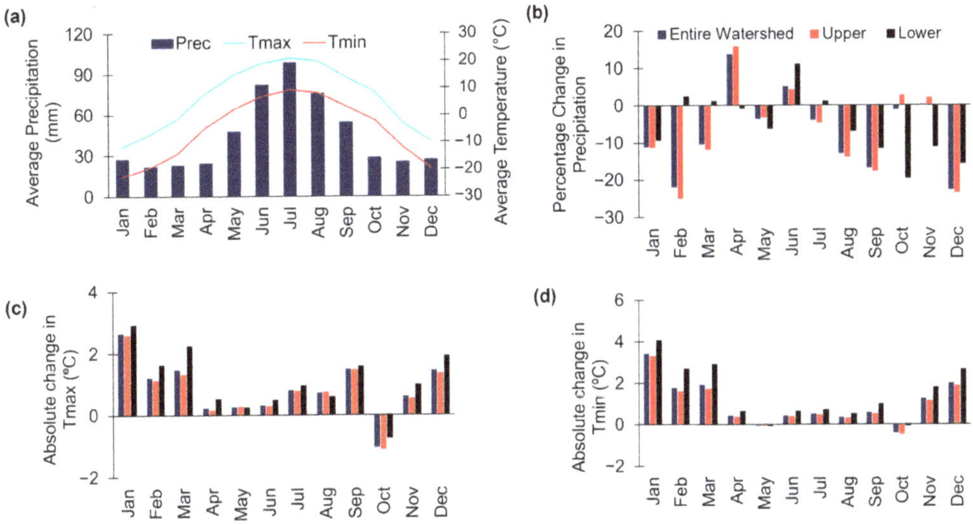

Figure 5. Seasonal distribution of precipitation, maximum and minimum temperature during the baseline period (1960–1989) for the entire watershed (**a**), percentage change in monthly precipitation (**b**), absolute change in monthly maximum (**c**) and minimum temperature (**d**) in the recent period (1990–2019) compared to baseline period for the upper, lower, and entire watershed.

However, sub-watershed analysis reveals that the precipitation changes in the lower part of the basin do not necessarily reflect the overall basin's changes. This is particularly visible during the late spring, late winter, and spring months, where the changes in lower region precipitation were opposite to that of the entire watershed and upper regions. For instance, the precipitation during October reduced slightly by 1 mm on average throughout the basin, but it declined significantly by almost 20 mm in the lower region. At the same time, precipitation increased by about 2 mm in the upper region, further highlighting the spatial variability of CC in the basin.

The seasonal changes in the two periods highlighted that temperature is increasing throughout the year, with only October experiencing a reduction (−0.74 °C and −0.06 °C in maximum and minimum temperature). The warming tendency was more profound in winter, with a rise of more than 3 °C observed during January for both maximum and minimum temperatures. Although the annual change in minimum temperature was higher than the maximum, summer months experienced a greater change in maximum temperature.

3.3. *Performance of the Hydrological Model*

The result of model calibration and validation against daily streamflow at three gauging stations is shown in Figure 6. The model performed well in capturing both high and low flows of the streamflow variability. During the calibration and validation periods, the NSE value for all three stations was greater than 0.70, indicating a very good performance [52]. The coefficient of determination was also above the acceptable range, ranging from 0.71 to 0.77 during calibration and 0.72 to 0.77 during validation. There was a slight overestimation bias at the Embarras station (negative PBIAS), while the Athabasca and Fort McMurray stations showed underestimation bias ranging from 0.34% to 11.48%. Overall, the model performed satisfactorily, and the model performance was comparable to other models previously applied in the basin [3,28,29].

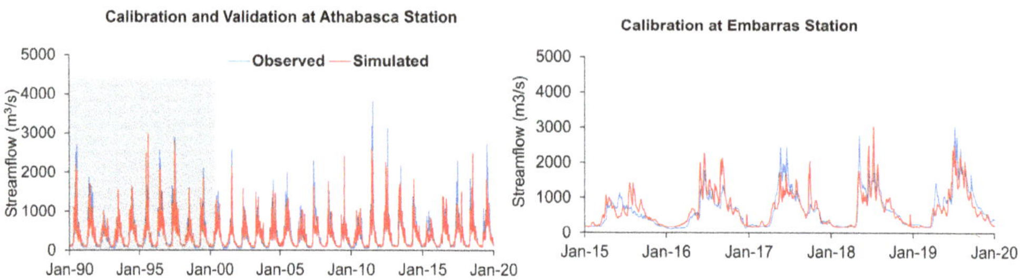

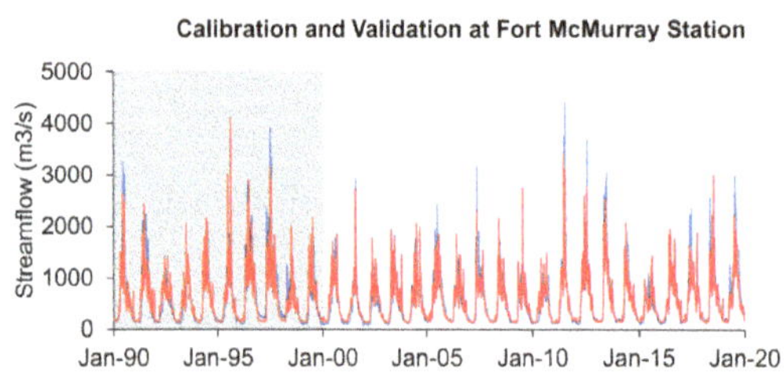

Location	Athabasca		Fort McMurray		Embarras
Parameter	Calibration	Validation	Calibration	Validation	Calibration
NSE	0.71	0.70	0.77	0.76	0.70
R^2	0.72	0.72	0.77	0.77	0.71
Pbias(%)	7.88	11.48	0.34	8.62	−1.33

Figure 6. Model performance during calibration (2000–2019) and validation (1990–1999) at Athabasca, Fort McMurray, and Embarras stream gauge stations (Calibration only from 2015–2019). Note: The grey shading indicates the validation period.

Among the selected parameters, the time constants for both interflow and baseflow reservoirs were found to be highly sensitive. These time constants directly govern the contribution towards streamflow in the form of interflow and baseflow in the linear reservoir method, resulting in higher sensitivity. Meanwhile, Manning's coefficient, degree day melting coefficient, and conductance between river and aquifer were found to be less sensitive compared to the saturated zone's parameter. The final calibrated values of parameters have been provided in Appendix C.

3.4. *Impacts of CC and LULCC on Hydrological Components*

The analysis of the annual change in hydrological indicators clearly shows that CC significantly influenced the hydrological cycle, surpassing the impact of LULCC (Table 3). For annual streamflow, CC caused an almost 40 times greater reduction than LULCC, with an 8% decrease versus a mere 0.2% decline. The upper and lower regions exhibited a more substantial impact of CC, with a higher impact observed in the former. On the other hand, seasonal variances in CC's impact on streamflow were observed (Figure 7). Spring months (March and April) experienced increased flow due to CC, while reductions were noted in other months, declining up to 17% in May. The changes due to CC were similar throughout the basin area. Meanwhile, LULCC had a clear semiannual pattern as it decreased from January to June (0.11 to 4.62%) while it increased in the latter half of calendar year (0.37 to 1.80%). The spatial variation in LULCC impact was evident, with the lower region showing inconsistent seasonal streamflow changes compared to the upper region.

Table 3. Average annual value of streamflow indicators, evapotranspiration, and groundwater recharge for all simulated scenarios with the contribution of CC, LULCC, and total change for the upper, lower, and entire watershed. Note. The values in parentheses represent percentage contributions.

Variables	Region	Mean Value in Each Scenario				Absolute Contribution of Each Factor (%)		
		S1	S2	S3	S4	CC	LULCC	Combined
Annual streamflow (m^3/s)	Entire	675.0	620.9	673.6	622.1	−52.9 (7.9%)	−1.5 (0.2%)	−54.1 (8.0%)
	Upper	574.9	527.6	573.5	528.7	−46.2 (8.0%)	−1.4 (0.2%)	−47.3 (8.2%)
	Lower	107.5	99.5	107.5	99.4	−8.1 (7.6%)	−0.1 (0.01%)	−8.0 (7.4%)
Annual 7-day minimum flow (m^3/s)	Entire	169.3	162.4	168.6	161.5	−7.8 (4.6%)	−0.6 (0.4%)	−6.8 (4.0%)
	Upper	143.1	135.9	142.5	136.4	−6.7 (4.7%)	−0.5 (0.4%)	−7.2 (5.0%)
	Lower	11.7	12.6	12.0	13.1	1.4 (12.4%)	0.3 (2.5%)	0.9 (7.6%)
Annual 7-day maximum flow (m^3/s)	Entire	2442.1	2042.7	2431.3	2088.4	−353.6 (14.5%)	−10.8 (0.5%)	−399.4 (16.4%)
	Upper	2160.3	1826.4	2151.1	1831.6	−328.7 (15.2%)	−9.2 (0.4%)	−333.9 (15.4%)
	Lower	511.3	426.1	510.0	425.4	−86.0 (16.8%)	−1.3 (0.3%)	−85.3 (16.6%)
Annual date of minimum flow (day)	Entire	63.0	58	64	57	−6.0 (9.5%)	1.0 (1.6%)	−5.0 (7.9%)
	Upper	63.0	58	64	57	−6.0 (9.5%)	1.0 (1.6%)	−5.0 (7.9%)
	Lower	172.0	173	173	171	−1.0 (0.6%)	1.0 (0.6%)	1.0(0.6%)
Annual date of maximum flow (day)	Entire	184.0	186.0	186.0	185	1.0 (0.5%)	2.0 (1.1%)	2.0 (1.1%)
	Upper	182.0	183.0	183.0	180	−2.0 (−1.1%)	1.0 (0.5%)	1.0 (0.5%)
	Lower	190.0	187.0	191.0	186	−4.0 (2.1%)	1.0 (0.5%)	−3.0 (1.6%)
Annual evapotranspiration (mm)	Entire	384	365.6	384.3	365.4	−18.6 (4.8%)	0.2 (0.0%)	−18.4 (4.8%)
	Upper	390.5	371.5	391.3	370.8	−19.7 (5.0%)	0.7 (0.2%)	−19.0 (4.9%)
	Lower	346.6	331.8	344.1	334.2	−12.4 (3.6%)	−2.5 (0.7%)	−14.8 (4.3%)
Annual groundwater recharge (mm)	Entire	144.3	133.8	143.7	134.3	−10.0 (6.9%)	−0.6 (0.4%)	−10.5 (7.3%)
	Upper	148.3	136.5	147.4	137.3	−11.0 (7.4%)	−0.9 (0.6%)	−11.8 (8.0%)
	Lower	121.4	118.2	122.5	117.1	−4.3 (3.6%)	1.1 (0.9%)	−3.2 (2.7%)

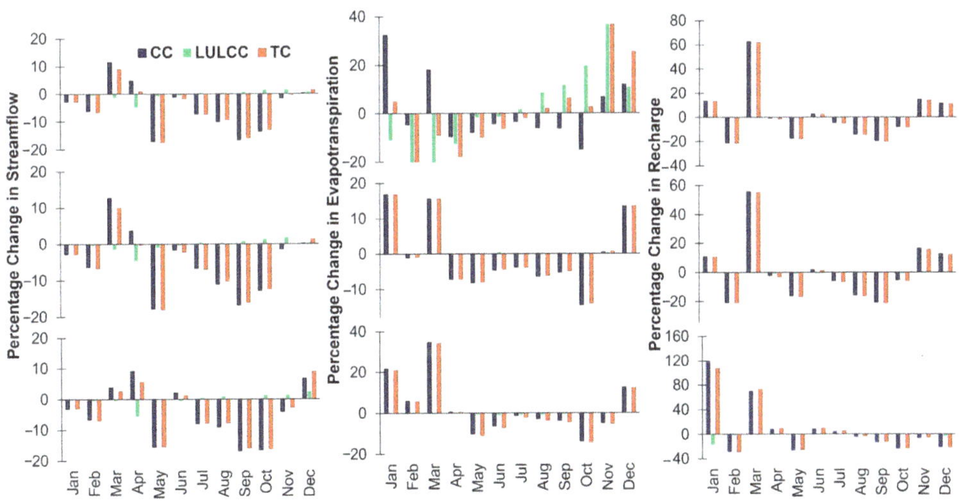

Figure 7. Percentage change under CC, under LULCC, and under both scenarios for monthly streamflow, evapotranspiration, and groundwater recharge for the entire watershed (1st row), upper (2nd row) and lower regions (3rd row).

Similar patterns are observed in annual extremes of streamflow. CC has contributed to a significant reduction in both the annual 7-day maximum (14.5%) and minimum (4.6%) flows, while LULCC had minimal impact. However, minimum flow in the lower region showed an opposite pattern, increasing due to CC and LULCC by approximately 1%. Notably, the reduction in minimum flow in the lower region due to LULCC was proportionally higher compared to other LULCC-induced changes. In the case of extreme timings, low flow was found to be shifting earlier due to CC (6 days) while it was delayed by one day due to LULCC. As a result, the low flow date was found to be shifted by five days ahead due to both stressors. Likewise, for maximum flow timing date was delayed by CC (1 day) and LULCC (2 days). However, spatial variation was observed in the case of CC's impact on high flow timings, as it was found to be delayed in upper (2 days) and lower (4 days) regions.

In contrast to streamflow, CC and LULCC had counteracting impacts on annual evapotranspiration. CC reduced evapotranspiration for the entire watershed by about 4.9%, while LULCC had a slight increasing effect of 0.1%. The sub-watershed analysis, however, shows that the changes observed are not consistent throughout the basin. Evapotranspiration has been found to be increasing in the upper region (0.2%) but declining in the lower region (0.7%) due to LULCC. The seasonal pattern of evapotranspiration follows a similar trend to precipitation, with higher rates during summer and lower rates during winter. A notable observation is the overall increase in evapotranspiration under CC from late fall to early spring, with the most significant percentage increment occurring in January, approximately 32%. On the other hand, LULCC caused a reduction in evapotranspiration in winter and spring months, while it increased in summer and fall. Sub-watershed analysis highlighted differences between the upper and lower regions' responses to LULCC impact, with evapotranspiration increasing throughout the year in the upper region and decreasing throughout the year in the lower region.

Both CC (6.9%) and LULCC (0.4%) caused a reduction of groundwater recharge for the entire watershed. The reducing impact of CC was consistent throughout the basin. As seen for other hydrological processes, the lower region had varying responses to LULCC as recharge was found to have increased by 0.9%. Seasonal changes showed reduced groundwater recharge due to CC during summer and fall in the upper region, which was consistent with the entire watershed. At the same time, CC contributed to increased recharge during late fall to early winter (12% to 14% per month) in the entire watershed. However, as with the case of other hydrological components, the lower region's response to CC showed a few differences with the upper region. For instance, the recharge in the lower region mostly increased in summer as opposed to declining summer recharge in the upper region due to CC. Meanwhile, LULCC resulted in decreased recharge (0.2% to 0.8% per month) throughout the year in the upper region, which was similar for the entire watershed. In contrast, the recharge rate in the lower region increased due to LULCC except during winter.

4. Discussion

4.1. Impact of CC and LULCC

CC has been found to be highly dominant in inducing the observed hydrological changes in the basin. This finding of a higher impact of CC aligns with various previous results from other regions around the globe [53–56]. For instance, a study by Ahmed et al. [54] found that CC caused an almost 98% increment in annual runoff while LULCC only contributed a 2% rise at the Yangtze River Basin, China. Similarly, Tan et al. [15] reported that CC had a higher impact on both annual streamflow and evaporation in the Johor River Basin, Malaysia. In our study, the primary influence of CC on streamflow has predominantly been caused by decreased precipitation, resulting in less water availability for other hydrological processes. This led to a reduction of flow in rivers and streams. Likewise, less water availability for evapotranspiration might have resulted in the decline of annual evapotranspiration. These findings are consistent with other studies in the ARB highlighting precipitation as the major driver of change [20,23,27]. The study by Bawden et al. [20] used a multiple linear regression (MLR) model to analyze the relationship between runoff, precipitation, and temperature in the ARB. They found that warm season precipitation trends explained over twice as much variation in runoff as temperature trends, with R^2 values of 62% and 30%, respectively. They also identified a significant correlation between precipitation and streamflow trends during the warm season (March–October) from 1976 to 2010, at a 10% significance level.

The seasonal impact of CC, however, shows that the impact is quite complex and dynamic in nature. The findings suggest that the governing climate variable varies with season. During the late winter and early spring, it has been seen that the temperature drives the hydrological processes as snowmelt forms a significant portion of the streamflow. In contrast, precipitation is the driving factor during the other periods of the year. This seasonal variability explains the increase in streamflow due to higher snowmelt from warming temperatures in March and April. These months are generally when spring freshet, a key event of cold region hydrology, is initiated in the watershed. The increasing annual 7-day minimum flow observed in the lower region can also be linked to this effect. As the lower region is warming at a higher rate, its impact has been directly reflected in increased low flows, which usually occur during winter and early spring. Likewise, the diminishing precipitation during the relatively warm season has direct implications for the streamflow generation, resulting in reduced flow. Furthermore, the warming tendency during late winter and early spring has likely contributed to increased rain-on-snow events. These events involve rainfall directly onto the snowpack, creating more overland flow, resulting in greater overland flow and consequently augmenting water availability in the river. Rising winter and spring temperatures also have cascading effects on summer at a lower elevation. The diminished snowpack due to snowpack sublimation in winter leads to reduced water storage, ultimately limiting the supply to the river in the summer.

The seasonal distribution and the governing mechanism of evapotranspiration are similar to those of streamflow. The reduced precipitation during late spring and fall has likely resulted in the reduction of ET during these months. However, during the winter and early spring, the increased temperature has accelerated the snowmelt, which in turn has allowed for increased evapotranspiration during this period. A similar finding of a possible increment in winter evapotranspiration due to rising temperatures in the future was also reported by Farjad et al. [10] in the Elbow watershed, Canada.

This study has shown that the impacts of LULCC in the hydrological processes have been manifested through reduced streamflow, alteration of peak flows, increased evapotranspiration, and reduction of groundwater recharge. During the specified period, there was an observed increase in the coverage of savannas and forests, resulting in a notable expansion of canopy areas conducive to evapotranspiration. This phenomenon likely contributed to the overall rise in annual evapotranspiration rates, while concurrently leading to a reduction in both annual streamflow and recharge. Despite minor increments in non-vegetated and urban lands, these changes were insufficient to counteract the prevailing trends. Notably, in the lower region, a significant decrease in deciduous forests was noted, partially explaining the increased evapotranspiration recharge in that area. Moreover, the increment in grasslands, non-vegetated lands, and developed areas introduced variable responses in terms of evapotranspiration and recharge dynamics. The landscape of the ARB is still undergoing changes at a decent rate [57]. The upper region of the basin experienced a reduction in agricultural lands and increased rural and urban settlements from 2011 to 2022. These changes are expected to continue in the future, which might have a significant impact on the basin's hydrology. Although the annual impact might not be significant, it can still significantly alter the timing and frequency of extreme flows. Moreover, changes in vegetation properties can have implications for the seasonal distribution of streamflow as well as other important hydrological processes.

The integrated impact of both stressors has resulted in seasonal disturbance in the hydrological cycle. These induced seasonal shifts have significant implications for aquatic ecosystems and water availability for various industrial activities, such as utilities, mining, and bitumen extraction in the lower part of the basin. The flow alteration in the ARB has cascading impacts on the downstream Peace-Athabasca Delta (PAD), which provides habitat for 45 mammal species, 214 bird species, and 20 fish species [58]. The flow shifts in the Athabasca River, alongside other nearby rivers, influence the formation of perched basins and their periodic flooding, processes vital for the survival of species such as muskrats, waterfowl, and various other wildlife. The lower Athabasca River supports year-round spawning, rearing, and feeding habitats for various fish species [35], with any changes in flow potentially impacting their distribution. For instance, a study by Morales-Marin et al. [59] found that reduced streamflow results in lower water velocities and depths, creating sub-optimal conditions for the Athabasca rainbow trout (ART), a species classified as "at risk".

Water consumption from the Athabasca River has surged significantly in recent decades [19], posing potential challenges for its users. This is particularly critical for the oil sands industry, which depends heavily on the river, accounting for over 70% of the total water usage [35]. To address this, a framework has been implemented to regulate cumulative water withdrawals, ensuring a balance between human and environmental demands [35]. The framework establishes weekly management triggers and withdrawal limits based on the seasonal flow variations. However, as climate and land use changes increasingly affect the flow regime, updates to the framework may be necessary to balance social, environmental, and economic priorities in the future. Different climate impact assessment studies carried out in the ARB have already reported that the flow regime is likely to experience shifts in the future in the form of increased annual flow, higher seasonal fluctuations, shifts in hydrography, and altered peak flow frequency, snow coverage, glacier contribution, and more which might result in spatiotemporal heterogeneity in water availability leading to water security challenges [3,28,29,60,61]. With anthropogenic activities expected to further alter the landscape properties, their integrated impact might lead to more uncertainty in the spatiotemporal water availability in the basin.

4.2. Uncertainties

Hydrological models have been widely recognized as significant contributors to uncertainties when simulating hydrologic processes. These uncertainties can arise from multiple sources, such as input data, parameter selection, the degree of process simplification, calibration, and structural uncertainty [62,63]. Furthermore, studies have observed that there are additional uncertainties associated with modeling at high altitudes [64] as well as the inadequate accounting of river ice cover dynamics [29] in most hydrological models. The hybrid climate datasets utilized in this study were developed using the REFerence Reliability Evaluation System (REFRES), which produces high-resolution hybrid climate data by mathematically ranking multiple datasets [37]. While these hybrid datasets offer extensive spatial coverage, they may not be as reliable as ground-based observational data. The authors also highlighted that the reliability of the climate data diminishes at higher altitudes, potentially due to factors such as topographic variability, orographic effects, and the limited availability of ground-based measurements.

Additionally, the selection of only two LULC maps, each representing one period of analysis, has essentially limited the temporal representation of the watershed's vegetation property. This limitation suggests the possibility of underestimating or overestimating

LULC features, leading to uncertainties in the simulated outcomes [2]. The model calibration was carried out by fine tuning only a few selected parameters, and ignoring other parameters may also induce uncertainties in the modeling results. Moreover, the calibration and validation of the model against daily streamflow are likely to introduce uncertainties in the model performance inherently. The calibration against daily streamflow may particularly induce uncertainties in a cold climate region like the ARB. During the winter season, significant ice coverage occurs in the river, which is not sufficiently addressed in MIKE SHE, potentially contributing to uncertainty. Additionally, the presence of ice coverage may have impacted the accuracy of the streamflow data used for calibration.

Future studies can address these uncertainties by adopting a more robust modeling framework. Employing physics-based methods to better capture snowmelt and thaw processes, incorporating river ice dynamics, and providing a more detailed representation of the underlying saturated zone could enhance model performance. Additionally, calibrating the model using multiple variables, such as soil moisture and groundwater levels, would improve its reliability and accuracy.

5. Conclusions

This study assessed the individual and integrated effects of CC and LULCC on the hydrological system of the ARB using the MIKE SHE-MIKE Hydro River model. The results showed that MIKE SHE can adequately represent hydrological conditions of cold climate regions like ARB and the model could be applied to other similar climatic regions with adequate calibration and validation. The findings clearly indicate that there have been considerable changes in climatic conditions and LULCC patterns in the ARB. These changes have altered the basin's hydrological system in which the CC impact emerged as a dominant factor of change.

The evaluation of CC revealed a warming and drying trend in the region, with the lower part of the basin witnessing a greater rise in temperature. The annual maximum and minimum temperatures have increased by 0.85 °C and 1.01 °C, while precipitation has decreased by 33 mm in the recent thirty-year period compared to the earlier baseline from 1960 to 1989. The findings suggest that CC alone would have been responsible for reducing annual flow along with its extremes, annual actual evapotranspiration, annual recharge, and annual baseflow. The hydrological response of the basin to climatic factors was found to be dynamic as temperature fluctuations had greater implications during winter, while precipitation fluctuations governed climatic impacts in other seasons. This also led to seasonal shifts in the streamflow, overland flow evapotranspiration, recharge, and baseflow of the basin.

At the same time, the basin was subjected to increasing anthropogenic activities that led to an increment in non-vegetated lands, forested areas, and developed areas at the expense of grasslands and shrublands. The modeling result suggests that these changes induced an increment in annual evapotranspiration and recharge. Sub-watershed analysis in the upper and lower regions highlighted the heterogeneity in the basin's response to these stressors, further complicated by seasonal variations in the dominant stressor. The lower region had different LULCC patterns, and this was also reflected in its hydrological response. These findings emphasize the complexity of the basin's hydrological system. The study demonstrated that both CC and LULCC have affected the hydrological processes of the Athabasca River Basin in various ways. Neglecting either of these factors in assessing the basin's current or future hydrological state could lead to an overestimation or underestimation of the hydrological response. Consequently, it is crucial to consider the potential changes in the hydrological system due to both stressors when formulating water management practices and policies for the foreseeable future.

Author Contributions: Conceptualization, S.A., B.F. and M.S.B.; methodology, S.A. and B.F.; software, S.A.; validation, S.A., B.F. and Q.K.H.; formal analysis, S.A.; investigation, S.A.; resources, B.F. and A.G.; data curation, S.A.; writing—original draft preparation, S.A.; writing—review and editing, D.K., A.G. and Q.K.H.; visualization, S.A.; supervision, M.S.B. All authors have read and agreed to the published version of the manuscript.

Funding: This work was funded under the Oil Sands Monitoring (OSM) Program. It is independent of any position of the OSM Program.

Data Availability Statement: The datasets generated during and/or analyzed during the current study are available from the corresponding author on reasonable request.

Conflicts of Interest: The authors declare no conflicts of interest.

Appendix A

Validation of Hargreaves–Samani (HS) Method

The daily potential evapotranspiration (ET) obtained from the HS method was compared with actual observed values at four locations from April to September 2019, and the root mean square error ranged from 0.73 to 0.81 as seen in Figure A1. Meanwhile, from October to March, calculations of ET at climate stations are based on the Penman–Montieth method, and validation with full annual data also showed satisfactory performance of the HS method (Figure A2).

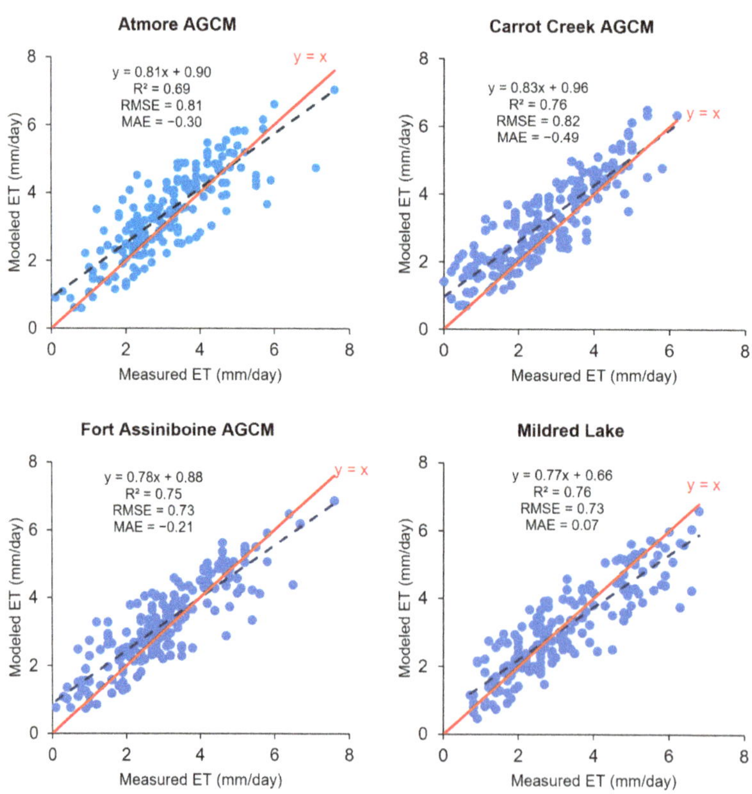

Figure A1. Validation of Hargreaves–Samani Method for Potential Evapotranspiration (ET) at four locations within Athabasca River Basin (April to September 2019). The dashed line represents the best fit line.

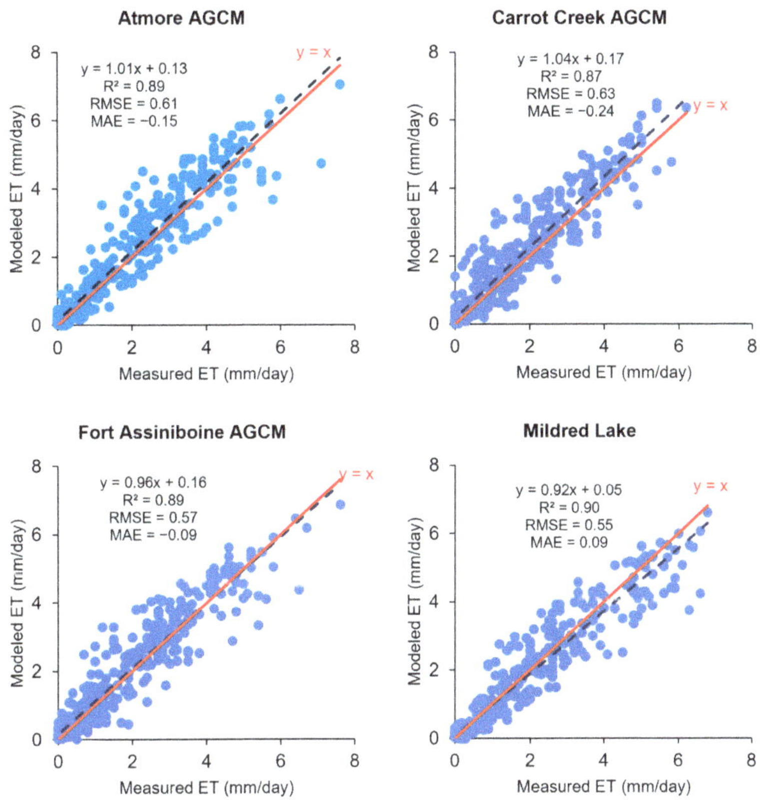

Figure A2. Validation of Hargreaves–Samani Method for Potential Evapotranspiration (ET) at four locations within Athabasca River Basin (2019). The dashed line represents the best fit line.

Appendix B

Population Changes in Athabasca River Basin

The population of the ARB was evaluated based on 2 improvement districts, 2 specialized municipalities, 10 towns, and 16 municipal districts lying fully and partially in the basin. The data were obtained from census data of the federal government [65] and municipal data from Alberta Municipal Affairs [66]. The results show higher growth in the human population from 2001 to 2011.

Table A1. Population changes in ARB from 1981 to 2021.

Year	Population	Percentage Change per 5 Years (%)
1981	167,261	--
1986	177,604	6.18
1991	187,590	5.62
1996	204,343	8.93
2001	230,356	12.73
2006	274,462	19.15
2011	307,958	12.20
2016	311,811	1.25
2021	304,604	−2.31

Appendix C

Calibrated Parameters

Table A2. Time varying degree day snow melting coefficient for each year.

Date	Degree Day Coefficient (mm/Deg C/day)	Date	Degree Day Coefficient (mm/Deg C/day)
1-Jan	0.2	10-Jul	3.96
11-Jan	0.28	20-Jul	3.92
21-Jan	0.36	30-Jul	3.88
31-Jan	0.44	9-Aug	3.84
10-Feb	0.52	19-Aug	3.8
20-Feb	0.6	29-Aug	2
2-Mar	0.68	8-Sep	1.16
12-Mar	0.76	18-Sep	1.08
22-Mar	0.84	28-Sep	1
1-Apr	0.92	8-Oct	0.92
11-Apr	1	18-Oct	0.84
21-Apr	1.08	28-Oct	0.76
1-May	1.16	7-Nov	0.68
11-May	2	17-Nov	0.6
21-May	4.8	27-Nov	0.52
31-May	5	7-Dec	0.44
10-Jun	4.8	17-Dec	0.36
20-Jun	4.6	27-Dec	0.2
30-Jun	4		

Table A3. Manning's M ($m^{1/3}/s$) for different land use classes.

Land Use Class	Range	Calibrated Value
Water	20.00–40.00	25.04
Developed Areas	80.00–100.00	90.90
Agriculture	20.00–40.00	28.57
Vegetated Open Land	20.00–40.00	33.33
Broadleaf Forests	6.25–10.00	10.00
Coniferous Forests	6.25–10.00	10.00
Mixed Forests	6.25–10.00	10.00
Wetlands	6.00–22.00	8.33
Rock/Rubble	33.00–44.00	40.00

Table A4. Parameters to define saturated zone.

Parameter	Value
Interflow Time Constant	16 days
Baseflow Time Constant for Reservoir 1	60 days
Baseflow Time Constant for Reservoir 2	3650 days

References

1. Rast, M.; Johannessen, J.; Mauser, W. Review of Understanding of Earth's Hydrological Cycle: Observations, Theory and Modelling. *Surv. Geophys.* **2014**, *35*, 491–513. [CrossRef]
2. Ji, L.; Duan, K. What Is the Main Driving Force of Hydrological Cycle Variations in the Semiarid and Semi-Humid Weihe River Basin, China? *Sci. Total Environ.* **2019**, *684*, 254–264. [CrossRef] [PubMed]
3. Dibike, Y.; Eum, H.-I.; Coulibaly, P.; Hartmann, J. Projected Changes in the Frequency of Peak Flows along the Athabasca River: Sensitivity of Results to Statistical Methods of Analysis. *Climate* **2019**, *7*, 88. [CrossRef]
4. Leong, D.N.S.; Donner, S.D. Climate Change Impacts on Streamflow Availability for the Athabasca Oil Sands. *Clim. Change* **2015**, *133*, 651–663. [CrossRef]

5. Saha, G.C.; Li, J.; Thring, R.W.; Hirshfield, F.; Paul, S.S. Temporal Dynamics of Groundwater-Surface Water Interaction under the Effects of Climate Change: A Case Study in the Kiskatinaw River Watershed, Canada. *J. Hydrol.* **2017**, *551*, 440–452. [CrossRef]
6. Scibek, J.; Allen, D.M.; Cannon, A.J.; Whitfield, P.H. Groundwater–Surface Water Interaction under Scenarios of Climate Change Using a High-Resolution Transient Groundwater Model. *J. Hydrol.* **2007**, *333*, 165–181. [CrossRef]
7. Zhou, Q.; Leng, G.; Su, J.; Ren, Y. Comparison of Urbanization and Climate Change Impacts on Urban Flood Volumes: Importance of Urban Planning and Drainage Adaptation. *Sci. Total Environ.* **2019**, *658*, 24–33. [CrossRef]
8. Gashaw, T.; Tulu, T.; Argaw, M.; Worqlul, A.W. Modeling the Hydrological Impacts of Land Use/Land Cover Changes in the Andassa Watershed, Blue Nile Basin, Ethiopia. *Sci. Total Environ.* **2018**, *619–620*, 1394–1408. [CrossRef]
9. Kayitesi, N.M.; Guzha, A.C.; Mariethoz, G. Impacts of Land Use Land Cover Change and Climate Change on River Hydro-Morphology- a Review of Research Studies in Tropical Regions. *J. Hydrol.* **2022**, *615*, 128702. [CrossRef]
10. Farjad, B.; Gupta, A.; Razavi, S.; Faramarzi, M.; Marceau, D.J. An Integrated Modelling System to Predict Hydrological Processes under Climate and Land-Use/Cover Change Scenarios. *Water* **2017**, *9*, 767. [CrossRef]
11. Lamichhane, S.; Shakya, N.M. Integrated Assessment of Climate Change and Land Use Change Impacts on Hydrology in the Kathmandu Valley Watershed, Central Nepal. *Water* **2019**, *11*, 2059. [CrossRef]
12. Marhaento, H.; Booij, M.J.; Hoekstra, A.Y. Hydrological Response to Future Land-Use Change and Climate Change in a Tropical Catchment. *Hydrol. Sci. J.* **2018**, *63*, 1368–1385. [CrossRef]
13. Dey, P.; Mishra, A. Separating the Impacts of Climate Change and Human Activities on Streamflow: A Review of Methodologies and Critical Assumptions. *J. Hydrol.* **2017**, *548*, 278–290. [CrossRef]
14. Jiang, C.; Xiong, L.; Wang, D.; Liu, P.; Guo, S.; Xu, C.-Y. Separating the Impacts of Climate Change and Human Activities on Runoff Using the Budyko-Type Equations with Time-Varying Parameters. *J. Hydrol.* **2015**, *522*, 326–338. [CrossRef]
15. Tan, M.L.; Ibrahim, A.L.; Yusop, Z.; Duan, Z.; Ling, L. Impacts of Land-Use and Climate Variability on Hydrological Components in the Johor River Basin, Malaysia. *Hydrol. Sci. J.* **2015**, *60*, 873–889. [CrossRef]
16. Marhaento, H.; Booij, M.J.; Ahmed, N. Quantifying Relative Contribution of Land Use Change and Climate Change to Streamflow Alteration in the Bengawan Solo River, Indonesia. *Hydrol. Sci. J.* **2021**, *66*, 1059–1068. [CrossRef]
17. Ramezani, M.R.; Helfer, F.; Yu, B. Individual and Combined Impacts of Urbanization and Climate Change on Catchment Runoff in Southeast Queensland, Australia. *Sci. Total Environ.* **2023**, *861*, 160528. [CrossRef]
18. Zeng, S.; Xia, J.; Du, H. Separating the Effects of Climate Change and Human Activities on Runoff over Different Time Scales in the Zhang River Basin. *Stoch. Environ. Res. Risk Assess.* **2014**, *28*, 401–413. [CrossRef]
19. Aryal, S.; Babel, M.S.; Gupta, A.; Farjad, B.; Khadka, D.; Hassan, Q.K. Assessment of Hydrological Baseline Condition and Its Alteration in Athabasca River Basin, Canada. *J. Hydrol. Reg. Stud.* **2024**, *53*, 101805. [CrossRef]
20. Bawden, A.J.; Linton, H.C.; Burn, D.H.; Prowse, T.D. A Spatiotemporal Analysis of Hydrological Trends and Variability in the Athabasca River Region, Canada. *J. Hydrol.* **2014**, *509*, 333–342. [CrossRef]
21. Peters, D.L.; Watt, D.; Devito, K.; Monk, W.A.; Shrestha, R.R.; Baird, D.J. Changes in Geographical Runoff Generation in Regions Affected by Climate and Resource Development: A Case Study of the Athabasca River. *J. Hydrol. Reg. Stud.* **2022**, *39*, 100981. [CrossRef]
22. Peters, D.L.; Atkinson, D.; Monk, W.A.; Tenenbaum, D.E.; Baird, D.J. A Multi-scale Hydroclimatic Analysis of Runoff Generation in the Athabasca River, Western Canada. *Hydrol. Process.* **2013**, *27*, 1915–1934. [CrossRef]
23. Zaghloul, M.S.; Ghaderpour, E.; Dastour, H.; Farjad, B.; Gupta, A.; Eum, H.; Achari, G.; Hassan, Q.K. Long Term Trend Analysis of River Flow and Climate in Northern Canada. *Hydrology* **2022**, *9*, 197. [CrossRef]
24. Chen, Z.; Grasby, S.E. Reconstructing River Discharge Trends from Climate Variables and Prediction of Future Trends. *J. Hydrol.* **2014**, *511*, 267–278. [CrossRef]
25. Rood, S.B.; Stupple, G.W.; Gill, K.M. Century-long Records Reveal Slight, Ecoregion-localized Changes in Athabasca River Flows. *Hydrol. Process.* **2015**, *29*, 805–816. [CrossRef]
26. Sauchyn, D.J.; St-Jacques, J.-M.; Luckman, B.H. Long-Term Reliability of the Athabasca River (Alberta, Canada) as the Water Source for Oil Sands Mining. *Proc. Natl. Acad. Sci. USA* **2015**, *112*, 12621–12626. [CrossRef]
27. Ghaderpour, E.; Zaghloul, M.S.; Dastour, H.; Gupta, A.; Achari, G.; Hassan, Q.K. Least-Squares Triple Cross-Wavelet and Multivariate Regression Analyses of Climate and River Flow in the Athabasca River Basin. *J. Hydrometeorol.* **2023**, *24*, 1883–1900. [CrossRef]
28. Shrestha, N.K.; Du, X.; Wang, J. Assessing Climate Change Impacts on Fresh Water Resources of the Athabasca River Basin, Canada. *Sci. Total Environ.* **2017**, *601–602*, 425–440. [CrossRef]
29. Eum, H.-I.; Dibike, Y.; Prowse, T. Climate-Induced Alteration of Hydrologic Indicators in the Athabasca River Basin, Alberta, Canada. *J. Hydrol.* **2017**, *544*, 327–342. [CrossRef]
30. Hwang, H.; Park, Y.; Sudicky, E.A.; Berg, S.J.; McLaughlin, R.; Jones, J.P. Understanding the Water Balance Paradox in the Athabasca River Basin, Canada. *Hydrol. Process.* **2018**, *32*, 729–746. [CrossRef]

31. Kerkhoven, E.; Gan, T.Y. Differences in the Potential Hydrologic Impact of Climate Change to the Athabasca and Fraser River Basins of Canada with and without Considering Shifts in Vegetation Patterns Induced by Climate Change. *J. Hydrometeorol.* **2013**, *14*, 963–976. [CrossRef]
32. DHI MIKE SHE: User Guide and Reference Manual. 2024. Available online: https://manuals.mikepoweredbydhi.help/latest/Water_Resources/MIKE_SHE_Print.pdf (accessed on 15 January 2024).
33. Keilholz, P.; Disse, M.; Halik, Ü. Effects of Land Use and Climate Change on Groundwater and Ecosystems at the Middle Reaches of the Tarim River Using the MIKE SHE Integrated Hydrological Model. *Water* **2015**, *7*, 3040–3056. [CrossRef]
34. Ramteke, G.; Singh, R.; Chatterjee, C. Assessing Impacts of Conservation Measures on Watershed Hydrology Using MIKE SHE Model in the Face of Climate Change. *Water Resour. Manag.* **2020**, *34*, 4233–4252. [CrossRef]
35. Government of Alberta. *Lower Athabasca Region: Surface Water Quantitiy Management Framework for the Lower Athabasca River*; Government of Alberta: Edmonton, AB, Canada, 2015; ISBN 978-1-4601-2173-3.
36. WaterSMART Solutions Ltd. *A Roadmap for Sustainable Water Management in the Athabasca River Basin*; WaterSMART Solutions Ltd. for Alberta Innovates: Calgary, AB, Canada, 2018; p. 247.
37. Eum, H.-I.; Gupta, A. Hybrid Climate Datasets from a Climate Data Evaluation System and Their Impacts on Hydrologic Simulations for the Athabasca River Basin in Canada. *Hydrol. Earth Syst. Sci.* **2019**, *23*, 5151–5173. [CrossRef]
38. Hargreaves, G.H.; Samani, Z.A. Reference Crop Evapotranspiration from Temperature. *Appl. Eng. Agric.* **1985**, *1*, 96–99. [CrossRef]
39. Wijesekara, G.N.; Farjad, B.; Gupta, A.; Qiao, Y.; Delaney, P.; Marceau, D.J. A Comprehensive Land-Use/Hydrological Modeling System for Scenario Simulations in the Elbow River Watershed, Alberta, Canada. *Environ. Manag.* **2014**, *53*, 357–381. [CrossRef]
40. Dastour, H.; Ghaderpour, E.; Zaghloul, M.S.; Farjad, B.; Gupta, A.; Eum, H.; Achari, G.; Hassan, Q.K. Wavelet-Based Spatiotemporal Analyses of Climate and Vegetation for the Athabasca River Basin in Canada. *Int. J. Appl. Earth Obs. Geoinf.* **2022**, *114*, 103044. [CrossRef]
41. Kalyanapu, A.J.; Burian, S.J.; McPherson, T.N. Effect of Land Use-Based Surface Roughness on Hydrologic Model Output. *J. Spat. Hydrol.* **2009**, *9*, 51.
42. Pradhan, N.R.; Downer, C.W.; Marchenko, S. Catchment Hydrological Modeling with Soil Thermal Dynamics during Seasonal Freeze-Thaw Cycles. *Water* **2019**, *11*, 116. [CrossRef]
43. Thompson, J.R.; Green, A.J.; Kingston, D.G.; Gosling, S.N. Assessment of Uncertainty in River Flow Projections for the Mekong River Using Multiple GCMs and Hydrological Models. *J. Hydrol.* **2013**, *486*, 1–30. [CrossRef]
44. Ma, L.; He, C.; Bian, H.; Sheng, L. MIKE SHE Modeling of Ecohydrological Processes: Merits, Applications, and Challenges. *Ecol. Eng.* **2016**, *96*, 137–149. [CrossRef]
45. Turgeon, F.; Larocque, M.; Meyzonnat, G.; Dorner, S.; Bourgault, M.-A. Examining the Challenges of Simulating Surface Water–Groundwater Interactions in a Post-Glacial Environment. *Can. Water Resour. J./Rev. Can. Des Ressour. Hydr.* **2018**, *43*, 262–280. [CrossRef]
46. Waseem, M.; Kachholz, F.; Klehr, W.; Tränckner, J. Suitability of a Coupled Hydrologic and Hydraulic Model to Simulate Surface Water and Groundwater Hydrology in a Typical North-Eastern Germany Lowland Catchment. *Appl. Sci.* **2020**, *10*, 1281. [CrossRef]
47. Dastour, H.; Hassan, Q.K. Quantifying the Influence of Climate Variables on Vegetation Through Remote Sensing and Multi-Dimensional Data Analysis. *Earth Syst. Environ.* **2024**, *8*, 165–180. [CrossRef]
48. Downing, D.J.; Pettapiece, W.W. *Natural Regions and Subregions of Alberta*; Natural Regions Committee: Edmonton, AB, Canada, 2006; ISBN 978-0-7785-4572-9.
49. Environment Canada. *Scientific Review for the Identification of Critical Habitat for Woodland Caribou (Rangifer Tarandus Caribou), Boreal Population, in Canada*; Environment Canada: Ottawa, ON, Canada, 2008.
50. Biodiversity in the Boreal Forest-Regional Aquatics Monitoring Program (RAMP). Available online: http://www.ramp-alberta.org/river/boreal/alberta.aspx (accessed on 3 December 2024).
51. Ahmed, M.R.; Hassan, Q.K.; Abdollahi, M.; Gupta, A. Introducing a New Remote Sensing-Based Model for Forecasting Forest Fire Danger Conditions at a Four-Day Scale. *Remote Sens.* **2019**, *11*, 2101. [CrossRef]
52. Moriasi, D.N.; Arnold, J.G.; Van Liew, M.W.; Bingner, R.L.; Harmel, R.D.; Veith, T.L. Model Evaluation Guidelines for Systematic Quantification of Accuracy in Watershed Simulations. *Trans. ASABE* **2007**, *50*, 885–900. [CrossRef]
53. Bi, H.; Liu, B.; Wu, J.; Yun, L.; Chen, Z.; Cui, Z. Effects of Precipitation and Landuse on Runoff during the Past 50 Years in a Typical Watershed in Loess Plateau, China. *Int. J. Sediment Res.* **2009**, *24*, 352–364. [CrossRef]
54. Ahmed, N.; Wang, G.; Booij, M.J.; Xiangyang, S.; Hussain, F.; Nabi, G. Separation of the Impact of Landuse/Landcover Change and Climate Change on Runoff in the Upstream Area of the Yangtze River, China. *Water Resour. Manag.* **2022**, *36*, 181–201. [CrossRef]
55. Akbari, S.; Reddy, M.J. Change Detection and Attribution of Flow Regime: A Case Study of Allegheny River Catchment, PA (US). *Sci. Total Environ.* **2019**, *662*, 192–204. [CrossRef]

56. Tong, S.T.Y.; Yang, H.; Chen, H.; Yang, J.Y. Hydrologic Impacts of Climate Change and Urbanization in the Las Vegas Wash Watershed, Nevada. *J. Water Clim. Change* **2016**, *7*, 598–620. [CrossRef]
57. Government of Alberta. *Annual Report 2022: Land Use Changes in Alberta*; Government of Alberta: Edmonton, AB, Canada, 2023.
58. Peace-Athabasca Delta Ecological Monitoring Program (PADEMP). Available online: https://web.archive.org/web/20190718032140/pademp.com/ (accessed on 25 December 2024).
59. Morales-Marín, L.A.; Rokaya, P.; Sanyal, P.R.; Sereda, J.; Lindenschmidt, K.E. Changes in Streamflow and Water Temperature Affect Fish Habitat in the Athabasca River Basin in the Context of Climate Change. *Ecol. Model.* **2019**, *407*, 108718. [CrossRef]
60. Chernos, M.; MacDonald, R.J.; Nemeth, M.W.; Craig, J.R. Current and Future Projections of Glacier Contribution to Streamflow in the Upper Athabasca River Basin. *Can. Water Resour. J./Rev. Can. Des Ressour. Hydr.* **2020**, *45*, 324–344. [CrossRef]
61. Dibike, Y.; Prowse, T.; Bonsal, B.; O'Neil, H. Implications of Future Climate on Water Availability in the Western Canadian River Basins. *Int. J. Climatol.* **2017**, *37*, 3247–3263. [CrossRef]
62. Karlsson, I.B.; Sonnenborg, T.O.; Refsgaard, J.C.; Trolle, D.; Børgesen, C.D.; Olesen, J.E.; Jeppesen, E.; Jensen, K.H. Combined Effects of Climate Models, Hydrological Model Structures and Land Use Scenarios on Hydrological Impacts of Climate Change. *J. Hydrol.* **2016**, *535*, 301–317. [CrossRef]
63. Moges, E.; Demissie, Y.; Larsen, L.; Yassin, F. Review: Sources of Hydrological Model Uncertainties and Advances in Their Analysis. *Water* **2020**, *13*, 28. [CrossRef]
64. Orth, R.; Staudinger, M.; Seneviratne, S.I.; Seibert, J.; Zappa, M. Does Model Performance Improve with Complexity? A Case Study with Three Hydrological Models. *J. Hydrol.* **2015**, *523*, 147–159. [CrossRef]
65. Government of Canada, S.C. Census Datasets. Available online: https://www12.statcan.gc.ca/datasets/index-eng.cfm?Temporal=2021 (accessed on 26 March 2024).
66. Alberta's Municipal Population Total from 1960 to 2023-Open Government. Available online: https://open.alberta.ca/publications/alberta-municipal-population-total (accessed on 26 March 2024).

hydrology

MDPI

Article

Development of Transient Hydrodynamic and Hydrodispesive Models in Semi-Arid Environments

Samir Hakimi [1,*], Mohamed Abdelbaset Hessane [1], Mohammed Bahir [2], Turki Kh. Faraj [3] and Paula M. Carreira [4]

1 Department of Geology, Faculty of Sciences Dhar Mahraz, Sidi Mohamed Ben Abdellah University, Fez 30000, Morocco; abdelbaset.hessane@usmba.ac.ma
2 High Energy and Astrophysics Laboratory, Faculty of Sciences Semlalia, Cadi Ayyad University, Marrakech 40000, Morocco; bahir@uca.ac.ma
3 Depatement of Soil Science, College of Food and Agricultural Sciences, King Saud University, Riyadh 11451, Saudi Arabia; talasiri@ksu.edu.sa
4 Campus Tecnológico e Nuclear, Instituto Superior Técnico, Universidade Técnica de Lisboa, 1049-001 Lisbon, Portugal; carreira@ctn.tecnico.ulisboa.pt
* Correspondence: samir.hakimi@usmba.ac.ma; Tel.: +212-6-54-8000-75

Abstract: The hydrogeological study of the Rharb coastal basin, located in the semi-arid northwest region of Morocco, focuses on its two aquifers: the Plio-Quaternary aquifer characterized by high-quality water with salt concentrations ranging from 0.4 to 2 g/L, and the Upper Quaternary aquifer, with lower water quality (2 to 6 g/L). The deep aquifer is overexploited for agricultural purposes. This overexploitation has led to declining piezometric levels and the worsening of the oceanic intrusion phenomenon. The study aims to develop a numerical model for a period of 15 years, from 1992/93 to 2006/07 for monitoring groundwater quantity and quality, considering recharge, exploitation, and basin characteristics. A hydrodynamic model based on storage coefficient calibration identifies overexploitation for irrigation, increasing from 93 Mm^3 in 1993 to 170 Mm^3 in 2007, as the primary driver of declining water levels. A hydrodispersive model highlights higher salt concentrations in the shallow aquifer (up to 6 g/L), high nitrate concentrations due to human activity, and pinpoints areas of seawater intrusion approximately 500 m from the shoreline. Although the deeper aquifer remains relatively preserved, negative hydraulic balances from -15.4 Mm^3 in 1993 to -36.6 Mm^3 in 2007 indicate an impending critical period.

Keywords: Rharb basin; semi-arid région; overexploitation; oceanic intrusion; hydrodynamic model; storage coefficient; hydrodispersive model

1. Introduction

Groundwater is a vital resource on a global scale [1]. It constitutes the primary source of drinking water in many countries due to its lower vulnerability to pollution compared to surface water, as well as its availability, accessibility, and relatively low cost [2]. However, excessive exploitation exacerbates the decline in piezometric levels and accelerates the advancement of saline intrusion [3], which is a global issue characterized by the landward displacement of the freshwater–saltwater interface due to density differences [4].

The study of aquifer systems, aimed at understanding the mechanisms that control groundwater flow regimes to preserve the resource, requires extensive information and data. The validity and reproducibility of aquifer analyses strongly depend on the availability of abundant and high-quality data. Organizing them into a coherent and logical structure, supported by appropriate software, ensures their validity and accessibility, thus providing a powerful tool for hydrogeological studies [5].

Hydrogeological modeling is increasingly used to verify data consistency, enhance understanding, and facilitate more effective analysis of complex hydrogeological contexts [6]. Furthermore, groundwater modeling allows for the assessment of aquifer responses to hydrological conditions and promotes sustainable groundwater resource management [7]. Various modeling approaches have been extensively explored in the literature [8–11]

Hydrogeological models are widely used worldwide for effective groundwater management. In Tunisia, these models help assess the impact of aquifer overexploitation on groundwater salinization [12]. In Australia, they are used to predict the effects of climate change on water availability in arid regions [4]. Similarly, in India, hydrogeological modeling is employed to manage overexploited aquifers by optimizing resource use for agriculture [13]. These tools provide tailored solutions to ensure sustainable water management, enabling informed decision-making and more resilient management strategies in the face of climatic and anthropogenic challenges.

In Morocco, irrigated agriculture is the primary consumer of groundwater [14], playing a crucial role in the national economy by contributing approximately 15% to the Gross domestic product (GDP) and employing nearly 43% of the active population [15]. The Rharb Plain, part of a series of Moroccan coastal plains [16], represents the downstream section of the Sebou hydraulic basin. This region is distinguished by its abundant agricultural production, particularly in cereals, vegetables, and irrigated crops, which are essential for food security and socio-economic development. Located in a semi-arid environment [16] characterized by irregular precipitation, high temperatures, and prolonged droughts, it heavily relies on irrigation to sustain agricultural production and mitigate economic losses associated with water shortages.

The Rharb basin comprises a Plio-Quaternary aquifer with coarse granulometry, facilitating the circulation of deep groundwater, as well as an Upper Quaternary aquifer with fine granulometry, where lower-quality water tends to stagnate. These aquifers are separated by a Quaternary aquitard [17], limiting direct exchanges between them. The expansion of irrigated areas, covering approximately 107,000 hectares [18], disrupts the equilibrium of the deep aquifer in the coastal region, leading to progressive overexploitation.

In this semi-arid context, excessive freshwater extraction from deep aquifers has resulted, since the early 1990s, in significant hydrological deficits and critical qualitative impacts such as contamination by pollutants and seawater intrusion into coastal aquifers. This phenomenon deteriorates water quality, jeopardizing both agriculture and human consumption.

To better understand these complex dynamics, transient hydrodynamic and hydrodispersive models have been developed for the Rharb basin. These models are based on a steady-state hydrodynamic model [18] and are reinforced by a database derived from a Geographic Information System (GIS) [17,19], enhancing analytical and decision-making capabilities. The transient hydrodynamic model, based on the Modular Groundwater Flow Model-Groundwater Modeling System (MODFLOW-GMS), integrates historical pumping [20] and recharge data while refining the storage coefficients of the Plio-Quaternary aquifer to better adapt them to actual conditions. This approach enabled a precise analysis of the piezometric decline of the deep aquifer between 1992–1993 and 2006–2007, identifying key factors influencing groundwater availability, including climate change and evolving agricultural practices.

Groundwater quality has a major impact on human health as well as agricultural and industrial activities. High concentrations of elements such as nitrates and other toxic substances can significantly degrade groundwater resources. In many cases, spatial variations in fluid density play a crucial role in groundwater flow patterns, particularly in coastal areas where density-dependent models are required to accurately simulate saline intrusion and submarine groundwater discharge [21]. Consequently, a hydrodispersive model was developed to monitor water quality changes in both aquifers and analyze the increase in oceanic intrusion into the Plio-Quaternary aquifer. Using the Three-Dimensional Multispecies Transport Model with Modular Solver (MT3DMS), this model simulated the spatio-temporal evolution of water concentrations, accounting for seasonal variations and pumping practices.

Given that the Saltwater Intrusion and Groundwater Flow Model (SEAWAT) is the most commonly used tool for characterizing and projecting the dynamic changes in groundwater flow and salinity [4,14], we employed it to further investigate oceanic intrusion and accurately simulate the advancement of saline water contaminating the deep aquifer in the coastal region.

The model highlights the impact of intensive groundwater exploitation on aquifer salinization. It introduces, for the first time, a multilayer model of the entire Rharb basin, accounting not only for the interactions between the Rharb aquifer system and its external environment but also for the interactions between the deep aquifer, the aquitard, and the shallow aquifer. Thus, developing an effective management model for precise groundwater level control is of paramount importance [22,23].

In light of the growing challenges posed by climate change and overexploitation, particularly in semi-arid regions, rational and effective interventions are necessary to promote sustainable development.

2. Geographical, Geological, and Hydrogeological Framework

The Rharb Basin is located in the northwest of Morocco and covers an area of 4000 km^2. It is bordered by the Atlantic Ocean to the west, the Drader-Souier plain to the north, the hills of the Rif mountains to the east, and the Maamora plateau to the south. With elevations ranging between 5 and 25 m, this area comprises a coastal zone, continental margins, and a central alluvial plain of the lower Sebou (Figure 1).

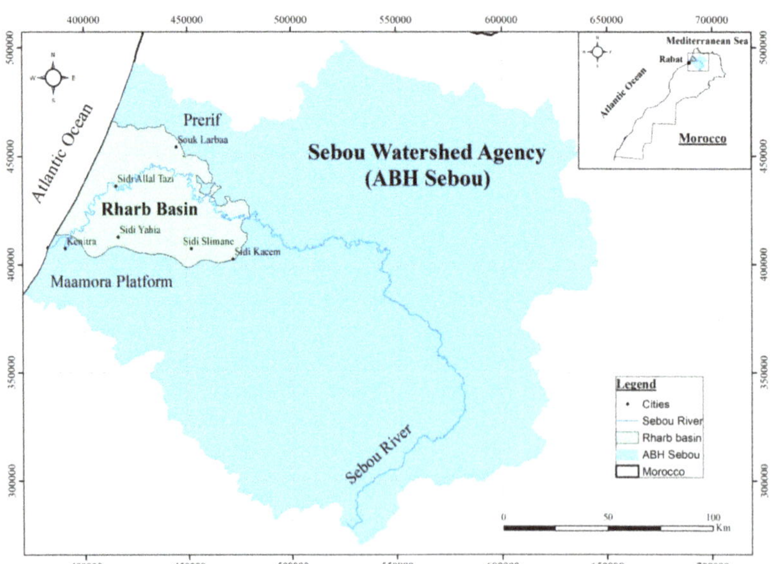

Figure 1. Location map of the Rharb basin.

From a geological standpoint, the Rharb Basin represents the foreland basin of the Rif mountain range, affected by continuous subsidence since the middle Vindobonian (a period during which the pre-Rif thrust sheets were formed). The Rharb, a subsiding basin since the Tortonian, is primarily structured by two main lineaments: the first, to the north, responsible for the structure of Lalla Zohra [24], and the second, to the south, represented by the Maamora front [24,25].

During the Plio-Villafranchian, the relief of the northern, southern, and eastern borders of the plain was uplifted in response to recent phases of Alpine compression, isostatic adjustments, and eustatic falls, compensated by the abundance of coarse continental inputs deposited by significant fluvio-estuarine systems [26]. In the central part of the basin [21] highlighted, within the Miocene substratum, three satellite basins largely controlled the subsequent formation of the aquifers during the Plio-Quaternary period. To the west of the basin, along the coastal deposits of the NW-SE fault corridors, is the growth of the Lalla Zahra front to the north and a localized folding structure around the city of Kenitra developed.

Furthermore, field data associated with electrofacies resistivity and high-resolution seismic refraction tomography indicate that this region is tectonically active and subjected to significant horizontal movements with a strike-slip component, as evidenced at the Dhar Doum-Lalla Zahra fault, where faults are associated with anticlines and synclines, and the structural setup resembles "fault-propagation folds" or propagation folds at the front of the Rif chain [27].

At the regional scale, new interpretations incorporate the Rharb Basin into a dynamic involving the lineaments of the Rif chain, notably the Jebha and Nekour faults [28,29]. The position of current seismic epicenters confirms the recent activity of some of these lineaments.

This basin is located between two major structural units that serve as sediment sources: the pre-Rif to the north and the Rifian ridges to the south, while to the south are the Maamora Plain and the Hercynian Meseta [30]. The lithostratigraphic succession of the basin ranges from the Miocene to the Upper Quaternary (Figure 2).

During the Lower Quaternary, two distinctive levels are sometimes distinguished within this detrital complex: the Tensiftien, characterized by the reappearance of coarse gravel and more or less clayey sands (up to a maximum thickness of 60 m), and the Amerian, with sediments consisting of clays and loams with limestone nodules (up to a maximum thickness of 50 m).

The Villafranchian (Moulouyan-Regreguian-Saletian) and the Lower Quaternary are difficult to distinguish due to similar depositional conditions. They contain significant spreads of pebbles and gravel found beneath the plain and its northern and eastern peripheries. To the south, the fine sands and gravel of the Maamora are connected to them.

The Pliocene is characterized by marine deposits of calcareous sandstones, sandstones, and sands located in the coastal zone. Regressive deposits [26] and conglomerates identified by their outcrops in the north of the basin, yellow sands in the east, and continental sediments, pebbles, and silts located upstream of the confluence zone of the Sebou and Ouergha rivers. The thickness of this level is low, and its hydrogeological interest is limited.

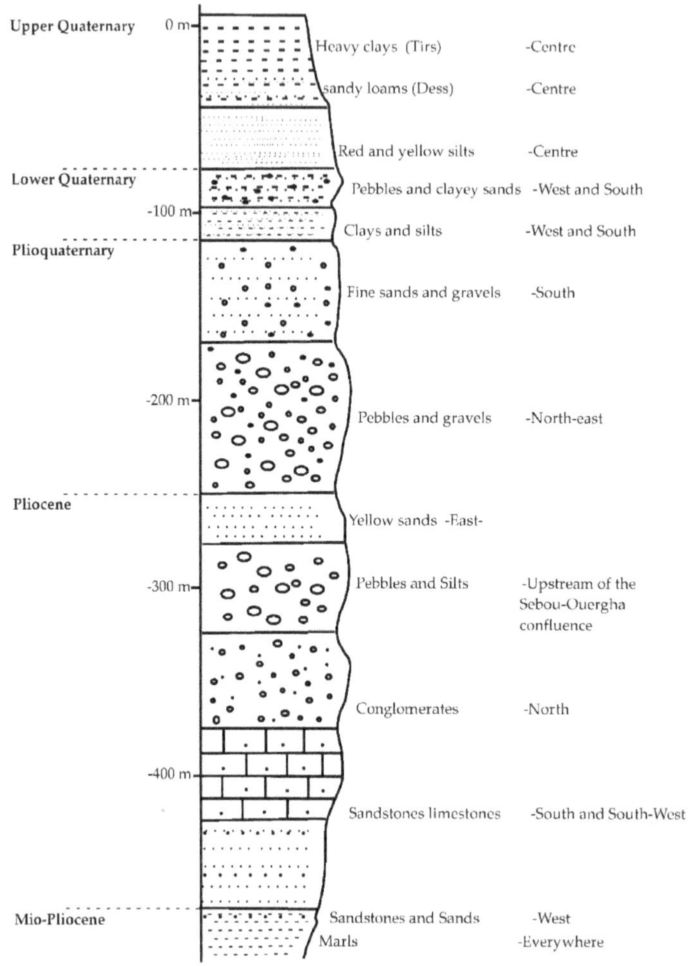

Figure 2. Schematic lithostratigraphic log of the Rharb basin.

The Mio-Pliocene is associated with the subsidence and sedimentary infill of the syntectonic foreland basin. It is characterized by a thick series, with a thickness that can reach 2000 m of gray or blue marls. Laterally, these marls transition to discordant molasses of the foreland, displaying chaotic facies where extensive submarine landslides can be observed. This forms the general substrate of the overlying terrains.

From a hydrogeological standpoint, this sedimentary succession includes two aquifers: one aquifer ranging in age from the Pliocene to the Lower Quaternary, composed of coarse sediments through which the waters of the deep aquifer flow, and another aquifer from the Upper Quaternary through which the waters of the shallow groundwater flow. The two aquifers are separated by an aquitard, consisting of compact Lower Quaternary clays, which pressurize the deep waters in the center of the basin. At the outcrops in the north, east, south, and west of the basin, the aquifer becomes unconfined (Figure 3).

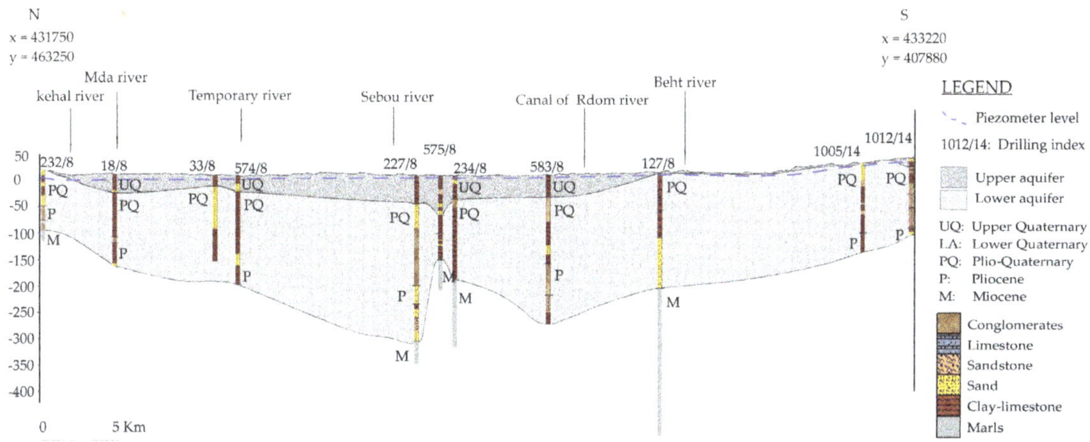

Figure 3. Schematic North-South section of the Rharb aquifer system. (Created from mechanical drilling data provided by the Directorate of Water Research and Planning, Ministry of Equipment and Water, Morocco).

3. Methods and Tools

3.1. Transient Hydrodynamic Model

For the transient hydrodynamic model, the working method involves using the numerical MODFLOW model, derived from the calibration of permeabilities in the steady-state system [17]. This model consists of a workspace containing the 3D block diagram of the basin (Figure 4a), built from mechanical boreholes and enhanced by data from geoelectrical profiles, as well as the layers of the conceptual model (Figure 4b). These data are stored in a geographic database created using ArcGIS (Figure 4c). New data, including storage coefficients to be calibrated (Appendix A), piezometric histories, and recharge and pumping histories [23], are input into the transient-state model.

The computational code MODFLOW, developed in Fortran by the USGS [31], is based on solving the general partial differential equation of diffusivity for flow, through a combination of the continuity equation and Darcy's law (Equation (1)).

$$\frac{\delta}{\delta x}\left(K_x\frac{\delta H}{\delta x}\right)+\frac{\delta}{\delta y}\left(K_y\frac{\delta H}{\delta y}\right)+\frac{\delta}{\delta z}\left(K_z\frac{\delta H}{\delta z}\right)-Q=S_S\frac{\delta H}{\delta t} \quad (1)$$

With H: hydraulic head [L], K: hydraulic conductivity [LT^{-1}], S_s: specific storage coefficient [L^{-1}], Q: injected or pumped water volume [L^3T^{-1}], t: time [T].

Subsequently, the MODFLOW calculation parameters are updated in transient mode to account for the temporal evolution of the groundwater system in response to external disturbances, primarily precipitation, and abstractions. Similar to the steady-state hydrodynamic model, the development of the transient model follows the model construction process [32] (Figure 5).

The main adjustments made to the MODFLOW tool include reconfiguring the model for transient state and defining initial conditions to consider during simulations. The objective of transient modeling is to calibrate the storage coefficients of the Plio-Quaternary aquifer by comparing the calculated piezometric levels, for each time interval, to measurements from the five pilot piezometers. The available measurements of this coefficient before calibration range from 0.01 to 0.2 for the unconfined aquifer and from 0.0001 to 0.001 for confined aquifers. A series of simulations is initiated with the aim of reducing the gap between calculations and measurements until the stage where the results are deemed satisfactory while adhering to the modeling cycle [33] (Figure 6).

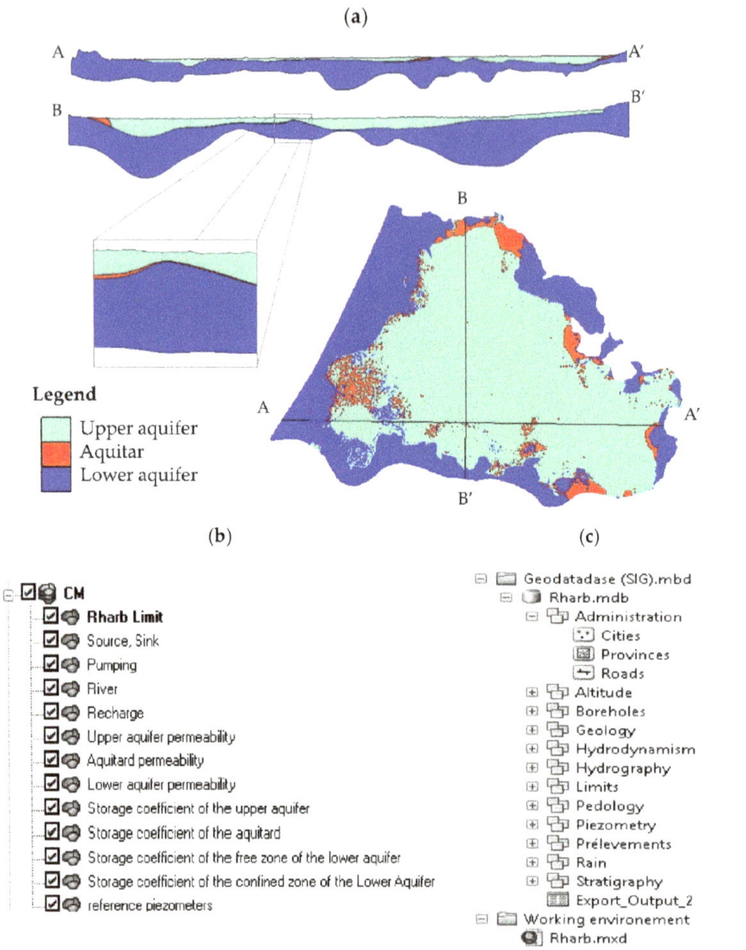

Figure 4. (**a**) Stratigraphic model of Rharb basin, AA′: East-West cross-section, BB′: North-South cross-section; (**b**) Conceptual model layers; (**c**) GIS Geodatabase.

The resolution of the first-order linear differential system to calculate the piezometric height (H) is done according to the following equation:

$$TH = Q + a^2 S\left(\frac{\delta H}{\delta t}\right) \quad (2)$$

With T: transmissivity (L^2T^{-1}), a: surface area of the cell (L^2), S: storage coefficient. Thus, the storage volume ΔV at each cell i, j, k level is expressed as follows:

$$\frac{\Delta V}{\Delta t} = S_{i,j,k}\left(\Delta R_j * \Delta C_i * \Delta V_k\right) * \frac{\left(H_{i,j,k}^{m} - H_{i,j,k}^{m-1}\right)}{\left(t^m - t^{m-1}\right)} \quad (3)$$

With $\Delta V/\Delta t$: variation of storage per unit time during the time step Δt, $S_{i,j,k}$: storage coefficient of cell i, j, k, ΔR_j: distance between neighboring nodes, ΔC_i*ΔV_k: surface area of the exchange facet, ${H^m}_{i,j,k}$ ${H^{m-1}}_{i,j,k}$: heads at nodes (i, j, k) and (i−1, j, k) at time step m, $t^m - t^{m-1}$: time interval separating time steps m and m − 1.

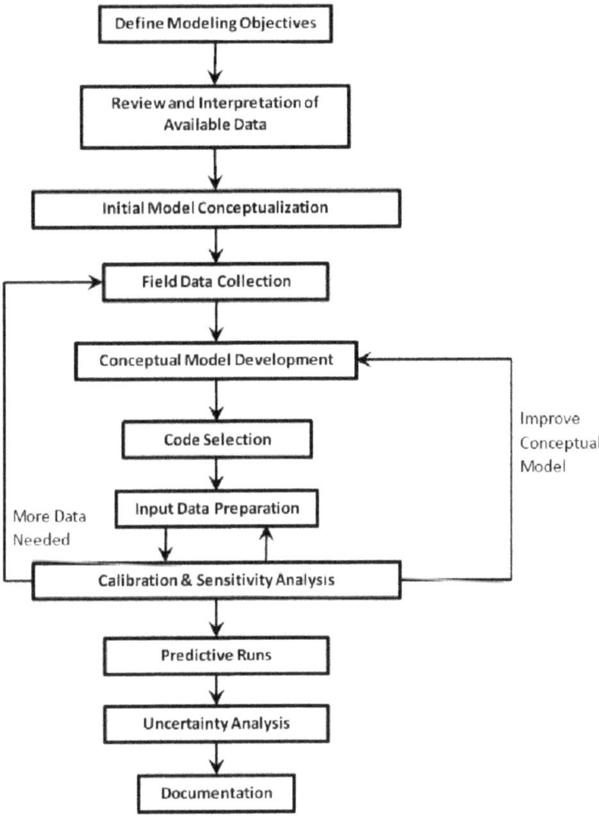

Figure 5. Model construction process.

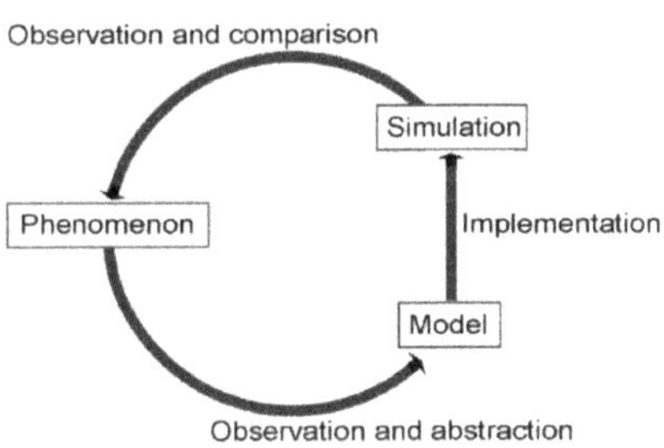

Figure 6. Modeling cycle.

3.2. *Hydrodispersive Model*

For the hydrodispersive model, and to replicate the history of the saltwater wedge advancement in the Plio-Quaternary aquifer, the MT3DMS module is utilized to study the dispersion of chemical solutes in both aquifer layers of the aquifer system, according to the finite-difference method [21] (Equation (4)).

$$\frac{\delta(\theta C_k)}{\delta t} = \frac{\delta\left(\frac{\theta D_{ij} * \delta C_k}{\delta x_j}\right)}{\delta x_i} - \frac{\delta(\theta \vartheta_i * C_k)}{\delta x_i} + qsCsk + \Sigma Rn \tag{4}$$

With C_k: dissolved concentration of species k [ML^{-3}], θ: porosity of the subsurface medium [$-$], t: time [T], x_i: distance along the respective Cartesian coordinate axis [L],

D_{ij}: hydrodynamic dispersion coefficient tensor [L^2T^{-1}]; v_i: seepage or linear pore water velocity [LT^{-1}], qs: volumetric flow rate per unit volume of aquifer representing fluid, Cs_k: concentration of the source or sink flux for species k [ML^{-3}], ΣRn: chemical reaction term [$ML^{-3}T^{-1}$].

The SAEWAT module is employed to investigate the inland advancement of the freshwater–saltwater interface. It is a 3D groundwater flow modeling application for variable-density flow coupled with multi-species solute transport. SEAWAT builds upon the MODFLOW and MT3DMS interfaces and utilizes them for boundary condition setting and post-processing. At the boundary conditions, it integrates flow equations, mass conservation, and solute transport equation.

3.3. Simulation Period, Spatio-Temporal Discretization, and Boundary Conditions

The Rharb aquifer system did not show significant declining trends until the early 1990s. Simulations are conducted from the beginning of this decade, where pumping activities experienced a notable increase, causing disturbance to the aquifer system equilibrium. We have considered a period of 15 intervals from 22 September 1992 to 22 September 2007, based on the availability of historical data on withdrawals for agricultural irrigation purposes. The discretization [28] of the study area respects the following two conditions: Horizontally, and to better target the most sensitive areas (aquifer–river interfaces, boundary conditions...), we opted for a refined mesh by assigning cell sizes ranging from 6.25×10^{-2} m^2 to 1 km^2 after utilizing the MODFLOW USG Quadtree Grid. This results in 46,748 cells to represent the entire system, with 14,252 for the upper aquifer, 12,259 for the aquitard, and 20,237 for the lower aquifer. Vertically, the model is three-layered, and the cells adhere to the actual thickness of the aquifer formations.

The boundary conditions of the transient model are as follows (Figure 7).

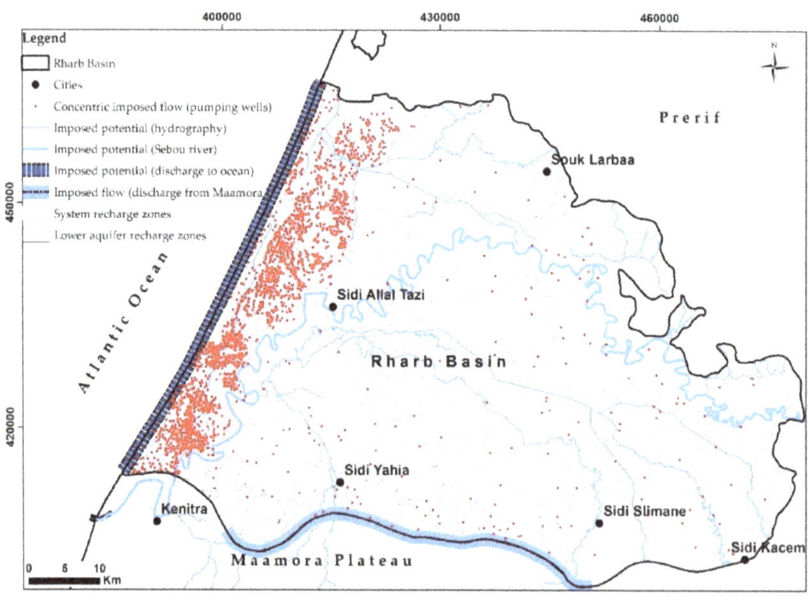

Figure 7. Boundary conditions.

The southern boundary is a constant head condition represented by the discharge of groundwater from the Maamora aquifer to the Rharb aquifer. The northern boundary is a no-flow condition represented by the divide between surface waters of the Drader-

Souier plain and the Rharb plain. The western boundary is an imposed potential condition representing the discharge of groundwater from the Rharb aquifer to the Atlantic Ocean. The eastern, southeast, and southwest boundaries are no-flow conditions defined by the extension limits of the hydrogeological basin. Rivers are represented as boundaries with imposed potential or Dirichlet conditions [34]. The main infiltration zones include coastal strips, eastern and southern areas where permeable outcrops allow the system to receive an average of 113.4 million cubic meters per year (0.4 mm/day).

3.4. Inputs and Outputs

For the transient hydrodynamic model, sampling points are represented as an imposed flux condition and are incorporated into the model with the indication of the pumping rate. These withdrawals are internal numerical data consisting of a 2368 daily historical value from three surveys conducted by COMBE [20], ENGEAMA (Inquiry 1992), and the Sebou Hydraulic Basin Agency [18] following the surveys of 2005 and 2008 (Table 1).

Table 1. Inventory of input and output data for the transient model of Rharb basin.

Data	Infiltration (Input)	Pumping Well (Output)	Storage Coefficients
Time series	22 September 1992–22 September 2007		-
Entity	156 polygons	2368 points	7 points (Appendix A)
Time (interval)	seasonal	annual	-

To simplify the task and since irrigation withdrawals constitute more than 90% of the total withdrawals, we only consider irrigation data. Data on drinking water reserved for consumption and industrial water are not precise, and some information such as coordinates is missing. The return flow from irrigation is also subtracted from the inputs to compensate for the portion of water supply for drinking and industrial purposes as outputs since their volumes are analogous. Recharge data are inserted based on the division of the Rharb plain into 156 parcels derived from the calibration of the steady-state model. Each group of parcels (cells with equal infiltration values) is assigned a series of 60 values corresponding to the seasonal distribution of recharge for the entire study period. This distribution depends on the recharge, which in turn is a function of rainfall and soil type (Figure 8).

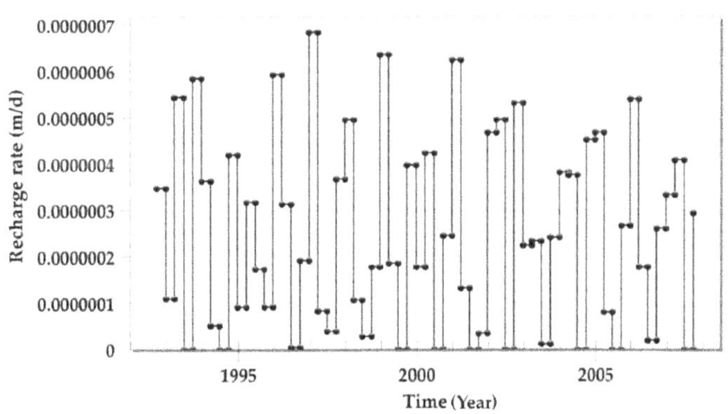

Figure 8. Example of interannual recharge in one of the recharge areas (Zone #1: Poorly developed soils).

For the hydrodispersive model, the first stage in building the Mass Transport model (MT3DMS) involves using input data from the hydrodynamic model, as well as data related to water chemistry [35] (Table 2).

Table 2. Inventory of hydrodispersive model data of the Rharb basin.

Data	Description
Geometry	Stratigraphic trilayer model.
Hydrogeological parameters	Hydraulic conductivities, Porosity, Dispersivity, etc.
External stresses and initial conditions	Initial water concentrations: rain, groundwater, etc.
Boundary conditions	Imposed volumic concentration of the nappe-ocean boundary.

The initial average porosities assigned to the three layers of the model are 0.4 to 0.45 for the upper aquifer, 0.35 to 0.45 for the aquitard, and 0.2 for the lower aquifer. The longitudinal dispersivity is approximately 20 m for all cells of the model. The groundwater concentration map of the Rharb basin established for the low-water period of 1964/1965 serves as the reference state (Figure 9). The dry residue at 180 °C is approximately 0.7 g/L for the Plio-Quaternary aquifer and ranges from 1 to over 6 g/L for the phreatic aquifer.

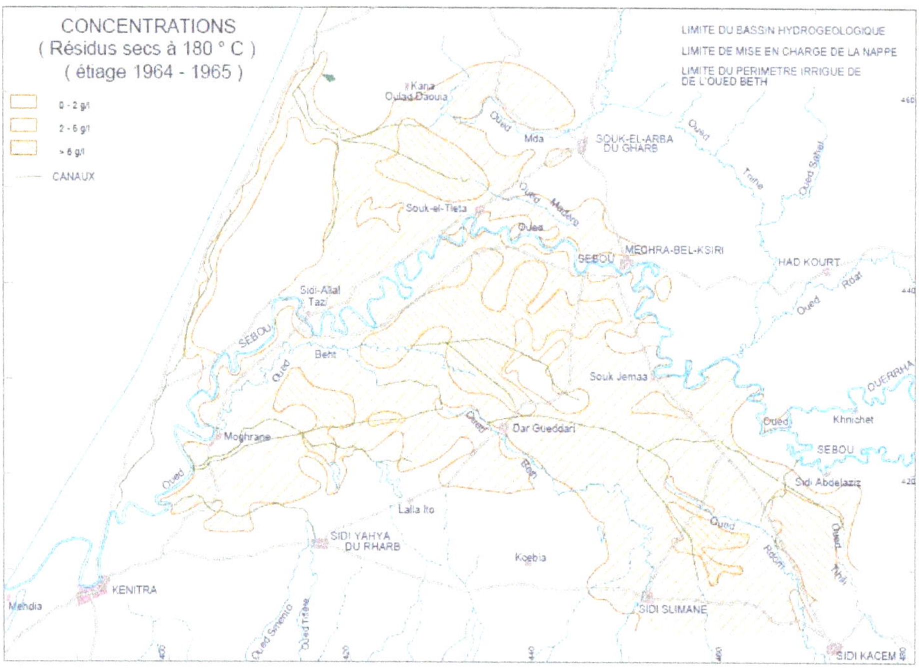

Figure 9. Reference map of Gharb groundwater concentrations [20].

The concentrations of rainwater range from 0.1 to 0.2 g/L and are assigned to recharge zones based on the conceptual model, except where cell-by-cell correction is necessary. The Advection, Dispersion, Sources/Sink Mixing, and Transport Observation packages are used when selecting MT3DMS packages. They facilitate the handling of solute propagation in the aquifer system, thereby adhering to the principles of hydrodispersive modeling and studying solute propagation.

Continuing with the hydrodispersive model, the second stage is building the Variable Density Flow Model (SEAWAT). The inputs of the VDF package enable control over density calculations (Table 3).

Table 3. Commonly used values for the Variable Density Flow package [36].

Variable	Reference Value	Slope
Density	1000 kg/m^3	n/a
Salt Concentration	0 mg/L	0.714
Temperature	25 °C	−0.375 kg/m^3/°C
Pressure Head	0 m	4.46×10^{-3} kg/m^4

Given that the concentration of seawater is 35.7 g/L compared to a reference concentration of 0 g/L, and its density is 1025 kg/m^3 compared to a reference density of 1000 kg/m^3, the DRHODC ratio (Density/concentration slope) is expressed as:

$$DRHODC = \frac{1025\left(\text{kg.m}^{-3}\right) - 1000\left(\text{kg.m}^{-3}\right)}{35\left(\text{kg.m}^{-3}\right) - 0\left(\text{kg.m}^{-3}\right)} = 0.7 \tag{5}$$

The coastline serves as a boundary condition of the imposed potential type. In the context above, this boundary condition sets a constant hydraulic head at 0 m, representing the natural condition of the groundwater base level. This boundary is considered during hydrodynamic calculations as a quantitative condition. At this stage, a value of 35.7 g/L is assigned to this boundary for qualitative characterization.

3.5. Water Quality and Géochimical Facies

To determine the hydrogeochemical facies of the shallow waters exploited in the basin, we analyzed a sampling network composed of 39 sampling points (the 2013 survey on the Rharb Basin, conducted by the Hydraulic Agency of the Sebou Basin). These water samples, taken from the groundwater, are considered representative of the entire aquifer (Figure 10).

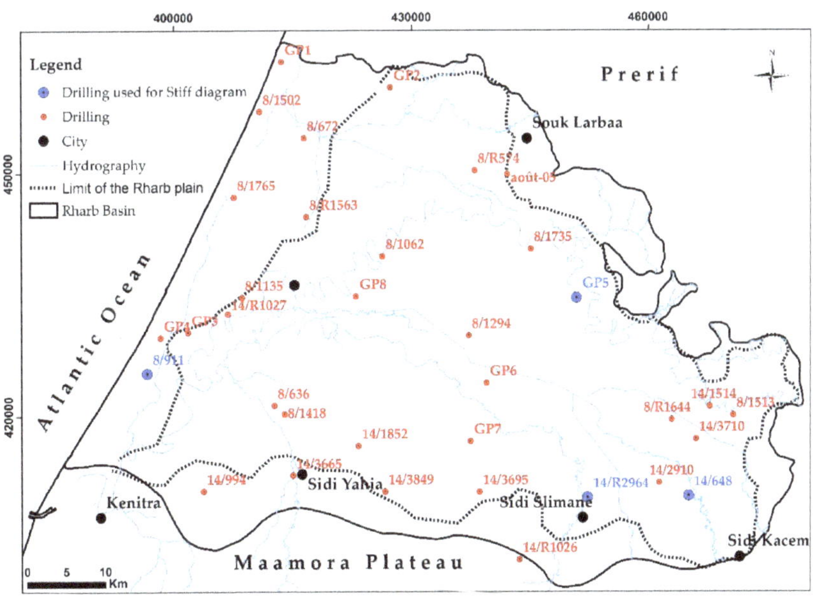

Figure 10. Location of sampling points for chemical analysis.

The analyses conducted are primarily mineralogical. Depending on the amount of solution sampled, the analyses are more or less complete, with priority given to the analysis of the nitrate ion, the most representative element of aquifer pollution. Other elements analyzed include major ions: Ca^{2+}, Mg^{2+}, Na^{+}, K^{+}, Cl^{-}, NO_3^{-}, and SO_4^{2-}; minor elements: SiO_2, and trace metals: Mn and Fe. Electrical conductivity, pH, total dissolved solids, and alkalimetric titration are also determined (Appendix C).

4. Results

The main objective in hydrodynamic modeling is the adjustment of the storage coefficient, which governs the hydraulic system's behavior under transient conditions. The calibration of the hydrodynamic characteristics is carried out as follows:

- Introduction of permeabilities obtained from calibration in steady-state conditions,
- Introduction of storage coefficients measured in the field,
- Refinement of the distribution of the storage coefficient to better match the piezometric history,
- Possible reassessment of the permeabilities (new steady-state calibration) if required.

The model calibration is considered successful after a series of simulations that result in a good match between the measured piezometric levels and the potentials calculated by the model. Furthermore, the quality of the model calibration can be assessed using the following criteria:

- Reconstruction of the historical data from monitoring piezometers,
- Distribution of the storage coefficient,
- Analysis of the balance from the calibration (inflows and outflows) for the different time intervals.

4.1. Storage Coefficients of the Plio-Quaternary Aquifer

Following the comparison of piezometric levels between the model and recordings from pilot piezometers (Table 4), adjusting storage coefficients enables a spatial distribution generalized at the basin scale.

Table 4. Lambert coordinates of pilot piezometers used to calibrate storage coefficients of Rharb basin.

Reference Piezometer	X (m)	Y (m)
540/8	421,070	420,144
698/8	397,061	425,246
1552/8	394,915	421,195
1562/8	416,958	448,815
1564/8	416,468	454,357
1765/8	407,665	447,225
1807/14	412,817	411,091

The reliability of the result is assessed through the measurement of the calibration uncertainty of the calculated and measured loads at the monitoring well levels (Figure 11a,b), and by establishing a measurement/calculation relationship (Figure 11c).

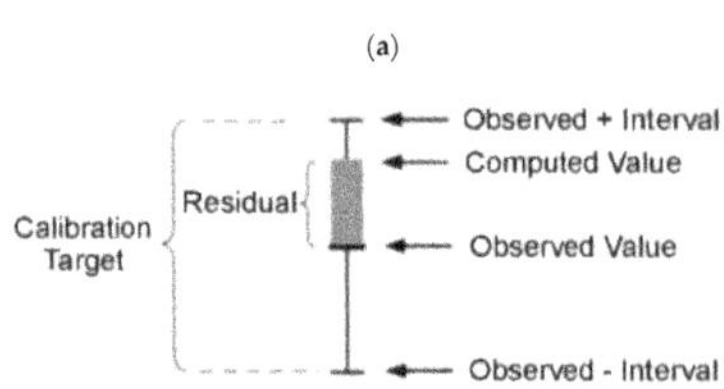

Figure 11. *Cont.*

(b)

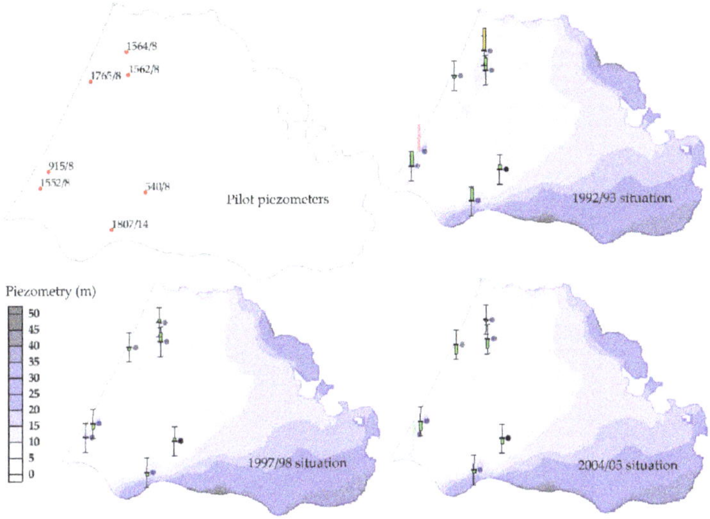

(c)

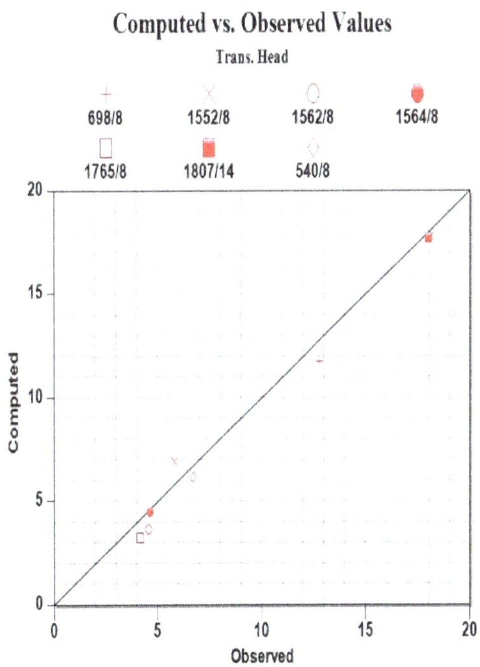

Figure 11. (**a**) Method for measuring calibration uncertainty. (**b**) Degree of correlation between observed and calculated piezometries. (**c**) Relationship between measured and calculated piezometric heights at pilot piezometers (example of September 2004 situation).

The pilot piezometers are selected based on their positions, which enable them to record piezometric fluctuations in rapid response to changes in recharge/pumping volumes (zones of outcrop in the West and South of the Plio-Quaternary aquifer). The storativity coefficients range from 0.01 to 0.3 along the coastal zone, which is influenced by the marine environment, where Quaternary formations are predominantly sandy. They range from 0.01 to 0.2 along the southern zone near the contact with the Maamora aquifer, an area influenced by continental factors, where formations are predominantly gravelly-sandy. In the outcrop areas of the Southeast and Northeast Plio-Villafranchian basin, where formations consist of coarse materials (pebbles, gravel, etc.), the coefficients range from 0.1 to 0.35.

Sometimes, lithological sequences of the Plio-Quaternary aquifer contain, in certain localities, compact clayey and marly layers ranging in thickness from 1 to 2 m, which act as impermeable screens (Appendix B). These clayey strata put pressure on the waters of the so-called free part of the deep aquifer and reduce the proportion of gravitational waters, resulting in storage coefficients ranging between 0.001 and 0.002, values relatively lower than the standards known for free aquifers.

The storage coefficients range from 5×10^{-6} to 1×10^{-5} in the central zone of the basin, where the deep aquifer is confined due to the impermeable formations of the Amerian and the Soltanian (Figure 12).

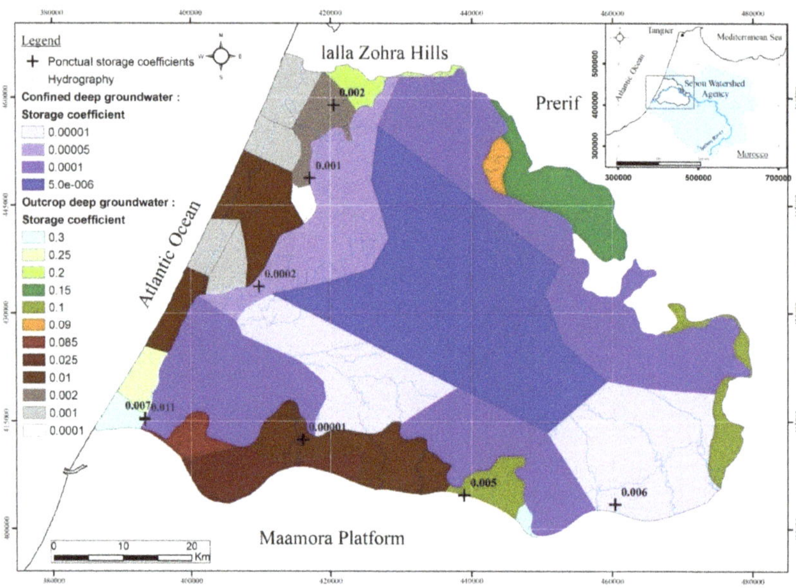

Figure 12. Spatial distribution of storage coefficients for the Plio-Quaternary reservoir.

4.2. Water Balances

The calibration of the transient model allows for the development of 15 balances relative to the selected intervals. According to the analysis of the balances, the system becomes increasingly deficient, transitioning from −15.4 million m^3/year in September 1993 to −36.6 million m^3/year in September 2007 (Table 5).

The volumes discharged to the ocean decrease over time, and the volumes drained by the hydrological network decrease in favor of infiltration (Figure 13). This is due to the decline in piezometric levels, caused by withdrawals which increased from 93 million cubic meters per year in September 1993 to 170 million cubic meters per year in September 2007. It is noted in this regard that the probable decrease in future precipitation may amplify the effect of exploitation on groundwater reserves, to discuss a combined effect of climate change and overexploitation.

Table 5. Hydrological balances for the periods 1992/93 and 2006/07 in the Rharb basin.

September 1993	In (m^3/d)	Out (m^3/d)	September 2007	In (m^3/d)	Out (m^3/d)
Recharge	346.217	0	Recharge	344.919	0
Maamora	182.220	0	Maamora	176.652	0
Ocean	0	−110.654	Ocean	0	−84.826
Rivers	126.829	−140.910	Rivers	157.537	−148.621
Drains	0	−80.831	Drains	0	−66.891
Well	0	−255.249	Well	0	−466.723
Storage	45	−109.767	Storage	6.611	−18.827
Total	655.311	−697.411	Total	685.719	−785.888
Deficit	−15.4 Mm^3/year		Deficit	−36.6 Mm^3/year	

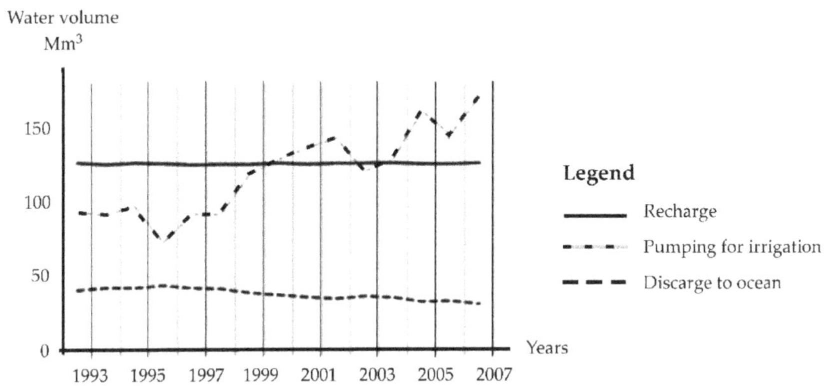

Figure 13. Evolution of recharge and discharge volumes.

The representation of the evolution of recharge/discharge shows that the fluctuation changes pace from one curve to another (Figure 13). During the study period, the two curves related to withdrawals and discharge exhibit significant variations compared to the rainfall curve. It is observed that despite rainfall remaining more or less constant over time, overexploitation of deep waters leads to a decrease in volumes discharged to the ocean.

This indicates that, despite the volumes of rainfall received by the basin during the study period, it is primarily the pumping of freshwater from the deep aquifer that plays the most impactful role in the Rharb aquifer system.

4.3. *Spatiotemporal Evolution of Salt Concentrations*

For the hydrodispersive model, the concentrations of the phreatic waters did not show remarkable changes between 1992/93 and 2006/07 because the geography of the iso-zones remained practically the same. However, the concentrations of waters evolved in certain areas at the level of the aquitard and the Plio-Quaternary aquifer.

The quality of phreatic waters deteriorates from the peripheries toward the center of the upper Quaternary aquifer (Figure 14a), with concentrations ranging from 2 to over 6 g/L. This is attributed to the poor lateral movement of solutes in the sediments of the upper Quaternary. The concentrations of water in the aquitard show an increase during the study period (Figure 14b). The impermeable aquitard acts as an obstacle that blocks the propagation of solutes toward the deep aquifer. The concentration of deep waters is less than 0.4 g/L in the peripheries of the basin, especially in the Mnasra area, thanks to the natural replenishment of waters during rainy events. This concentration increases from 0.7 g/L (reference value) to 2 g/L in some localities at the center of the basin due to exchanges with mediocre phreatic waters where the aquitard is locally absent (Figure 14c).

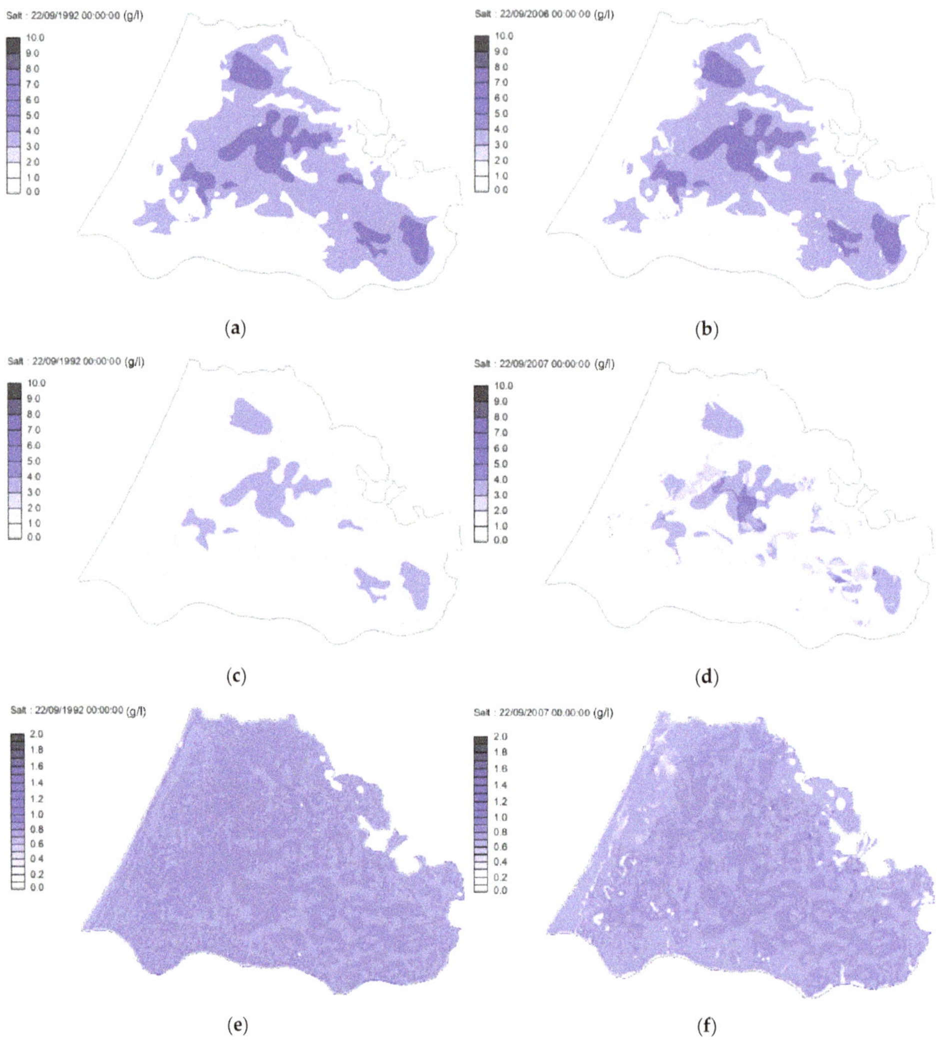

Figure 14. Spatial distribution of concentrations (g/L) between 1992–1993 and 2006–2007. (**a**) Upper Aquifer on 22 September 1992. (**b**) Upper Aquifer on 22 September 2007. (**c**) Aquitard on 22 September 1992. (**d**) Aquitard on 22 September 2007. (**e**) Lower aquifer on 22 September 1992. (**f**) Lower aquifer on 22 September 2007.

The curve showing the evolution of salt concentration in the deep aquifer at the outcrops of the Plio-Quaternary aquifer (cell n° 273,532) records a decrease, from 0.7 g/L (the initial value assigned to the waters of this aquifer) to 0.2 g/L (Figure 15). This implies that over the years, groundwater tends to reach chemical equilibrium with values close to those of rainwater; hence, the softest waters of the deep aquifer are found near recharge zones. Confined waters may register concentration values of up to 2 g/L in some areas (cell n° 295,951) due to exchange by drainage with underlying wet environments.

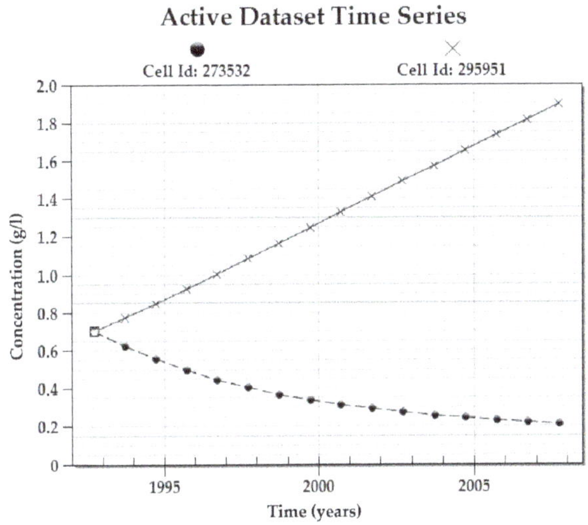

Figure 15. Examples of changes in water concentrations in the deep aquifer. Cell n° 273,532: free zones; cell n° 295,951: confined zones.

4.4. Oceanic Intrusion

For the Variable Density Model (SEAWAT), the iso-concentration curves of the water are parallel to the coastline, responding to the advance of oceanic waters due to the lowering of the piezometric level of the Plio-Quaternary aquifer and the resulting loss of head. The level of contamination of the aquifer by oceanic waters is not the same along the coastal boundary of the basin. In the south at latitude 416,000 m, oceanic waters penetrate more than 500 m into the aquifer, with a transition zone of 250 m (total of 750 m) (Figure 16). However, in the north, this interface is close to the shore with an average transition zone of 200 m.

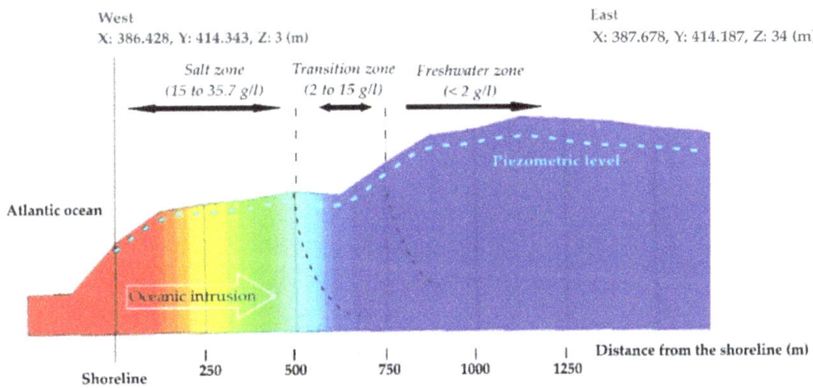

Figure 16. Salt concentration along a cross-section for the September 2007 situation.

The cross-sectional tomographic profile, perpendicular to the coastline at the location 'Ouled Berjal Oued 1′, southwest of the basin (West X: 386.428, Y: 414.343, Z: 3 m; East X: 387.678, Y: 414.187, Z: 34 m), is one of the measurements taken during a 2008 survey conducted by the Sebou Hydraulic Basin Agency (ABHS). In accordance with the norms predefined by the ABHS, the electrical tomography profiles were carried out using a

SYSCAL PRO system with 5 cables, each 500 m long, and an inter-electrode spacing of 10 m. The maximum electrode separation for the current electrodes is 560 m.

Oulad Berjal-1 profile serves as a reference for comparison to confirm SEAWAT results. It reveals the presence of saline waters extending 350 m from the shoreline, with electrical resistivities below 10 Ω.m. Between 350 and 500 m, freshwater from the Plio-Quaternary aquifer appears on the profile, giving way to the mediocre waters of the phreatic aquifer beyond 500 m from the shoreline (Figure 17).

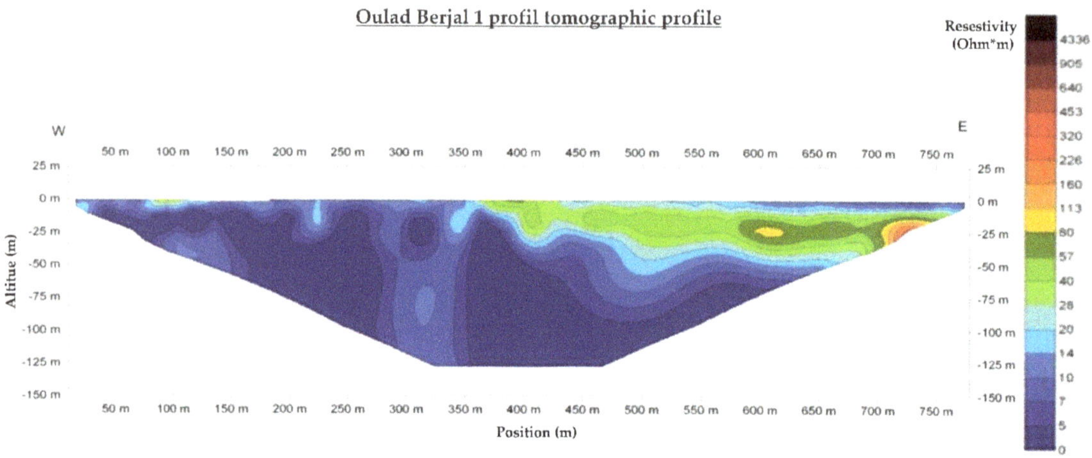

Figure 17. Example of a freshwater-saltwater contact tomographic profile [18].

4.5. Water Quality and Géochimical Facies

The presence of nitrates in the aquifer indicates an exogenous influence on the natural conditions of the environment. These high nitrate concentrations, their random spatial distribution, and significant local variation signal the existence of localized and important anthropogenic activity.

Stiff diagrams are selected to graphically represent the concentrations of major elements, excluding nitrates, expressed in meq/L. These concentrations are plotted on either side of a vertical axis. This type of representation has two main advantages: different water types produce distinct diagram shapes that can be recognized at a glance, and the total concentrations are easily perceived by the elongation of the diagram. An improvement was made by adding lines connecting the $Na^+ + K^+$ poles to Mg^{2+} and Cl^- to SO_4^{2-} [37]. (Figure 18).

In general, the chemical facies of the groundwater in the Rharb aquifer varies between two extreme types: calcium bicarbonate on one hand, and sodium and/or calcium chloride on the other. The calcium bicarbonate type characterizes the water in the dune cordon areas (F8/911), while the sodium and/or calcium chloride type is found between the cities of Sidi Slimane and Sidi Kacem (F14/648, GP5, and F14/R2964). These variations are not uniform across the entire aquifer but depend on the water table level and, especially, the geological context of the sampling point. Typically, the water transitions through an intermediate type, such as calcium chloride or sodium bicarbonate, indicating the presence of chemical reactions between the water and its environment (dissolution-precipitation reactions, cation exchange reactions).

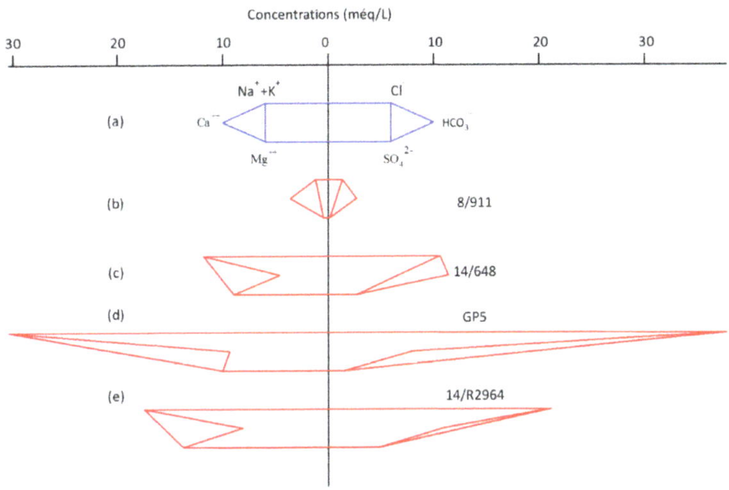

Figure 18. Stiff diagrams illustrating the chemical facies (signature) of the water for the second campaign of 2013. (**a**) Hypothetical water showing the location of different elements, (**b**) Water at sampling point 8/911 (calcium bicarbonate water), (**c**) Water at sampling point 14/648 (calcium bicarbonate water), (**d**) and (**e**) show the very high mineralization of water in wells GP5 and 8/R2964; sodium chloride type.

5. Discussion

The first limitation of the hydrodispersive model is the representation of the displacement of the sea wedge. The geometry of the freshwater–saltwater interface along the coastal zone is not vertical as theoretically indicated in the model, but rather inclined from top to bottom toward the onshore direction (Figure 19).

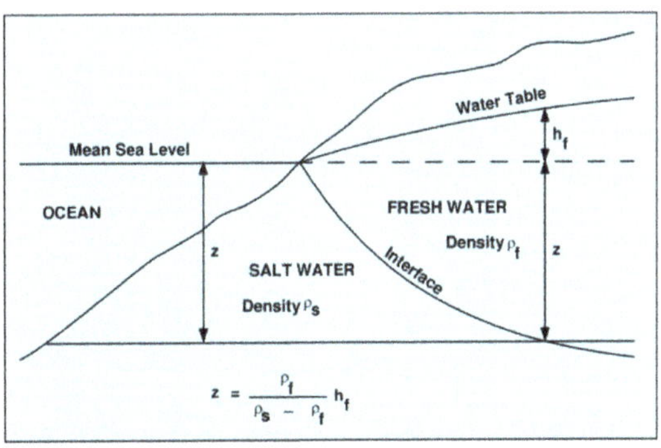

Figure 19. Geometry of the freshwater-saltwater contact [33].

The model calculates the movement of the saltwater wedge toward the continent, cell by cell (Figure 16). Each cell maintains the same concentration throughout its entirety (35.7 g/L) just a few meters from the oceanic boundary where saltwater reaches the water table surface. Once the water at the water-table surface becomes fresh, the application of the Ghyben–Herzberg principle (Equation (6)) becomes the sole means to locate the depth of the freshwater–saltwater interface. SEAWAT provides the location perpendicular to the ocean coastline of this interface, and it's up to the user to calculate the depth.

$$Z = 40 * h \quad (6)$$

With Z: depth of the freshwater–saltwater interface relative to sea level (m), h: piezometric height (m).

Another limitation of the model developed in this study is the choice of the period 1992/93–2006/07. This choice is based on the availability of agricultural withdrawal data collected during the 2005 and 2008 campaigns by the Sebou Hydraulic Basin Agency. We are considering improving the work by extending the simulation beyond 2007, within the framework of a new collaboration with the relevant authorities. Once this is done, we will be able to propose forecasting scenarios to assess the effects of climate change and demographic growth.

To now discuss future scenarios for water resource management at the Rharb Basin, we first wish to mention that the Rharb plain plays a key role in Moroccan agriculture, with significant contributions to national cereal production (73.6% of the cultivated area), as well as substantial areas dedicated to beetroot (17,300 ha) and citrus (129,288 ha) cultivation. However, this intensive agricultural activity, coupled with decreasing rainfall and climate change, places considerable pressure on local groundwater resources. The overexploitation of the Rharb coastal aquifer exacerbates the situation, necessitating urgent management measures to preserve water quality and availability.

To mitigate these impacts and safeguard the environment, a range of strategies must be adopted. These include the development of new dams to reduce withdrawals from groundwater, facilitate aquifer replenishment through controlled water releases, and accelerate artificial recharge techniques, such as treated water injection. Additionally, rationalizing fertilizer use and implementing wastewater treatment plants in rural centers are crucial steps to prevent pollution from agricultural and industrial activities [38,39].

Furthermore, the implementation of monitoring systems is essential to track real-time water quality and levels, providing early warning alerts in case of degradation. An integrated water resources management (IWRM) approach, involving all stakeholders, will help ensure the sustainable management of water resources, balancing demand with the need to protect the Rharb aquifer. The prudent use of these resources should be a top priority to address both qualitative and quantitative groundwater challenges in arid and semi-arid regions [40].

6. Conclusions

This study, focusing on the Rharb basin, a region situated within a semi-arid environment, enables the development of a transient hydrodynamic model and a hydrodispersive model using the MODFLOW, MT3DMS, and SEAWAT computational codes.

The development of a hydrodynamic model in a transient regime, by adjusting the storage coefficients for the period September 1992/September 2007, reveals a declining water situation marked by increasing deficit balances, due to irrigation pumpings. The justification of the results through calibration and correlation with the monitoring wells allows for the attribution of zoning of storage coefficients on the basin scale, ranging from 5.10^{-6} to 1.10^{-4} in the deep confined zone and from 1.10^{-3} to $3.5.10^{-1}$ in the free zones.

The development of the hydrodispersive model allows for tracking the spatial distribution of salt concentrations during the same period, taking into account various hydrogeological parameters and the initial concentrations assigned to different layers of the model.

The quality of groundwater is poor over the 15 study intervals, with concentrations ranging from 2 to 6 g/L, and the distribution of iso-concentrations has not evolved over time due to poor water circulation in the upper Quaternary aquifer. Limited water circulation, long residence time, and significant evaporation in the surface aquifer restrict the spread of ionic species, leading to notable local enrichment in major elements. In the Plio-Quaternary aquifer, the water quality is generally good (ranging from 0.4 to 2 g/L). In recharge zones, the concentrations of deep groundwater do not exceed 0.4 g/L, due to water renewal during rainy episodes, while locally in the center of the basin, these values can reach 2 g/L due to exchanges with overlying environments. The aquiclude acts as a barrier to the chemical contamination of the deep aquifer waters, thanks to the properties of the lower Quaternary clays that slow down the propagation of dissolved elements.

The development of the hydrodispersive model also allows for locating the freshwater–saltwater interface in September 2007 at 500 m from the oceanic shoreline, with a transition zone of 250 m thickness to the southwest of the basin. This interface gradually approaches the shoreline in a northward direction.

Modeling work over a geographical extent of 4000 km^2 represents a significant challenge, aimed at providing for the first time a multilayer model of the entire Rharb basin. This model takes into account not only the interaction between the aquifer system of the Rharb and the external environment but also the interaction between the deep aquifer, the aquitard, and the shallow aquifer. Nevertheless, it is recommended to use this work as preliminary guidance for pumping tests to ensure greater accuracy and provide new data, which in turn can enrich such modeling efforts.

In summary, this research highlights the crucial importance of understanding the hydrogeological mechanisms at work in the coastal basin of Rharb, especially as groundwater aquifers in Morocco's coastal regions are under serious threat due to climate change [41]. It underscores the need to develop tools and numerical models for sustainable management of groundwater resources, preventing the harmful consequences of overexploitation, and ensuring a balanced and responsible use of these precious resources for future generations, thus contributing to environmental sustainability.

Author Contributions: Conceptualization: M.A.H. and S.H.; In-field work: S.H.; Methodology: S.H.; Software: S.H., Formal analysis: M.B. and T.K.F.; Data interpretation: M.B., T.K.F. and P.M.C.; Writing—original draft preparation: S.H.; Writing—review and editing: S.H. and M.B.; Supervision: M.A.H. and M.B. All authors have read and agreed to the published version of the manuscript.

Funding: This research was funded by the Researchers Supporting Project at King Saud University, Riyadh, Saudi Arabia (Fund no. RSP2025R487).

Data Availability Statement: Data are included in the manuscript.

Acknowledgments: The authors would like to thank the anonymous reviewers and editors, and express their sincere gratitude to the Laboratory of "Geodynamics and Natural Resources", Reception Structure, Faculty of Sciences Dhar Mahraz, Fez, for its assistance through the provision of equipment useful for the work. We also wish to thank the staff of the Directorate of Water Research and Development, Rabat, for their collaboration; the staff of the Water Management and Planning Department at the Sebou Water Basin Agency, Fez, for their support through the provision of available data useful for the work; and the staff of the Western Regional Directorate, "National Office of Electricity and Drinking Water/Water Branch", Kenitra, for providing relevant data essential to the work. We extend our appreciation to the Researchers Supporting Project at King Saud University, Riyadh, Saudi Arabia, for funding this research project, (Fund no. RSP2025R487).

Conflicts of Interest: The authors declare no conflicts of interest.

Appendix A

Table A1. Storage coefficients (S) of the Plio-Quaternary aquifer, Rharb basin.

IRE	X (m)	Y (m)	S
1521	417,025	448,800	0.001
1541	420,400	458,950	0.002
1716	438,900	404,600	0.005
1819	409,800	433,700	0.0002
3312	460,350	403,350	0.006
3415	393,270	415,270	0.007
3416	393,242	415,040	0.11

Appendix B

Table A2. Lithostratigraphic column from borehole 253/8 (UQ: Upper Quaternary, PQ: Plio-Quaternary).

From	To (m)	Lithology	Stratigraphy	Z (m)
0	3	Topsoil	UQ	5.32
3	9	Yellow clayey sand	PQ	2.32
9	11	Compact clay	PQ	−3.68
11	24	Yellow sandy clay	PQ	−5.68
24	30	Yellow clayey sand	PQ	−18.68
30	33	Sandstone and Yellow Sand	PQ	−24.68
33	35	Sandy clay	PQ	−27.68
35	43	Sandstone and Yellow Sand	PQ	−29.68
43	45	Red sandy clay	PQ	−37.68
45	46	Compact marl	PQ	−39.68
46	54	Clay	PQ	−40.68
54	61	Sandstone and Sand	PQ	−48.68
61	73	Red Clay and Sand	PQ	−55.68
73	76	Clay	PQ	−67.68
76	81	Clay and Sand	PQ	−70.68
81	84	Yellow clay	PQ	−75.68
84	88	Yellow sandstone	PQ	−78.68
88	94	Clay	PQ	−82.68
94	118	Yellow Sands and Sandstone	PQ	−88.68
118	127	Red sand	PQ	−112.68
127	151	Sandstone, Yellow Clay and Sand	PQ	−121.68
151	153	Compact marl	PQ	−145.68
153	164	Sandstone and Sand	PQ	−147.68
		-	-	−158.68

Appendix C

Table A3. Results of chemical analyses at different sampling points.

Sample							Cations (mg/L)							Anions (mg/L)				
	T (°C)	COND. (uS/cm)	pH	TH (°F)	TAC (°F)	RS (mg/L)	NH_4^+	Na^+	K^+	Ca^{2+}	Mg^{2+}	Fe^{2+}	Mn^{2+}	Cl^-	NO_2^-	NO_3^-	HCO_3^-	SO_4^{2-}
8/911	19.2	550	7.2	20.495	13.5	386	0.051	27.44	0.41	73.1	5.46	*	*	49	*	55.3	164.7	8.66
8/1135	21.5	1105	7.4	37.509	17.85	783	0.019	71.92	1.47	136.3	8.51	*	*	161	0.038	42	217.7	64
8/1418	22.4	1470	7.4	49.097	17.9	1104	0.009	102	1.36	144	32	*	*	315	*	18.23	218.3	20
14/R2964	21.5	4050	7.3	110.054	55	2702	0.032	326	132	164	168	*	*	750	0.182	160	671	237
14/1852	23.5	1390	7.5	30.536	15	643	0.022	79.37	2.42	96	16	*	*	204	*	7.04	184	13
14/1025	19	405	7.9	9.833	11	231	0.019	36	1.47	31	5.1	*	*	27	0.008	9.56	134.2	9
14/R1026	20	1090	7.4	29.956	26.1	585	0.139	118	1.82	88	19.44	*	0.118	153	0.084	25	318	35
14/3665	21	870	7.6	26.723	19.1	578	0.036	64	21.68	72.1	21.2	*	*	70	0.053	87	233	61
14/3710	23.4	1440	7.6	31.457	29.5	818	0.027	168	1.75	70	34	*	*	212	*	6.2	359.9	54
14/2910	19.5	1961	7.2	37.989	41.5	1101	0.014	282	1	50	62	*	0.013	350	0.025	5.41	506.3	76
14/648	18.7	2560	7.3	68.06	57	1476	0.03	272	0.63	93	109	*	*	376	0.094	49.08	695.4	132
14/3849	20	700	7.4	20.467	16.7	405	0.054	59.49	1.79	66	9.72	*	*	73		11	203.7	27.5
14/994	20.5	1010	7.4	31.962	22.2	605	0.077	83	1.31	103.2	15.1	*	*	138	*	32.24	271	31
14/3695	20.5	770	7.5	22.963	15.5	498	0.013	60	1.59	70	13.36	*	*	105	*	2.7	189	23
8/R574	19.5	3073	7.1	79.789	35.75	2194	0.037	447	6.7	173	89	*	*	685	0.649	252	439	143
8/1294	22.9	925	7.4	20.482	26.9	514	0.014	117.08	1.82	59	14	*	*	110	*	0.03	328	17
8/636	20.2	1110	7.4	28.206	18.3	709		107	1.41	86	16.4	*	*	210	*	5.3	223.2	17
8/R1644	20	3450	7.7	74.462	26.9	2480	0.008	383	2.11	150	90	*	*	802	*	3.57	326.9	188
8/1502	20.5	2090	7.1	53.614	20.4	1450	0.069	203	2.61	172	26	*	*	426	0.033	60	249	117
8/R1563	20.8	2080	7.2	63.756	20	1605	0.076	141	6.53	211	27	*	*	259	0.007	302	244	130
8/1765	21	2330	7.1	47.775	36.5	1336	0.055	262	27	142	30	*	*	366	0.02	46	445.3	152
8/672	21.2	1810	7.3	51.132	12.7	1406	0.05	111	42	167	23	*	*	263	0.033	258	156	76
8/1513	21	3680	7.6	68.263	48.8	2286	0.013	556	3.25	140	81	*	*	642	0.006	44.81	595.3	318
14/1514	17.5	2850	7.8	73.924	37.3	1970	0.012	292	1.29	133	99	*	0.136	518	0.133	29.87	455	227
8/1062	21.7	4200	6.7	139.831	51.6	3276	0.021	351	11.38	268.5	177	*	0.691	798	0.136	57.62	629.5	385
8/1735	22	1680	7.5	32	24.9	925	0.026	169	0.85	80.2	29.16	*	0.024	302	*	0.19	303.7	12
Août*05	20.5	2390	7.4	44.593	20.8	1437	0.022	258	6.92	121	35	*	0.016	539	0.03	19.65	254	32
14/R1027	19.5	1360	7.3	26.954	21.6	729	0.018	166	0.62	92	9.72	*	0.06	246	0.038	8	263.5	45
GP1	21.4	790	7.6	22.99	20	462	0.071	63	1.82	74	11	*	*	66.6	0.014	32	244	36.3
GP2	21	2760	7	60.934	27	1793	0.046	287	1.86	170	45	*	*	635	0.007	5.15	329.4	11
GP3	22.5	965	7.2	32	14.45	696	0.022	65.27	0.74	104.2	14.58	*	*	160	*	54	176.2	29
GP4	19.3	845	7.5	29.999	14.4	637	0.014	48.23	1.41	105.2	9.11	*	*	92	0.152	115	175.6	24
GP5	24	9890	7.1	271.955	33.9	6885	0.028	1161	0.82	549	328	*	1.545	2734	0.068	93.52	413.5	893
GP6	23	4900	7.2	152.032	32.7	3630	0.021	654.59	3.26	197	250	*	0.148	1422	0.028	5.67	399	482
GP7	22.5	2780	7.4	67.946	10	1847	0.012	170.27	2.81	208	39	*	*	631	*	5.69	116	32
GP8	22	5120	6.8	97.112	40	3490	0.527	697	5.35	188	122	*	0.385	1344	0.015	0.29	488	73
P1	22	875	7.1	11.482	8.95	498	0.089	143.9	10.2	35	6.68	*	*	206	0.02	18.2	109	26
P2	22	420	6.9	13.99	1.2	293	0.014	23.59	6.76	37.6	11.2	*	*	47.6	0.005	67.42	73.2	1.82
P3	24.5	460	7.8	19.97	14.9	285	0.171	20.4	4	60	12.15	*	*	40	*	50	181.8	4.69

* Undefined.

References

1. Bouras, E.; Jarlan, L.; Khabba, S.; Er-Raki, S.; Dezetter, A.; Sghir, F.; Tramblay, Y. Assessing the impact of global climate changes on irrigated wheat yields and water requirements in a semi-arid environment of Morocco. *Sci. Rep.* **2019**, *9*, 19142. [CrossRef]
2. Brouziyne, Y.; Abouabdillah, A.; Hirich, A.; Bouabid, R.; Zaaboul, R.; Benaabidate, L. Modeling sustainable adaptation strategies toward a climate-smart agriculture in a Mediterranean watershed under projected climate change scenarios. *Agric. Syst.* **2018**, *162*, 154–163. [CrossRef]
3. Tao, J.; Huang, L.; Xiong, S.; Li, L.-X.; Wang, L.L.; Xu, L. Two-dimensional TMDs/MN (M = Al, Ga) van der Waals heterojunction photocatalyst: A first-principles study. *J. Mater. Sci.* **2023**, *58*, 14080–14095. [CrossRef]
4. Werner, A.D.; Bakker, M.; Post, V.E.A.; Vandenbohede, A.; Lu, C.; Ataie-Ashtiani, B.; Simmons, C.T.; Barry, D.A. Seawater intrusion processes, investigation and management: Recent advances and future challenges. *Adv. Water Resour.* **2013**, *51*, 3–26. [CrossRef]
5. Radu, G.; Carabin, G.; Hallet, V.; Peters, V.; Dassargues, A. GIS-based hydrogeological databases and groundwater modelling. *Hydrogeol. J.* **2001**, *9*, 555–569. [CrossRef]
6. Hamdi, M.; GoitA, K.; Zagrarni, M.F. Integrated 3D geological and hydrodynamic modeling for the water budget estimation: Sisseb El Alem Nadhour Saouaf case study. *Int. J. Innov. Sci. Res.* **2021**, *52*, 185–192.
7. Mary, X.A.; Rose, L.; Rajasekaran, K. Continuous and Remote Monitoring of Ground Water Level Measurement in a Well. *Int. J. Water* **2018**, *12*, 356–369. [CrossRef]
8. Alkhatib, J.; Engelhardt, I.; Ribbe, L.; Sauter, M. An integrated approach for choosing suitable pumping strategies for a semi-arid region in Jordan using a groundwater model coupled with analytical hierarchy techniques. *Hydrogeol. J.* **2019**, *27*, 1143–1157. [CrossRef]
9. Dehghanipour, A.; Morid, S.; Delavar, M.; Ahmadalipour, A. A WEAP-MODFLOW surface water-groundwater model for the irrigated Miyandoab plain, Urmia lake basin, Iran: Multi-objective calibration and quantification of historical drought impacts. *J. Hydrol. Reg. Stud.* **2019**, *22*, 100590. [CrossRef]
10. Radulovic, M.M.; Wong, H.; Al Tarawneh, M. Groundwater flow modelling for an assessment of abstraction and climate change impacts on groundwater resources—An example from the Wadi Araba Basin (Jordan). *Arab. J. Geosci.* **2016**, *13*, 1142. [CrossRef]
11. Salameh, E.; Abdallat, G.; van der Valk, M. Planning Considerations of Managed Aquifer Recharge (MAR) Projects in Jordan. *Water* **2019**, *11*, 182. [CrossRef]
12. Hamdi, M.; Goïta, K.; Karaouli, F.; Zagrarni, M.F. Hydrodynamic groundwater modeling and hydrochemical conceptualization of the mining area of Moulares Redeyef (southwestern Tunisia): New local insights. *Phys. Chem. Earth* **2021**, *121*, 102974. [CrossRef]
13. Motevalli, A.; Moradi, H.R.; Javadi, S. A comprehensive evaluation of groundwater vulnerability to saltwater up-coning and sea water intrusion in a coastal aquifer (Case Study: Ghaemshahr-Juybar Aquifer). *J. Hydrol.* **2018**, *557*, 753–773. [CrossRef]
14. Rasmussen, P.; Sonnenborg, T.O.; Goncear, G.; Hinsby, K. Assessing impacts of climate change, sea level rise, and drainage canals on saltwater intrusion to coastal aquifer. *Hydrol. Earth Syst. Sci.* **2013**, *17*, 421–443. [CrossRef]
15. Busico, G.; Kazakis, N.; Cuoco, E.; Colombani, N.; Tedesco, D.; Voudouris, K.; Mastrocicco, M. A novel hybrid method of specific vulnerability to anthropogenic pollution using multivariate statistical and regression analyses. *Water Res.* **2020**, *171*, 171–115386. [CrossRef] [PubMed]
16. Jahromi, M.N.; Gomeh, Z.; Busico, G.; Barzegar, R.; Samany, N.N.; Aalami, M.T.; Tedesco, D.; Mastrocicco, M.; Kazakis, N. Developing a SINTACS based method to map groundwater multi-pollutant vulnerability using evolutionary algorithms. *Environ. Sci. Pollut. Res.* **2020**, *28*, 28–7854. [CrossRef] [PubMed]
17. Hakimi, S.; Hessane, M.A.; Lyazidi, R. Elaboration of a steady-state hydrodynamic model: The case of the Gharb basin, Morocco. *Comun. Geológicas* **2020**, *106*, 11–27.
18. ABHS. *Etude D'actualisation du Plan Directeur D'aménagement Intégré des Ressources en Eau du Bassin Hydraulique de Sebou*; Agence du Bassin Hydraulique de Sebou: Fès, Morocco, 2011; p. 103.
19. Lyazidi, R.; Hessane, M.A.; Filali, J.M.; Bahir, M.; Ouhamdouch, S. Système d'aide à la décision pour la gestion de la ressource en eau en zones semi-arides—Cas de l'aquifère Gareb-Bouareg (Rif, Maroc). *Comun. Geológicas* **2017**, *104*, 43–52.
20. Combe, M. *Carte Hydrogéologique de la Plaine du Gharb au 1/100 000. Notes et Mémoires*; Service Géologique du Maroc: Rabat, Morocco, 1969; p. 221 bis.
21. Zheng, C.; Wang, P. *MT3DMS: A Modular Three-Dimensional Multispecies Transport Model for Simulation of Advection, Dispersion, and Chemical Reactions of Contaminants in Groundwater Systems*; Documentation and User's Guide; USA Army Corps of Engineers: Washington, DC, USA, 1999.
22. Raju, K.R.S.R.; Varma, G.H.K. Knowledge-Based Real-Time Monitoring System for Aquaculture Using IoT. In Proceedings of the 2017 IEEE 7th International Advance Computing Conference (IACC), Hyderabad, India, 5–7 January 2017; pp. 318–321. [CrossRef]
23. Kili, M.; El Mansouri, B.; Taky, A.; Chao, J. Nouvelle approche d'estimation des prélèvements d'eau d'irrigation à partir des ressources souterraines: Cas de la nappe côtière du Gharb. *Bull. Inst. Sci. Rabat Sci. Terre* **2006**, *28*, 31–39.

24. Flinch, J.F. Tectonic Evolution of the Gibraltar Arc. Ph.D. Thesis, Rice University, Houston, TX, USA, 1993; p. 212.
25. Zouhri, L.; Gorini, C.; Mania, J.; Deffontaines, B.; El, A.; Zerouali, H. Spatial distribution of resistivity in the hydro-geological systems, and identification of the catchment area in the Rharb basin, Morocco. *Hydrol. Sci. J.* **2004**, *49*, 3. [CrossRef]
26. Cirac, P. Le Bassin Sud Rifain Occidental au Néogène Supérieur. Evolution de la Dynamique Sédimentaire et de la Paléographie au Cours D'une Phase de Comblement. Ph.D. Thesis, d'État, Univ. Bordeaux, Bordeaux, France, 1985; p. 285.
27. Toto, E.A.; El Miloudi, A.; El Basri, M.; Hafid, L.; Zouhri, L.; El Mouraouah, S.; Benammi, M.; El Mouraouah, A.; Iben-Brahim, A.; Birouk, A.; et al. New geophysical and geological evidence for the present-day southernmost active deformational front of the Rif thrust and fold belt and the oceanic accretionary prism of Cadiz: The Dhar Doume Lalla Zahra fault, Northwestern Atlantic Coastal Morocco. *Environ. Earth Sci.* **2012**, *67*, 2411–2422. [CrossRef]
28. Tahayt, A.; Mourabit, T.; Rigo, A.; Feigl, K.L.; Fadil, A.; McClusky, S.; Reilinger, R.; Seroukh, M.; Ouazzani-Touhami, A.; BenSari, D.; et al. Mouvements actuels des blocs tectoniques dans l'arc Bético-Rifain à partir des mesures GPS entre 1999 et 2005. *Comptes Rendus Geosci.* **2008**, *340*, 400–413. [CrossRef]
29. Koulali, A.; Ouazar, D.; Tahayt, A.; King, R.W.; Vernant, P.; Reilinger, R.E.; McClusky, S.; Mourabit, T.; Davila, J.M.; Amraoui, N. New GPS constraints on active deformation along the Africa-Iberia plate boundary. *Earth Planet. Sci. Lett.* **2011**, *308*, 211–217. [CrossRef]
30. Platt, J.P.; Whitehouse, M.J.; Kelley, S.P.; Carter, A.; Hollick, L. The ultimate arc: Differential displacement, oroclinal bending, and vertical axis rotation in the External Betic-Rif arc. *Tectonics* **2003**, *22*, 3. [CrossRef]
31. McDonald, M.G. *A Modular Three-Dimensional Finite-Differences Ground-Water Flow Model*; USA Geological Survey: Dallas, TX, USA, 1984.
32. Wels, C.; Mackie, D.; Scibek, J. *Guidelines for Groundwater Modelling to Assess Impacts of Proposed Natural Resource Development Activities*, British Columbia Ministry of Environment: Victoria, BC, Canada, 2012; p. 289.
33. Tranouez, P. Contribution à la Modélisation et à la Prise en Compte Informatique de Niveaux de Descriptions Multiples. Application aux Écosystèmes Aquatiques. Ph.D. Thesis, Univ. du Havre, Le Havre, France, 2005; p. 191.
34. De Marsily, G. *Cours d'Hydrogéologie*; Université Pierre et Marie Curie: Paris, France, 2004; p. 226.
35. Dassargues, A. Modèles mathématiques en hydrogéologie et paramétrisation des aquifères. *Rev. D'hydrogéologie* **2007**, *15*, 45–67.
36. Langevin, C.D.; Guo, W. MODFLOW/MT3DMS–Based Simulation of Variable-Density Ground Water Flow and Transport. *Groundwater* **2006**, *44*, 339–351. [CrossRef] [PubMed]
37. Habot, J. Diagrammes trilineaires et diagrammes rectangulaires pour l'analyse des eaux. *Hydrogeologie* **1987**, *1*, 57–63.
38. Ministère de l'Agriculture, de la Pêche Maritime, du Développement Rural et des Eaux et Forêts. *Situation de L'agriculture Marocaine n° 15*; Recherche et Développement-Innovation Agricole Nationale 2030: Feuille de route; Ministère de l'Agriculture, de la Pêche Maritime, du Développement Rural et des Eaux et Forêts: Rabat, Maroc. 2023. Available online: https://www.onicl.org.ma/portail/wp-content/uploads/2023/02/rapport-situation-agriculture-marocaine-n%C2%B015-2019-2020.pdf (accessed on 31 December 2024).
39. Haut Commissariat au Plan. *Monographie régionale de la région Rharb-Chrarda-Béni Hssen*; Haut Commissariat au Plan: Rabat, Maroc, 2013; Available online: https://www.hcp.ma/region-kenitra/attachment/647182/ (accessed on 31 December 2024).
40. El Mountassir, O.; Bahir, M.; Hakimi, S.; Faraj, T.K.; Carreira, P.M. Assessment of Groundwater Quality in the Semi-Arid Environment: Implications of Climate Change. *Limnol. Rev.* **2024**, *24*, 520–542. [CrossRef]
41. El Mountassir, O.; Mohammed, B. The Assessment of the Groundwater Quality in the Coastal Aquifers of the Essaouira Basin, Southwestern Morocco, Using Hydrogeochemistry and Isotopic Signatures. *Water* **2023**, *9*(15), 1769. [CrossRef]

hydrology

MDPI

Article

Assessing the Effects of Bioenergy Cropping Scenarios on the Surface Water and Groundwater of an Intensively Agricultural Basin in Central Greece

Maria Sismanidi [1], Lamprini Kokkinaki [1], Sofia Kavalieratou [1], Haralampos Georgoussis [1], Kyriakos D. Giannoulis [2], Elias Dimitriou [3] and Yiannis Panagopoulos [1,*]

[1] Department of Hydraulics, Soil Science and Agricultural Engineering, School of Agriculture, Aristotle University of Thessaloniki, 54124 Thessaloniki, Greece; msismanid@agro.auth.gr (M.S.); lakokkin@agro.auth.gr (L.K.); kavalier@agro.auth.gr (S.K.); harisg@agro.auth.gr (H.G.)

[2] Laboratory of Agronomy and Applied Crop Physiology, Department of Agriculture, Crop Production & Rural Environment, University of Thessaly, Fytokoy Str., 38446 Volos, Greece; kgiannoulis@uth.gr

[3] Hellenic Centre for Marine Research, Institute of Marine Biological Resources and Inland Waters, 46.7 km Athens—Sounio Ave., 19013 Anavyssos, Attica, Greece; elias@hcmr.gr

* Correspondence: ypanag@agro.auth.gr; Tel.: +30-2310998708

Abstract: Pinios river basin constitutes the most important agricultural production area in Greece but contributes to the degradation of the quality and quantity of surface water and groundwater bodies. Bioenergy crops implemented as part of the existing cropping systems could be a novel and efficient mitigation strategy against water degradation, contributing to the production of energy through renewable sources. This study uses the Soil and Water Assessment Tool (SWAT) to first develop a representative model of Pinios river basin and evaluate its current state with respect to water availability and nitrate water pollution. A low-input perennial bioenergy crop, switchgrass, is then simulated closely to the Greek conditions to investigate its potential effects on water in three implementation scenarios: the installation and growth of switchgrass in the entire irrigated cropland, exclusively in irrigated sloping (slopes > 1.5%) cropland, and exclusively in irrigated non-sloping cropland. The simulated results demonstrate that under all scenarios, the water quality improvements with respect to the nitrate loads entering surface water and groundwater bodies were significant, with their reduction being directly affected by the extent to which switchgrass replaced resource-demanding conventional crops. Specifically, the reduction in the annual nitrate loads in the surface water under these three scenarios varied from 7% to 18% at the river basin scale, while in certain cropland areas, the respective reduction even exceeded a level of 80%. The potential to improve the water status was also considerable, as the implementation of the bioenergy crop reduced the irrigation water used annually in the basin by 10% (64 Mm^3) when switchgrass replaced the conventional crops only on the sloping land and by almost 30% (187 Mm^3) when it replaced them throughout the irrigated land. At the same time, significant biomass production above 18 t/ha/y applied in all of the simulations. This study also highlights the contribution of the bioenergy crop to the rehabilitation of the groundwater levels across the basin, with the possibility of increasing them by >50% compared to the baseline, implying that the adoption of switchgrass could be a promising means against water scarcity.

Keywords: bioenergy crop scenarios; biomass production; nitrate pollution; Pinios river basin; SWAT; switchgrass; water quality; water quantity

1. Introduction

Water degradation is often caused by the intensification of agricultural activities. Especially around the rural landscape of the Mediterranean region, agriculture has led to the overexploitation of water reserves for irrigation and to a decline in water quality due to non-point source pollutants like fertilizers and pesticide residues entering water bodies [1–5]. While intensive agriculture is crucial for food, feed, and fiber production, achieving a balance between water quality and quantity is a challenge. Therefore, appropriate management practices are required to accomplish these conflicting objectives and meet the objective of a good ecological status required by the Water Framework Directive (WFD) (Directive 2000/60/EC, 2000) and the Nitrates Directive (Directive 91/676/EEC) [6].

In addition, agriculture is being promoted globally as a significant method for sustainable energy production through the cultivation of bioenergy crops for renewable energy. The European Union (EU) has implemented measures to advance biofuels through the Renewable Energy Directive (Directive 2018/2001), which is part of the European Green Deal aiming for a climate-neutral EU by 2050. Biomass and biofuels are seen as promising energy sources within the current energy crisis in the EU, as they can be produced domestically, thereby reducing the dependence on energy imports. The need to increase bioenergy production in Greece has been recognized in recent decades [7]. Efforts to increase the combined heat and power production from biomass and to replace portions of gasoline and diesel with liquid biofuel have been reported from more than a decade [8]. However, the national use of renewable energy sources remains low. As an EU member state, Greece must comply with the European Directive, which, for example, mandates that biofuels should contribute to >10% of the total transport energy consumption by 2030 [9].

Besides energy targets, bioenergy production can, however, have positive environmental impacts when it comes from bioenergy crops that are incorporated within existing resource-demanding cropping systems. Specifically, perennial bioenergy crops are considered a viable option since they have the capacity to substantially reduce nitrate losses and subsequently reduce non-point source pollution [10]. In addition, bioenergy crops like sunflower and sorghum require less irrigation water than most irrigated crops, enhancing groundwater conservation [11]. In particular, the perennial bioenergy crop switchgrass, introduced from the US and characterized as low-input, is expected to exhibit similar behavior. Nevertheless, producing bioenergy requires extensive agricultural areas, often leading to significant land use changes. Therefore, it is essential to assess a bioenergy crop's socio-economic and environmental footprint before selecting it for regional planning. In Greece, no such plan has yet been proposed for any agricultural region to ensure the continuous, uninterrupted production of bioenergy products.

The Thessaly plain, located in the center of Greece, represents the most important agricultural producer in the country. The largest river basin in the area is Pinios river basin (PRB), covering roughly 11,000 km^2. It is characterized by a Mediterranean climate, fertile soils, and a mild topography and is considered ideal for a variety of crop cultivation. These characteristics have caused the intensification of irrigated agriculture in the region, with water quantity and quality implications [5,12]. This region is included in the list of areas characterized as nitrate-vulnerable zones, with a negative water balance [13].

For the assessment of land use changes at the river basin scale, including changes in cropping patterns and their effects on water resources, advanced hydrological models that can provide time-efficient water quantity and quality predictions are widely utilized. Specifically, the process-based Soil and Water Assessment Tool (SWAT) model has been used worldwide in a wide range of studies, among which one can find a few involving bioenergy crop simulations [14–16]. Of substantial interest to the present paper are SWAT studies on bioenergy crops that have highlighted significant nitrate loss reductions and considerable biomass accumulation for potential energy production, even when they are grown in low-productivity areas [17,18]. Moreover, Gassman et al. [19] found that the realistic, limited inclusion of bioenergy crops within agricultural systems can result in a considerable reduction in the loss of nitrates from land to water.

Recent publications on SWAT from Greece that can be considered relevant to the present study are only studies that have investigated the impacts of water management practices on the hydrological status of rivers or the water budgets of agricultural areas, as well as studies that have dealt with land use changes and their effects on pollution loads [20–22]. To the best of our knowledge, no studies dealing with bioenergy crop simulation in Greece exist in the literature.

The present study aims to build a dynamic SWAT hydrological and management model of the intensively managed PRB that considers climate, land use, and agricultural practices to estimate the crop yields and evaluate the resulting quantity and quality of the surface water and groundwater. Using this model, this study is the first to attempt a realistic simulation of the switchgrass growth cycle in Greece and execute rough scenarios of its implementation across large agricultural areas in order to explore the potential upper levels of water improvements and biomass production at the river basin scale.

2. Materials and Methods

2.1. The Study Area

Pinios river basin covers the river basin district (RBD) of Thessaly almost entirely, which is located in central Greece, as can be observed in Figure 1. Specifically, it covers an area of about 11,000 km^2 and is the most important agricultural producer in the country. Agriculture is associated with almost 45% of the total basin area [23]. According to the Hellenic Statistical Authority's Agriculture, Livestock, and Fishery sector, for the year 2022, crops on arable land covered 80% of the total cultivated agricultural and fallow land [24]. The main crops cultivated in the PRB are cotton and wheat, followed by much smaller areas of corn and alfalfa. The average annual precipitation in the basin is estimated at 700 mm [25,26], and the average annual river flow beyond "Tempi" and near the outlet (Figure 1) is roughly 80 m^3/s, as reported by Panagopoulos et al. [12].

The region has been classified as a nitrate-vulnerable zone (Nitrates Directive) [27], mostly due to its intensive agricultural activities, particularly related to the growth of wheat, cotton, and corn. Irrigation accounts for 94% of the total water usage according to the Hellenic Ministry for the Environment, Energy and Climate Change's Special Secretariat for Water [28]. Since groundwater currently supplies more than 65% of the region's total water usage, groundwater is crucial to the region's sustainability [23]. Two of the twenty-seven groundwater bodies recognized in the PRB, covering 2184 km^2, have been classified as "bad"-quality, while nine are recorded to have a "bad" quantity status according to the most recent regional water management plan [29]. In addition, three reservoirs operate in the area, contributing to a lower extent as irrigation sources (Figure 1). Specifically, "Girtoni" reservoir is located in the northeastern part of the basin, "Karla" is the reservoir in subbasin 41 in the southeastern part (Figure 1), and "Smokovo" is the reservoir in the south. These three reservoirs irrigate approximately 13,500 ha, 7600 ha, and 11,500 ha, respectively. Additionally, 9600 ha of cropland in the southwestern part of the basin (subbasins 39, 43, and 47) is irrigated by an outside source, "Plastiras" lake, as shown in Figure 1.

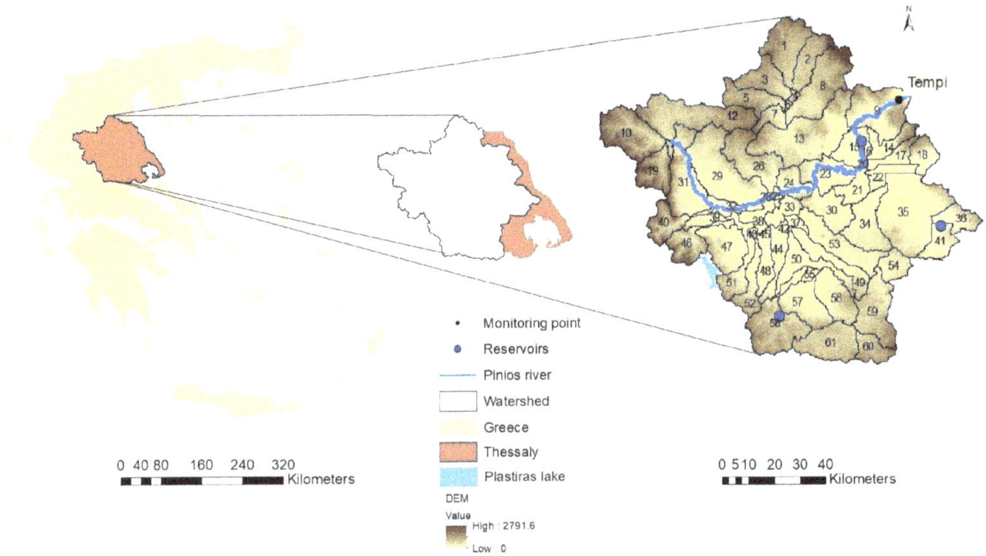

Figure 1. The location of Pinios river basin in Thessaly, central Greece, and a Digital Elevation Model of the basin with monitoring points and reservoirs.

2.2. Description of the SWAT Model

SWAT was developed by the US Department of Agriculture in collaboration with Texas A&M University [30] for use in complex agricultural landscapes. It is a semi-distributed, process-based, small watershed- to river-basin-scale model used for the simulation of hydrologic, sediment, and nutrient processes, as well as the crop growth, management practices, and land use changes, in a watershed using the water balance approach [31]. The watershed in the GIS-based SWAT is divided into subbasins, and one reach is associated with each subbasin. The subbasins are then divided into Hydrologic Response Units (HRUs), which represent a unique combination of land use, soil type, and topography.

The runoff and loadings such as nutrients, sediments, and pollutants transported by runoff are calculated separately in each HRU and then summed together to define the total loadings from the subbasin. Crop growth is also modeled at the HRU scale. Water balance is the driving force behind the water quality processes and crop growth rates occurring in the basin, including the simulation of surface runoff/infiltration, evapotranspiration, lateral flow, percolation, and return flow. Surface runoff is estimated at the HRU level by using a modification of the Soil Conservation Service (SCS) curve number (CN) method [32]. Groundwater aquifers are separated into two systems in SWAT: a shallow unconfined aquifer, which governs the amount of return flow to streams, and a deep confined aquifer, which does not contribute to the return flow but can act as a source or sink. The soil erosion caused by rainfall is calculated by the Modified Universal Soil Loss Equation (MUSLE) [33], which is a modified version of the Universal Soil Loss Equation (USLE) developed by Wischmeier and Smith [34].

The crop growth component of SWAT is a simplified version of the Erosion Productivity Impact Calculator (EPIC) model, which is capable of simulating a wide range of crop rotation, pastureland, and trees. Agricultural management practices are defined in SWAT by the specific management operations [12,32,35] affecting every cropping and livestock system, including bioenergy crops [36,37], by defining planting, harvesting, tillage passes, irrigation, grazing, and nutrient and pesticide applications. In order to simulate the PRB, the latest available version of ArcSWAT 2012, Release 687 compatible with ArcGIS 10.8.2, was implemented.

2.3. The Model Parameterization for Pinios River Basin in SWAT

A terrain layer was inserted to start the modeling setup for the PRB. An available 25 × 25 Digital Elevation Model (DEM) for the study area [12] was used to delineate the study area of PRB, covering a large part of the river basin district of Thessaly in central Greece (Figure 1). The area of the basin delineated was 10,623 km^2, covering roughly 75% of the Thessaly region. The elevation ranges significantly in this region, with the maximum values around 2000 m near the borders of the basin. The basin was then divided into 61 subbasins based on the natural hydrologic pathways defined by the surface topography. For delineation of the stream, a 10,000 ha upstream drainage area was decided (1% of the total basin's area), defining the minimum drainage area required to form the origin of a stream. The river locations with accessible measured flows, the meteorological stations of the area, and the existing dam locations were also taken into account in the definition of the subbasins.

A land cover layer derived from Corine (CLC) 2012 [38] was used in the modeling. The use of a land use map from 2012 to the present was facilitated by a direct comparison with CLC 2018, which showed insignificant changes in the land cover types in PRB, but also by comparing the current crop patterns in the agricultural land with those from previous decades. So, both recent and older crop allocation data from the Hellenic Statistical Authority were analyzed and confirmed the rather stable crop types and irrigated land in PRB in the last two decades [24]. In the final land use layer for SWAT, only crops covering a significant area (>1% of the total agricultural land) were considered. Specifically, the 10 land use classes included in the final map were as follows: pasture, forested areas, wetlands, urban lands, orchard trees, fallow areas, and wheat, corn, cotton, and alfalfa cultivation. Table 1 presents the basin areas assigned to the crops that were considered representative from 2010 onwards in PRB. Corn was the most resource-demanding crop in this study, receiving 630 mm of irrigation water and 364 kg N/ha annually. In addition, cotton was fertilized with 185 kg N/ha and was irrigated with 420 mm every year. Wheat, which occupied the largest area, as a non-irrigated crop, received 169 kg N/ha. Lastly, alfalfa, as a perennial crop with significant irrigation water consumption, needed >600 mm of irrigation water annually, while its fertilization did not include the N inputs.

Table 1. Distribution of cultivated crops in Pinios river basin.

Land Use	Area (ha)	Basin Percentage (%)
Alfalfa	15,287	1.4
Corn	22,908	2.0
Cotton	164,497	15.5
Fallow areas	84,159	7.9
Wheat	165,636	15.6

A soil map for PRB was produced using data obtained from the European Soil Database (ESDB) [39,40]. Table 2 presents some of the soil parameters required by SWAT and the corresponding parameters of the datasets based on ESDB. From the ESDB shapefile, the polygons located in the PRB were extracted, resulting in a shapefile with 19 different soil types. All of these 19 soils had two layers, with the first layer always being 300 mm in depth. Most of the soils belonged to hydrological category B, having moderate infiltration rates and water capacities. Specifically, among the most important parameters used by SWAT, the available water capacities of the soils ranged between 0.14 and 0.17 mm water/mm soil, and their saturated hydraulic conductivity ranged between 9.98 and 21.69 mm/h.

Table 2. Soil parameters in SWAT Pinios river basin model extracted from the European Soil Database [39,40].

SWAT Soil Layer Parameters	SWAT Parameter Definition	Parameters of Datasets Based on ESDB	Dataset
SOL_BD1 (g/cm^3)	Moist bulk density in the first soil layer	STU_EU_T_BD	ESDB-derived data
SOL_AWC1 (mm/mm)	Available water capacity of the first soil layer	(=FC-WP)	3D soil hydraulic DB
SOL_CBN1 (% wt.)	Organic carbon content in the first soil layer	STU_EU_T_OC	ESDB-derived data
SOL_K1 (mm/h)	Saturated hydraulic conductivity in the first soil layer	KS	3D soil hydraulic DB
CLAY1 (% wt.)	Clay content in the first soil layer	STU_EU_T_CLAY	ESDB-derived data
SILT1 (% wt.)	Silt content in the first soil layer	STU_EU_T_SILT	
SAND1 (% wt.)	Sand content in the first soil layer	STU_EU_T_SAND	
ROCK1 (% wt.)	Roch Fragment content in the first soil layer	STU_EU_T_GRAVEL	
USLE_K1	USLE equation soil erodibility (K) factor in the first soil layer	K_factor_soiltexture_Wischmeier	Global soil erodibility

The study area was divided into two slope classes. Land with a higher slope was considered more erosive, and therefore, for the needs of this study, a threshold to distinguish it from lowland was required. From the initial model executions and simulated erosion rates per HRU, a threshold of 1.5% was considered critical for the significant occurrence of erosion. A similar approach was followed in another published study, with the critical threshold being 2%, very close to our selection [17]. A map of the irrigated cropland in the basin with distinctions between slopes above and below 1.5% is shown later in Section 2.8 of this manuscript. The above layers of the land use, soils, and slope classes were overlayed, creating 1850 HRUs. The average HRU area in PRB was 5.7 km^2.

Additional N pressures in the basin are added by livestock. The parameterization of the model took into consideration the recently reported numbers from the Hellenic Statistical Authority [41]. These are 108,363 cattle, 1,410,538 sheep, and 936,838 poultry during the years 2011 to 2021. These numbers corresponded to the three distinctive regions where livestock operations occur—the prefectures of Larisa, Trikala, and Karditsa—and the animals were spatially allocated during the setup of the model to match the respective number for each district. The pressures of livestock activity were assessed based on the total number of animals within each subarea and the types of manure associated with each animal category [32]. Grazing in the region occurs from April to October for cattle and goats, whereas during the remaining wet months of the year, it was assumed that the animals were confined to indoor or restricted open spaces, with their manure being collected into manure heaps and then deposited onto grassland areas on a regular weekly basis. Grazing and manure deposition in the model were simulated in pastureland. Due to insufficient data regarding poultry in the PRB, a unified approach was adopted, involving weekly spreading of poultry manure across the grassland areas (pastures) for the entire number of poultry in the basin (both domestic and those raised in organized poultry farms) throughout the year.

Finally, the nitrogen (N) pressures from the waste water treatment installations (WWTPs) of the two biggest cities within the PRB, "Larisa" and "Trikala", were added to the model as point source discharges into the rivers. Data were collected from the official website of the Ministry of Environment and Energy [42], as well as from the approved River Basin Management Plans of the 1st Revision, and specifically from detailed documentation of an analysis of anthropogenic pressures and their impacts on the surface water and groundwater systems of the Thessaly Water District [43]. In Larisa, the calculated daily N load was 175.2 kg, corresponding to 63,948 kg N/year, whereas in Trikala, the respective loads were 104.5 kg N/day or 38,142.5 kg N/year.

All of the datasets that were used in the development and subsequent evaluation of the model are summarized comprehensively in Table 3 below.

Table 3. Summarized information on the data used for the schematization, parameterization, and performance evaluation of the Pinios river basin SWAT model.

Dataset	Description of the Data
Topography	25 × 25 m Digital Elevation Model (DEM) data used to delineate the area and the river network
Land use	1:100,000 land cover layer derived from Corine 2012; crop patterns evaluated from the Hellenic Statistical Authority (ELSTAT) for the years 2012–2021
Soil	1 × 1 km soil map produced using information provided by the European Soil Database (ESDB)
Daily climate	Daily precipitation, maximum and minimum temperature, relative humidity, and wind speed data from 31 stations during the years 2016–2023
Crop parameters	Plant parameters needed for the SWAT plant database were provided by assistant professor at the School of Agriculture of the University of Thessaly Kyriakos Giannoulis (co-author of this article)
Point sources	Point source data were collected from the official website of the Ministry of Environment and Energy for the operation of the WWTPs, as well as from the Approved River Basin Management Plan of the Thessaly Water District (EL08)
Major dams/reservoirs	Key reservoirs for main channels of the Pinios river and its tributaries (Smokovo, Girtoni, and Karla) and Plastiras reservoir located outside the basin
River flow	Daily measured river flow data (in m^3/s) accessible from the automatic monitoring network of hydrological stations of the Inland Water Department of the Hellenic Centre for Marine Research (HCMR) for the years 2016–2023
River water quality	Observed seasonal data related to N species concentrations in rivers acquired by the national monitoring program for the implementation of the WFD in Greece conducted by HCMR for the years 2018–2022
Crop yields	Reported values provided by the Hellenic Statistical Authority for the years 2011–2021

2.4. Hydrological Evaluation of the Pinios River Basin Model

Our SWAT model of PRB was developed using meteorological data for 8 years (2016–2023). Measured river flow data were also accessible from the automatic monitoring network of hydrological stations of the Inland Water Department of the Hellenic Centre for Marine Research (HCMR) for almost the same period [44]. Flow data from the station "Tempi" near the outlet of PRB (Figure 1) were used to evaluate the flow simulations. A warm-up period of two years was also considered necessary with this rather short simulation period in order to eliminate initial uncertainties [45]. To deal with the problem of limitations in the flow data and accelerate the representation of the hydrological baseline using SWAT, we took advantage of previous detailed studies conducted using SWAT for the same study area [12,46]. Even with small differences in the land use types and crop growth schedules, transferring some of the parameters to a new model of the same area was considered realistic for simulating hydrology. Therefore, the default values for certain parameters, related mostly to the soil and groundwater modules in SWAT, were modified by adopting the respective well-calibrated parameters from these previous studies, which evaluated simulated flows in a location on Pinios river upstream of the "Tempi" station but also in several other nested subbasins within the model's domain. Therefore, the parameter values used for hydrology in the present study were quite representative of a reliable simulation of the monthly river flows on a multi-site basis, as recommended to achieve the required efficiency in the model's performance across the entire landscape [47]. The statistical results produced from the comparison of past measured flows with the respective simulations were sufficient [12], with the Nash–Sutcliffe Efficiencies (NSEs) and coefficients of determination (R^2) being close to "1", quite above the minimum thresholds required for a satisfactory performance [48]. The final parameter values used in the present SWAT modeling study were thus based on the reported optimum values, and some small further adjustments were made in the present work (Table 4) with the purpose of optimizing the convergence of the monthly flows simulated for "Tempi" station with the most recently available measured flows. Special attention was given to the parameters related to groundwater due to their important sensitivity [49].

Table 4. Calibrated parameters used consistently in the Pinios river basin model.

ID	Parameter	Optimum Value
1	GWQMIN (mm)	1000
2	GW_REVAP	0.02
3	REVAPMN (mm)	750
4	GW_DELAY (days)	31
5	CN2	Ranging from 65.05 to 82.65 depending on the land use and soil type
6	SOL_AWC (mm)	0.11–0.17 water/mm soil for each layer depending on the soil type
7	SOL_K (mm/h)	8.2–21.69 for each layer depending on the soil type
8	SHALLST (mm)	0

GWQMIN: threshold for the depth of water in a shallow aquifer required for return flow to occur (mm); GW_REVAP: groundwater "revap" coefficient; REVAPMN: threshold for the depth of water in a shallow aquifer for "revap" or percolation into a deep aquifer to occur (mm); GW_DELAY: groundwater delay (days); CN2: initial SCS runoff curve number for moisture condition I; SOL_AWC: available water capacity of the soil layer (mm); SOL_K: saturated hydraulic conductivity (mm/h); SHALLST: initial depth of the water in a shallow aquifer (mm).

It should be noted that the zero values selected for the last parameter in Table 4 had a critical role in modeling the PRB. This parameter represents the initial water content in the shallow aquifers in the basin, with significant values allowing for the exploitation of groundwater regardless of the magnitude of the occurrence of natural recharge. In the PRB, however, groundwater has been overexploited over the years. As a result, pumping from permanent reserves, which are usually found at high depths, is not economically and technically feasible nowadays. To allow the model to exploit renewable groundwater reserves only, namely the water resources that entered the aquifers each year, and in this way mimic the actual situation across the intensively irrigated cropland of the PRB, a zero value was assigned to this parameter for all HRUs at the beginning of the simulation (1 January 2016).

The hydrological performance of the model for the recent period of 2016–2023, with the available discharge data from HCMR, is shown in Figure 2. This figure depicts a comparison of the simulated versus the observed river flows at the flow station "Tempi", near the outlet of the basin (Figure 1). The available observed data covered a period of almost 4 years (2019 to 2023), with a few other stations having much fewer data available and therefore not allowing for a reliable comparison. The accuracy of the simulation was measured with the use of the NSE and R^2, which were calculated as 0.59 and 0.65, respectively. The former parameter was not at the highest levels, and this can be attributed to the inability of the model to sufficiently predict one flow peak in 2020 and two flow peaks in 2022, as shown in Figure 2. There are two possible explanations for this mismatch, which was disproportionally reflected in the statistics. Possible limitations in the precipitation input data which may have inadequately represented the extremity of the three indicated events are the first explanation. The second may possibly be related to the measuring process for the flows, which may have led to consistent overestimations of high-river-flow events. Nevertheless, as the graphical comparison (Figure 2) between the simulated and observed flows for the whole 4 y period with the available data revealed no bias or significant differences in the timing or magnitude of the peaks or the shape of the recession curves, the current overall match and the associated statistics were considered acceptable for the purposes of this study.

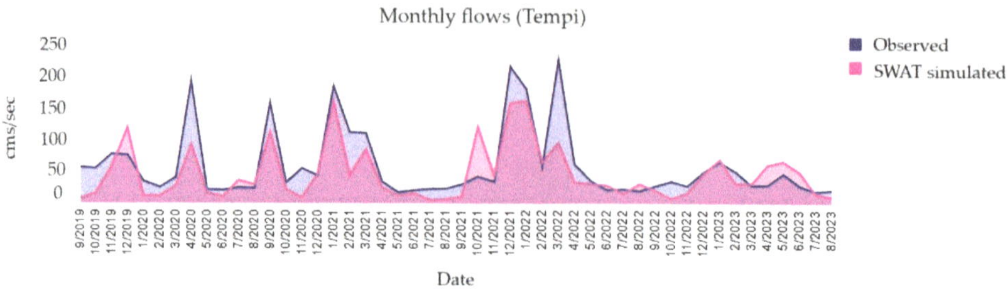

Figure 2. Simulated vs. observed river flows (2019–2023) at Tempi station.

2.5. Water Quality Evaluation of the Pinios River Basin Model

Due to the insufficient observed data on the river loads of nitrate-nitrogen (N-NO_3) or total nitrogen (TN) to conduct a detailed comparison with simulated loads, we adjusted a few empirical parameters. In Table 5, the most important calibrated parameters are shown with their final values. The very sensitive N percolation coefficient (NPERCO) was adjusted at the entire basin level based on data found in the literature in order to depict a simulated value that was more realistic to Greek conditions [50]. Moreover, the parameters CDN and SDNCO, which governed soil denitrification, were adjusted to achieve a reasonable simulated denitrification rate. The adjustment was terminated when the simulated denitrification from the soil was within 10–20% of the N fertilization [32].

Table 5. The main optimum nitrogen parameters of the SWAT Pinios River basin model.

Variable Name	Description	Normal Range	Final Value
CDN	Denitrification exponential rate coefficient	0.0–3.0	0.1
SDNCO	Denitrification threshold water content	0.1–1.1	0.997
NPERCO	N-NO_3 percolation coefficient	0.0–1.0	0.2
RCN	Concentration of nitrogen in rainfall (mg N/L)	-	1
RCN_SUB	Atmospheric deposition of nitrate (mg/L)	-	1

The river water quality observations related to N species were acquired by the national monitoring program for the implementation of the WFD in Greece conducted by HCMR [51]. These observed data comprised scarce samples within the five-year period of 2018–2022, evenly covering the wet (November–April) and dry (May–September) periods. Due to the limited and irregular frequency of the grab samples for measuring N-NO_3 and TN, these specific measurements could not be compared with simulated data reliably. Therefore, the observed sparsely sampled concentrations were averaged across the wet and dry periods for each year and then compared to the respective simulated seasonal concentrations. In situations like this where a complete measured time series is unavailable, the data are insufficient for an analysis using the recommended statistics mentioned in the hydrological evaluation in Section 2.4. Therefore, the model outputs for N-NO_3 and TN were compared graphically with the observations to ensure that the calibrated outputs were within the range of the seasonal nutrient levels in the river [50]. Figure 3 presents these graphic comparisons for the "Tempi" station located close to the basin outlet, which allowed for an adequate model evaluation of the N species at large time steps.

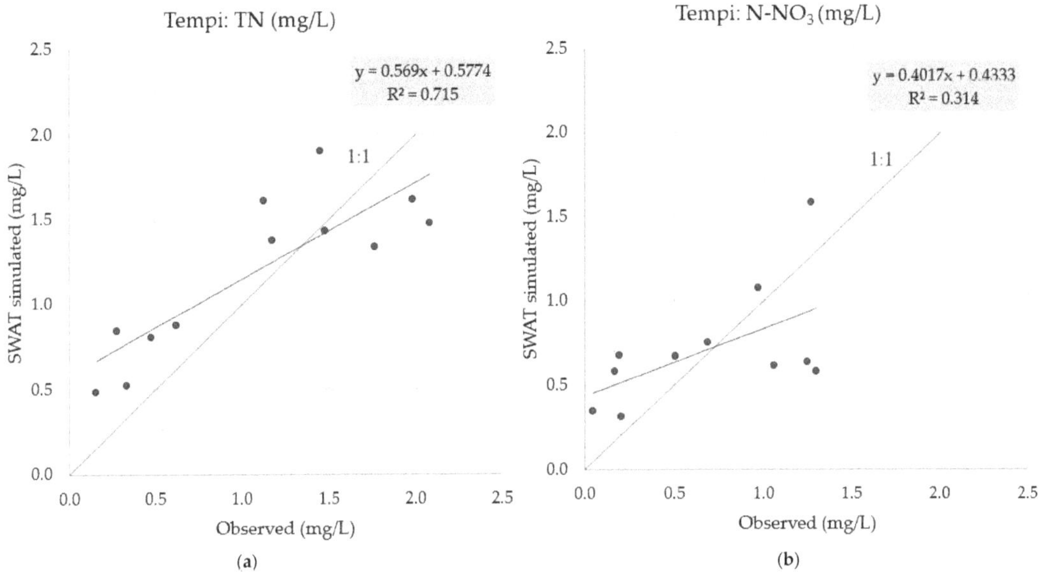

Figure 3. Total nitrogen (**a**) and N-NO_3 (**b**) comparisons at the basin outlet Tempi.

Upon initial observation, the scattergrams demonstrate a positive correlation and a comparable magnitude between the observed and simulated seasonal concentrations. In Figure 3a, the simulated TN concentrations are either above or below the 1:1 line, indicating no systematic over- or under-estimation by the model on a seasonal basis. The N-NO_3 concentrations were slightly overestimated, but the data pairs were not far from the 1:1 line, suggesting that the model predicted a magnitude comparable to that in the observed data. The correlation coefficient for TN was 0.84 (R^2 = 0.71), indicating a strong positive correlation between the simulated and observed values (Figure 3). The correlation coefficient for N-NO_3 was 0.56 (R^2 = 0.31), which was not as strong as that for TN. Nevertheless, even the current degree of coincidence between the seasonal observed and simulated values allowed for quite a realistic aggregation of the seasonal concentrations, which led to reasonable estimations of the TN and N-NO_3 river loads at large time steps (annual, mean annual). Specifically, the mean annual (2018–2022) observed TN and N-NO_3 concentrations were 1.37 and 0.89, with the respective simulated values being 1.97 and 0.85.

2.6. Plant Growth Evaluation

Additionally, the simulated crop yields for all four conventional crops in the area served as evaluation indicators for this study. In order to accurately simulate crop growth, several plant parameters were adjusted slightly in the SWAT plant database according to local knowledge and the convergence of the simulated yields with the reported ones. Table 6 summarizes the adjusted parameters and their final values for each land cover type.

The simulated mean annual yields were compared with the observed and reported values provided by the Hellenic Statistical Authority [24], and the comparison proved that with the above crop parameter values, the model simulated the annual production of the cultivated crops with acceptable deviations. Specifically, the simulated values for corn, cotton, alfalfa, and wheat were 11.45 t/ha, 3.11 t/ha, 12.38 t/ha, and 3.84 t/ha, with their respective observed production yields being 12.24 t/ha, 3.45 t/ha, 11.88 t/ha, and 3.64 t/ha. Upon further analysis of the available observed data, it was determined that the simulated spatial variability in the yields produced was also reasonable.

Table 6. Adjusted plant parameters for the four conventional crops in the Pinios river basin.

Variable Name	Definition	Cotton	Wheat	Corn	Alfalfa
HVSTI	Harvest index for optimal growing conditions [(kg/ha)/(kg/ha)]	0.5	0.4	0.6	0.8
WSYF	Lower limit of harvest index [(kg/ha)/(kg/ha)]	0.5	0.2	0.5	0.6
BLAI	Maximum potential leaf area index (LAI)	5	5	8	4
CHTMX	Maximum canopy height (m)	1	0.9	3	0.9
RDMX	Maximum root depth	1.5	1	1.5	3
T_OPT	Optimum temperature	28	20	26	25
T_BASE	Minimum (base) temperature for plant growth (°C)	14	0	9	2
ALAI_MIN	Minimum leaf area index for plants during the dormant period (m^2/m^2)	0	0	0	0.5
C_USLE	Minimum value of the USLE C-factor for the water erosion applicable to the land cover/plant	0.2	0.03	0.2	0.03
HEAT UNITS	Total heat units for the cover/plant to reach maturity	1700	1342	2700	1264

2.7. Baseline Simulation

The validated PRB model produced the baseline results with respect to hydrology, water quality, and crop production with reference to the period 2018–2023 (the first two years 2016 and 2017 were disregarded as warm-up years). As explained above, the initial groundwater tables at the beginning of the simulation were set as equal to zero in order to simulate only the exploitation of renewable groundwater reserves for irrigation. This representation was considered realistic due to the overexploitation of groundwater in recent decades and the resulting decrease in the level of groundwater tables at rather inaccessible levels.

The average annual surface runoff at the entire basin level was calculated to be 96 mm. The average annual evapotranspiration for the basin was estimated at 447.3 mm for the same period, with a total rainfall of 700 mm. Areas where the annual abstractions covered the needs of the crops fully or partly were also detected. The irrigation deficit varied across the basin, with a maximum at 90% but with irrigated HRUs that saw a severe water scarcity of $\geq$50% covering a rather small part of the irrigated land. At the baseline, during the period of 2018–2023, the total amount of water consumption per year for irrigation was roughly equal to 680 $\times$ 10^6 m^3. Irrigation from shallow aquifers was the main source of irrigation in the region, as 79% of the total arable land, without including fallow land, was irrigated by this source. The remaining 21% was irrigated by surface water, including the three reservoirs located within the basin (Figure 1) and an outside source, "Plastiras lake", which satisfied the irrigation needs of approximately 10,000 ha of cropland in the southwestern part of the basin. In order to examine the extent of groundwater exploitation, the amount of water stored in shallow aquifers at the end of the simulation period (the SWAT parameter SA_ST) was used. In the most intensively managed areas, the groundwater content at the end of the simulation varied between 0 and 200 mm.

Regarding water quality, the analysis of the baseline results shows that the area is indeed sensitive to N-NO_3 pollution, as its concentration exceeds the threshold of 0.6 mg/L, which, according to the Greek classification system for rivers [52], has been set as the upper limit of a "good" physicochemical status for water bodies. Specifically, the simulated mean annual N-NO_3 concentration at the basin's outlet was 1.68 mg/L. In addition, the mean annual TN loss from the entire land of the PRB to its surface waters was calculated to be 4.65 kg/ha. In the groundwater, N-NO_3 pollution was caused by the accumulation of nitrates, which in some cases reached levels that were prohibitive for the use of water for water supply purposes. The concentrations of N-NO_3 leaching, used as a pollution indicator in this work, were directly or indirectly governed by several N-NO_3 pathways, such as the N applied in fertilizers in the topsoil and the N uptake of different crops, as well as the N that was transported directly into streams. The average annual N-NO_3 leached concentration in the basin was calculated at 21.2 mg/L.

2.8. Bioenergy Crop Simulation

In this study, the implementation of a perennial bioenergy crop, switchgrass, in the PRB was examined through three what-if scenarios. The developed scenarios are listed in Table 7, together with the baseline scenario, which was used for comparison purposes.

Table 7. Switchgrass scenarios applied across the cropland of the Pinios river basin.

Scenario No.	Description
1	Baseline scenario
2	Switchgrass in the entire irrigated cropland
3	Switchgrass in irrigated sloping cropland
4	Switchgrass in irrigated non-sloping cropland

Sloping land was taken into consideration in the switchgrass implementation separately because grasses and perennial vegetation with persistent roots theoretically have the capacity to adapt to these areas of potentially lower productivity, and it was thus of interest to compare the results produced there with those in non-sloping areas. The term "sloping land" refers to areas of land that are not generally considered very well suited to intensive agricultural use due to the land slope, which limits the use of machinery and several practices compared to lowland areas. These areas may also be subject to soil degradation due to increased soil erosion, which causes the loss of valuable topsoil, and are usually described as erosive, droughty, and nutrient-poor, which may lead to reduced productivity [17]. In the PRB model of the present study, crop areas with slopes > 1.5% were considered sloping land based on the definition of the HRUs. The exact decision on this threshold level was taken after analyzing the baseline scenario simulations. It was concluded that the occurrence of soil erosion in areas with slopes higher than 1.5% was considerably higher compared to that in the HRUs with lower slopes. Moreover, it was chosen not to extend the switchgrass implementation scenarios to pastureland or semi-forested regions but only to irrigated cropland (sloping, non-sloping) areas. Non-irrigated cropland was excluded too, as no additional water pressure from switchgrass irrigation would be realistic in the water-scarce PRB.

Figure 4 illustrates the sloping and the non-sloping irrigated cropland areas of the PRB at the HRU level. Sloping irrigated cropland (presented in a red color in the map) represented 5.8% of the whole basin. Specifically, 219 out of 1850 HRUs represented sloping irrigated cropland in the PRB, accounting for 30% of the irrigated cropland. The remaining 70% represented non-sloping irrigated cropland (illustrated in a green color in the map in Figure 4). In numbers, from the 202,692 ha of the total irrigated cropland, 61,156 ha represented the irrigated sloping cropland and 141,537 ha the irrigated non-sloping cropland, as presented in Table 8.

None of the developed scenarios can be considered very realistic, mostly due to the extensive implementation of the bioenergy crop in the sloping and non-sloping irrigated landscapes or total irrigated land. However, our focus in developing these initial scenarios was to test the behavior of our PRB model across a large variety of types of land with switchgrass established and to explore the potential effects on water under significant changes to the existing irrigated cropping system.

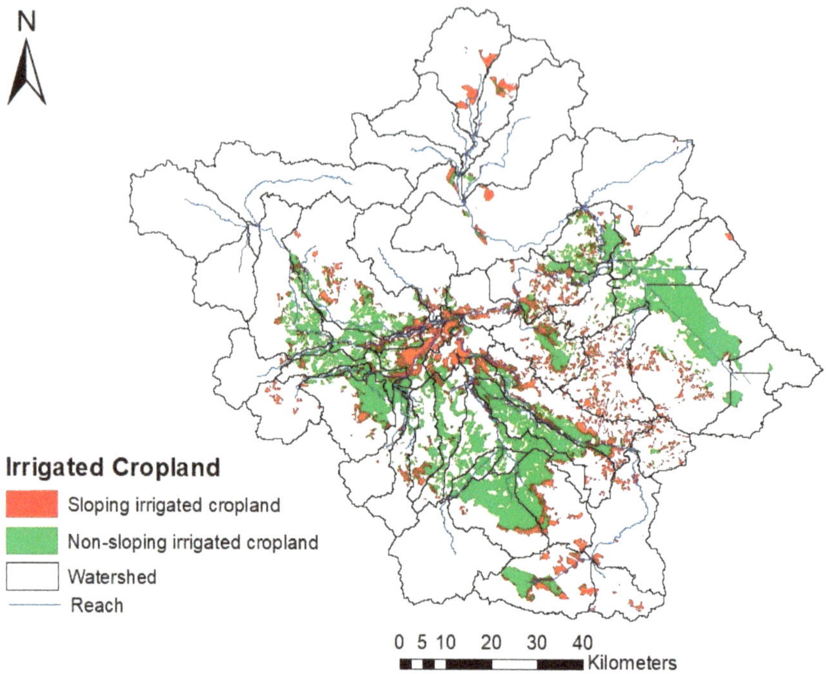

Figure 4. Sloping (>1.5%) and non-sloping irrigated cropland map of Pinios river basin.

Table 8. Sloping and non-sloping irrigated cropland distribution in Pinios river basin.

Category	Area (ha)	% Basin	% Cropland	% Irrigated Cropland
Basin	1,062,270	-	-	-
Cropland	368,328	34.7	-	-
Irrigated cropland	202,692	19.0	55.0	-
Irrigated sloping land	61,156	5.8	16.6	30.0
Irrigated non-sloping land	141,537	13.2	38.4	70.0

To simulate switchgrass closely to the Greek conditions, several plant parameters were adjusted in the SWAT plant database according to the recommendations of experts from the University of Thessaly and their relevant published works [53,54]. The parameters as well as their final adjusted values are presented in Table 9.

Switchgrass was simulated with a sowing date in May of the first year of simulation (2016) and harvest and kill operations in October of the last year of simulation (2023), with annual harvests in all intermediate years [54]. The results produced from its growth referred to the last 6 years (2018–2023) of the simulation period due to the 2-year warm up period that was neglected. The irrigation amount applied annually during the dry period (May–September) was 250 mm [54]. Only N fertilization was considered in this study since the requirements of switchgrass for P are minimal, with P application being only necessary in cases of significant soil deficiency [55]. The usual N fertilization rate for Greek conditions ranged between 80 and 160 kg per ha and was considered adequate to satisfy the plant's needs to result in a satisfactory amount of biomass being produced [53]. In this study, 150 kg N per ha during each year was applied with an additional 30 kg of N per ha right after seeding. To harvest switchgrass, the one-cut system was selected as the optimal system for the study area.

In all of the developed scenarios, switchgrass replaced all of the irrigated crops existing in the respective land in each scenario, namely cotton, corn, and alfalfa. When cotton was replaced with switchgrass, almost 20% less N was applied to the respective HRUs, while there was no P application. Moreover, this replacement resulted in 40% less irrigation water being used (if the groundwater availability allowed for the optimum irrigation). Similarly, when corn was replaced, there was a 59% reduction in the N applied with fertilization and a 60% theoretical reduction in the water consumed for irrigation. On the other hand, only when alfalfa was replaced with switchgrass did the amount of N applied increase (since alfalfa did not receive N at all), but there was a significant theoretical reduction of 58% in the amount of irrigation water applied.

Table 9. Plant parameters adjusted for Alamo switchgrass.

Variable Name	Definition	Switchgrass
HVSTI	Harvest index for optimal growing conditions [(kg/ha)/(kg/ha)]	0.95
BLAI	Maximum potential leaf area index (LAI)	9
CHTMX	Maximum canopy height (m)	2.5
RDMX	Maximum root depth	3
T_OPT	Optimum temperature	27
T_BASE	Minimum (base) temperature for plant growth (°C)	10
WSYF	Lower limit of harvest index [(kg/ha)/(kg/ha)]	0.95
HEATUNITS	Total heat units for cover/plant to reach maturity	2500

3. Results

The three switchgrass implementation scenarios in Table 7 were tested one by one in the Pinios river basin. The baseline scenario, which simulated an 8-year period from 2016 to 2023 by providing results for the 6-year period of 2018–2023 due to the first 2-year warm-up period, was used as a test bed for evaluating these alternative switchgrass implementation scenarios. The results of the baseline scenario (scenario 1) proved that the study area was vulnerable to N-NO_3. The mean annual effects in all four scenarios on the surface runoff, ET, groundwater content, total irrigation, N-NO_3 loads, N leached concentrations, and switchgrass biomass production at the entire basin scale were calculated and are listed in Table 10. The following three subsections analyze these results and focus more on the most interesting simulations related to water quantity, water quality, and bioenergy crop production, showing the spatial differentiation of the model predictions across a map of the Pinios basin.

Table 10. SWAT estimates of water, N, and biomass production in Pinios river basin during 2018–2023 under the implementation of the four scenarios.

Parameters	Results			
	Scenario 1	Scenario 2	Scenario 3	Scenario 4
Average annual surface runoff (mm)	157	133	150	140
Average annual ET (mm)	461	470	465	466
Groundwater content at the end of the simulation period (10^6 m^3)	5610	6180	5817	5914
Total annual irrigation water used (10^6 m^3)	680	413	616	493
N-NO_3 loss (kg/ha)	1.48	1.21	1.37	1.38
N-leached (mg/L)	21.2	11.9	16.4	15.9
Average annual switchgrass biomass production (t/ha)	-	18.6	18.2	18.4

Scenario 1: baseline scenario; Scenario 2: switchgrass implementation in the entire irrigated cropland; Scenario 3: switchgrass implementation in irrigated sloping cropland; and Scenario 4: switchgrass implementation in irrigated non-sloping cropland.

3.1. Results on Hydrology and Water Quantity

Switchgrass, as a perennial crop with a significantly high biomass production, had increased evapotranspiration values compared to those of the other crops. Specifically, 2.2%, 1%, and 1.5% increases in the mean annual baseline ET (461 mm in Table 10) of the entire basin were observed in the second, third, and fourth scenarios, respectively, compared to that in the first scenario. This effect was much more pronounced at the HRU level, with ET increases of up to 10% on an annual basis where cotton was the baseline crop. The average increase in the HRUs where corn was replaced with switchgrass was almost 2.5%, while the replacement of alfalfa with switchgrass resulted in almost the same ET values, as both were perennial crops with similar simulated biomass yields.

The surface runoff in the switchgrass implementation scenarios was expected to be reduced compared to the baseline mean annual 157 mm that was simulated at the entire basin level. As presented in Table 10, from that value, mean annual reductions of 15%, 4%, and 11% occurred in the second, third, and fourth scenarios, respectively.

Figure 5 illustrates the effect of switchgrass installation on the groundwater content of the basin. Figure 5a represents the first scenario (the baseline), while Figure 5b depicts the impact of the switchgrass implementation in the second scenario with switchgrass simulated in the entire irrigated land. The data used for the creation of the figure maps represent the values simulated by SWAT on the last day of the simulation (31 December 2023), allowing for an assessment of the long-term effect of switchgrass on groundwater exploitation. Thus, the map on the left, referring to the first scenario, presents the groundwater content in mm at the end of 2023, while the map on the right depicts the increase in groundwater (in mm) per subbasin from the baseline that was caused by the second scenario (switchgrass in all of the irrigated cropland).

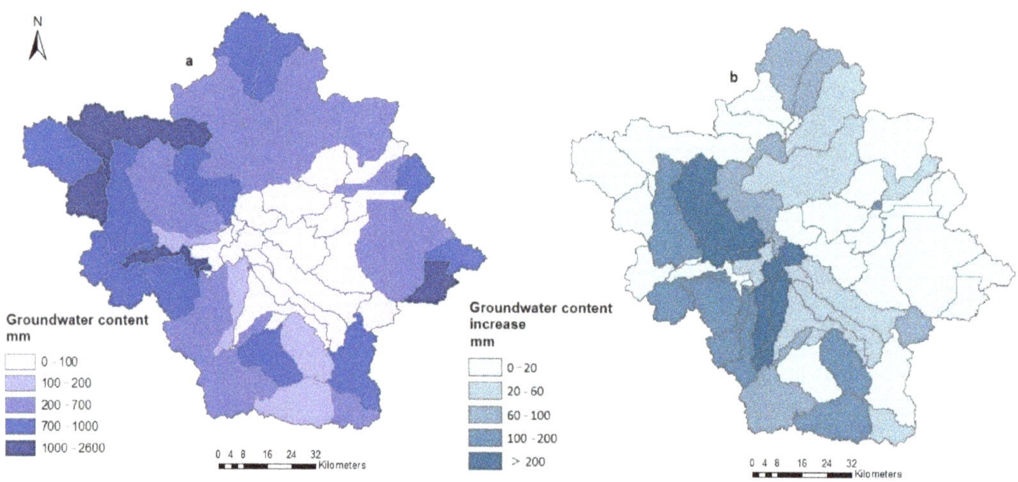

Figure 5. Groundwater content in the Pinios river basin at the end of the simulation period (end of 2023), expressed in mm of water for the baseline (**a**) and as the increase from the baseline (in mm) in the second scenario (**b**).

As can be observed in Figure 5, switchgrass resulted in significant increases in the groundwater content, with almost half of the PRB's area gaining >60 mm (or 600 m^3/ha) of water in the aquifers when switchgrass replaced more water-demanding conventional crops. The majority of the subbasins presented an increase in the water stored in shallow aquifers. The average increase observed among the subbasins was 68 mm, while 22 out of the 61 subbasins met >50% increases in the groundwater content from the baseline. As shown by the results in Table 10, the total effect of the switchgrass implementation scenarios on the amounts of groundwater stored in the aquifers of the entire PRB at the end of the simulation was positive in all cases. Specifically, the 5.61 billion m^3 of stored groundwater at the end of the simulation period in scenario 1 was increased by 10%, 4%, and 5% under scenarios No. 2, 3, and 4, respectively.

The irrigation water abstracted from shallow aquifers with the switchgrass implementation was considerably reduced as well. As stated in Section 2.8, according to the theoretical irrigation water needs of the crops, 40%, 60%, and 58% less water consumption was expected by replacing cotton, corn, and alfalfa with switchgrass, respectively. According to the results in Table 10, the total 680 hm^3 of water that was actually abstracted from the shallow aquifers of PRB for irrigation purposes on a mean annual basis in the baseline scenario was reduced by 40%, 10%, and 28% due to the implementation of the second, third, and fourth scenarios, respectively. Obviously, these numbers resulted from the extent of the irrigated areas and the irrigated conventional crops that were replaced by switchgrass in each scenario.

3.2. Results on Water Quality Due to Nitrate Loading in the Water

The impacts on N-NO_3 were quite positive too, with the predicted N-NO_3 loss reductions being quite or very significant at the entire basin level depending on the extent of perennial crop installation. Figure 6 illustrates the effect of the N-NO_3 loss in the basin in the first (Figure 6a) and second scenarios (Figure 6b). The data used to create the maps represent the average year for each scenario, calculated from the model's output data for the 6-year period of interest (2018–2023).

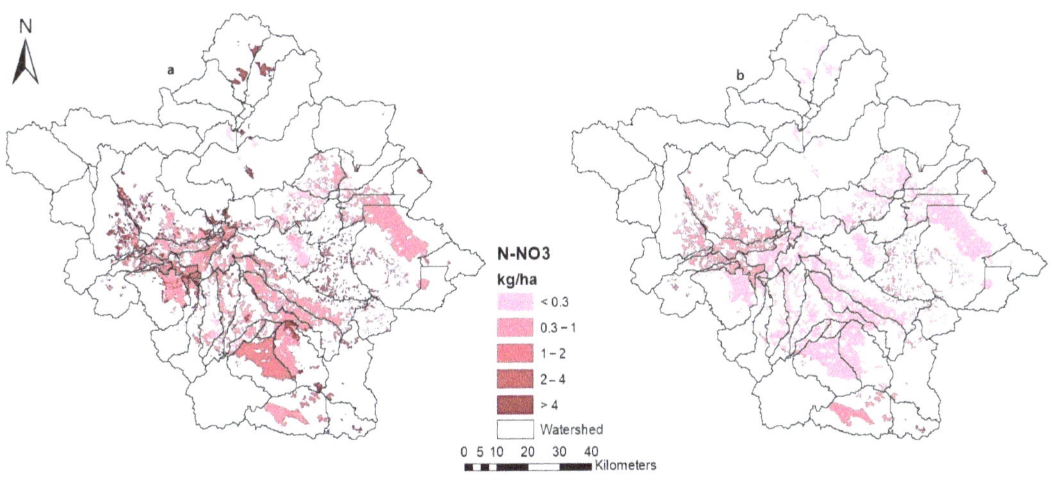

Figure 6. Mean annual (2018–2023) N-NO_3 load of the irrigated cropland within Pinios river basin under (**a**) the baseline scenario and (**b**) the switchgrass implementation in all irrigated cropland scenarios.

It is evident that there was an important effect of reduced N-NO_3 under the switchgrass implementation in parts of the PRB's cropland. Specifically, the average percentage reduction observed in the HRUs in the most lowland areas of the Thessaly plain, covering the central part of the basin, was 80%, with the reduction in certain HRUs approaching 95%. As can be observed in Figure 6, in the central, southern, and western parts of the PRB, the annual N-NO_3 loads in many irrigated areas were reduced to the level of <0.3 kg/ha, with the respective baseline values being greater than 1 kg/ha. The maximum annual N-NO_3 load in the HRUs simulated in the first baseline scenario was 19.8 kg/ha, whereas in the second extensive switchgrass implementation scenario, the corresponding value was reduced to 11.1 kg/ha. The mean annual N-NO_3 loads at the entire basin level were calculated for all switchgrass scenarios (Table 10), and precisely, from the mean annual 1.48 kg of lost N-NO_3 per ha of PRB land into streams, an 18%, a 7%, and again a 7% reduction was simulated for the second, third, and fourth implementation scenarios, respectively.

Another important output is the mean annual N-NO_3 leached concentrations in the baseline and three switchgrass scenarios, which are included in Table 10 as well. From the mean concentrations calculated at the basin level, 44%, 23%, and 25% reductions in the baseline N-NO_3 leached concentration of 21.2 mg/L were predicted on a mean annual basis in the second, third, and fourth scenarios, respectively.

3.3. Results on Switchgrass Biomass Production

Figure 7 illustrates the mean annual biomass production of switchgrass under the second simulated scenario, with switchgrass implemented in the entire irrigated land. The presented map was created by using the simulated mean annual (2018–2023) data on switchgrass biomass production.

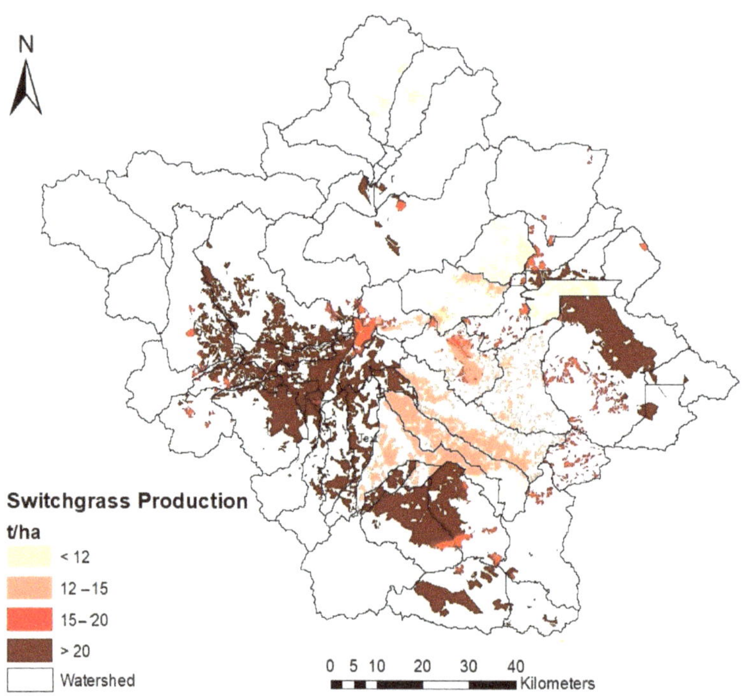

Figure 7. Mean annual (2018–2023) switchgrass biomass production in t/ha in the second scenario (switchgrass implementation in the entire cropland).

The simulated production of switchgrass, both the mean annual production (2018–2023) and that averaged across space, was calculated at 18.6 t/ha (Table 10), with the maximum biomass simulated for a single HRU being 22 t/ha. Published data on the actual growth of switchgrass within the study area [53] reported a typical range from 15 t/ha to 20 t/ha of biomass production on an annual basis. According to this evidence, the simulated biomass yields that belonged to the first class in the map in Figure 7 (<12 t/ha/y) were considered very low, while those of the second (12–15 t/ha/y) and third (15–20 t/ha/y) classes were rather low and normal, respectively, and the rest (>20 t/ha/y) were near-optimum. It is remarkable that according to the last row of Table 10, in the third scenario, when switchgrass was only implemented in the sloping irrigated land, its mean annual biomass production was simulated to be equal to 18.2 t/ha, while in the fourth scenario (only in non-sloping cropland), it was just a little higher and equal to 18.4 t/ha. These almost identical production levels per unit area have shown that in our model, growing switchgrass in areas with a milder topography yielded similar production levels to those when growing it in steeper areas. Thus, the selection of sloping area was not a limiting factor for biomass production.

4. Discussion

The PRB SWAT model developed in this study realistically simulated the hydrological and N cycle processes occurring at the river basin scale. Even though the N model predictions were only compared with river observations made at the basin's outlet, the available data allowed for an adequate evaluation by the model at large time steps and for the entire basin. For the acceptance of the model predictions at nested sites of the basin, we had to rely on robust spatial parameterization of the model with respect to the land use, soil, and management.

This was the case for the crop yields and switchgrass biomass production in particular, as well as for the impact of switchgrass on water quality and quantity, for which it was essential to rely on other research studies with which to compare our model's predictions. The reduction in surface runoff caused by the switchgrass implementation was the combined effect of (a) the more dense land coverage that characterizes most perennial crops compared to that of annual crops, resulting in reductions in surface runoff [56]; (b) the reduced amount of water that switchgrass received as irrigation, which also resulted in the production of less water capable of becoming runoff; and (c) the slightly increased ET, leaving a reduced amount of water to be lost via runoff [57].

As has been stated, switchgrass received notably less irrigation water compared to that received by the existing irrigated crops in the PRB, which led to the assumption that its adoption could increase groundwater reserves. Indeed, apart from the reduced direct abstraction of water, in the majority of the HRUs, switchgrass was also able to reduce the surface runoff and increase the soil's water content and percolation, the water that moves below the root zone over time [58]. While this was a fair observation at the entire basin level, there were certain HRUs where the opposite was observed due to less water being applied to the topsoil via irrigation. Overall, despite the fact that the areas that were available for the installation of switchgrass covered small percentages of the entire basin (19% of the basin represented the total irrigated cropland, 5.8% was the sloping irrigated cropland, and 13.2% was the non-sloping irrigated cropland), the individual effects in the subbasins were anything but trivial, as they proved to have the potential to improve the water availability considerably because of the reduced abstractions.

Regarding the total irrigation water used in the PRB, as expected, the effect of the fourth scenario (switchgrass in non-sloping irrigated cropland) was greater than the effect of the third scenario (sloping irrigated cropland), and this difference was attributed to the significantly greater area of non-sloping irrigated land that was available for the installation of switchgrass in the fourth scenario. This resulted in the replacement of more crops with greater irrigation water needs. Based on all of these results concerning the hydrology of the area in all of the developed scenarios, it can be concluded that the perennial crop could be effective in alleviating the high water abstraction from the shallow aquifers of the PRB and in this manner enhancing their water balance.

The agricultural use of N fertilizers has been a major contributor to the N export into the water from agricultural areas. Specifically, in PRB, which is highly vulnerable to N-NO_3 pollution, N-NO_3 reduction was considered to be of great importance. The trend in the N loss reduction was correlated with the N transported in dissolved form via runoff (mainly N-NO_3). In general, the decrease in N-NO_3 loading was mainly because switchgrass required a lower N fertilizer input in comparison to that for the baseline crops. The model simulations indicated that the adoption of the biofuel crop could reduce the N losses compared to those in the current cropping systems used in PRB, as observed in other studies as well [17,19,59,60]. The level of the reduction varied between scenarios, with the areal extent of the deployment of the bioenergy crop being the key factor for the greatest reduction being caused by the second scenario (with all irrigated crops replaced with switchgrass). However, the third and the fourth scenarios had the same overall effect at the basin level, resulting in average N-NO_3 loss rates of 1.37 and 1.38 kg/ha, even though a lower rate was expected in the fourth scenario with switchgrass implemented in double the area compared to that in the third scenario. The almost identical N-NO_3 reduction is obviously attributed to the greatest effect of the perennial crop in sloping areas where an increased reduction in runoff led to a more pronounced N-NO_3 reduction as well. The rather low N-NO_3 percentage reduction of 7% (1.37 and 1.38 from 1.48 kg/ha as shown in Table 10) was considerable for both scenarios since the conventional crop replacement represented a small percentage of land cover change in the entire basin (5.8% for the sloping irrigated cropland scenario and 13% for the non-sloping irrigated cropland scenario), and additionally, in both scenarios, there were extensive areas of non-irrigated wheat, not replaced by switchgrass, that still contributed greatly to N pollution (wheat HRUs contributed 47% of the total N-NO_3 pollution from the cropland in the first baseline scenario). So, even though the effect was not as significant at the entire basin level as in the irrigated HRUs, if the perennial crop was planned to be installed in the entire sloping cropland (irrigated and non-irrigated), the N-NO_3 could be reduced substantially, probably leading to impressive water quality improvements.

Nitrate pollution in PRB had anthropogenic activities as its main sources, mostly crop cultivation and secondly livestock farming. In groundwater, nitrate pollution is generally caused by the accumulation of nitrates, which in some cases can reach levels that are prohibitive to the use of water for water supply purposes. This means that the magnitude of N-NO_3 leaching is directly or indirectly governed by N fertilization, N crop uptake, and the N lost from the soil that is transported into streams. Additionally, the increased concentration of leaching nitrates can be attributed to the high irrigation rates in these areas [61,62]. Thus, the reduced leached N-NO_3 concentrations due to the switchgrass implementation were expected, as the perennial crop could reduce the N-NO_3 leaching from the agricultural lands in our study area significantly due to its lower irrigation and N fertilization needs.

One limitation of SWAT in the simulation of crop growth is its inability to account for crop diseases in specific areas and years that may have led to relatively higher simulated yields in certain years than those reported. Hence, it is always preferred to discuss the simulated average annual yields, and this was the case for the switchgrass biomass production simulated under these scenarios. By calculating the biomass yields in sloping areas only, it could be concluded that some of the highest yields occurred there. Evidently, switchgrass had great potential to grow in these areas. The average biomass production calculated in the sloping land scenario was 18.2 t/ha, and the maximum production observed in a single HRU was 22 t/ha. The areas where inadequate production (of less than 12 t/ha) was observed (Figure 7) could be related to irrigation water deficits in these areas. In the center of the basin where the main part of the Thessaly plain (lowland areas) is located, there were constant biomass production yields of less than 15 t/ha (Figure 7). This can be also attributed to the insufficient groundwater content of the aquifers in this subregion due to the lower precipitation levels in this area, which has subsequently led to less percolation and aquifer recharge compared to those in other areas in the western parts of the basin. The deficit irrigation observed further proved that irrigation water was the main limiting parameter for switchgrass crop growth, as was stated by Giannoulis et al. [63] about this particular subregion of the PRB. It should be also noted that all of the irrigated areas served by reservoirs in this study had no problems with irrigation water shortages. A representative example of the positive role of reservoirs in meeting irrigation water needs in the PRB is the large area irrigated by the Karla reservoir in the most eastern part, where despite low precipitation and groundwater recharge, the switchgrass biomass production reached its maximum values of >20 t/ha/y (Figure 7).

Concerning the conventional irrigated crops in the PRB, their spatially averaged simulated yield rates per ha of cropland in the last two scenarios, when only part of the irrigated land was replaced with switchgrass, did not practically change from their average (per unit area) simulated yields at the baseline. However, an added benefit that was observed at the very local level under the partial replacement of irrigated conventional crops with switchgrass in the third and fourth scenarios was the increased availability of groundwater due to the reduced overall abstraction. The implementation of switchgrass solely in sloping/non-sloping irrigated areas offered the potential for the production yields of the adjacent conventional crops in non-sloping/sloping areas that were irrigated inadequately before to be improved due to the increased amounts of irrigation applied from groundwater.

5. Conclusions

The purpose of this study was to provide estimates of the effects that switchgrass cultivation could have on the water quantity and quality when implemented in large areas of a Greek agricultural basin using SWAT. The bioenergy crop proved to have great potential to grow in sloping areas and produce sufficient amounts of biomass, even though these areas were expected to have a relatively lower productivity due to their higher susceptibility to soil and nutrient losses. Specifically, the N-NO_3 reductions at the basin level varied between 7 and 18% depending on the extent of deployment of the bioenergy crop, while there were individual HRUs with switchgrass installed where these reductions were greater than 80%. Implementing switchgrass on sloping land proved to be very effective in reducing the N-NO_3 loads. Furthermore, the cultivation of the bioenergy crop could significantly alleviate the irrigation water abstraction by 10–40% depending on the replaced crops and the areal extent of the bioenergy crop's installation. As a result, switchgrass increased the groundwater availability at several subbasins; specifically, for groundwater bodies in the most intensively irrigated central part of the basin, this increase was locally even higher than 50%.

Indeed, this study supports the progressive substitution of parts of the cropland containing conventional crops with bioenergy crops, promoting them as an effective restoration measure for water bodies, according to the WFD requirements, that could be also adopted in the next phases of regional WFD management plans. Ongoing research activities such as an economic analysis and the development of an efficient optimization tool could add value to the preliminary results presented in this paper, as they will allow additional criteria to be taken into consideration when selecting the locations for switchgrass implementation. It is believed that this current study on switchgrass simulation using SWAT, along with future directions, constitutes a useful approach to hydrological, water quality, and bioenergy crop simulations that could help decision-makers and water managers incorporate bioenergy crops into their combined efforts towards renewable energy production and water protection.

Author Contributions: Conceptualization: M.S., L.K. and Y.P. Methodology: M.S., L.K., K.D.G. and Y.P. Software: M.S., L.K., S.K. and H.G. Validation: M.S., L.K., S.K., H.G. and K.D.G. Formal analysis: M.S., L.K. and Y.P. Investigation: M.S., L.K., S.K., H.G. and E.D. Resources: M.S., L.K. and Y.P. Data curation: M.S., L.K., K.D.G., E.D. and Y.P. Writing—original draft preparation: M.S. and L.K. Writing—review and editing: M.S., L.K., S.K., H.G., K.D.G., E.D. and Y.P. Visualization: M.S., L.K., S.K., H.G. and Y.P. Supervision: Y.P. Project administration: Y.P. Funding acquisition: Y.P. All authors have read and agreed to the published version of the manuscript.

Funding: This research study was carried out within the framework of the National Recovery and Resilience Plan Greece 2.0, funded by the European Union—NextGenerationEU (Implementation body: HFRI). More specifically, this research was supported under the Basic Research Financing Action "Horizontal support of all sciences", Sub-action 1 (Project Number: 16425; project title: BIOGRASS).

Data Availability Statement: The data used in this study are available from the authors of this paper upon request.

Acknowledgments: We would like to thank the reviewers for their constructive and insightful comments that helped us improve our article.

Conflicts of Interest: The authors declare no conflicts of interest.

References

1. Claro, A.M.; Fonseca, A.; Fraga, H.; Santos, J.A. Future Agricultural Water Availability in Mediterranean Countries under Climate Change: A Systematic Review. *Water* **2024**, *16*, 2484. [CrossRef]
2. Panagopoulos, Y.; Karpouzos, D.; Georgiou, P.; Papamichail, D. Ecosystem Services Evaluation from Sustainable Water Management in Agriculture: An Example from An Intensely Irrigated Area in Central Greece. In Proceedings of the 7th International Electronic Conference on Water Sciences, Online, 15–30 March 2023. [CrossRef]
3. Malamataris, D.; Tsavdaridou, A.I.; Banti, D.; Malliaras, A.; Karalis, A.; Theocharis, N.; Michos, C.; Mazaris, A. Water-Ecosystems-Food nexus security achievement in the context of climate change: The case study of an agricultural Mediterranean Basin, Greece. *Res. Ideas Outcomes* **2023**, *9*, e116070. [CrossRef]
4. De Girolamo, A.M.; Lo Porto, A. Source Apportionment of Nutrient Loads to a Mediterranean River and Potential Mitigation Measures. *Water* **2020**, *12*, 577. [CrossRef]
5. Psomas, A.; Dagalaki, V.; Panagopoulos, Y.; Konsta, D.; Mimikou, M. Sustainable agricultural water management in Pinios river basin using remote sensing and hydrologic modeling. *Procedia Eng.* **2016**, *162*, 277–283. [CrossRef]
6. Hellenic Ministry for the Environment, Energy and Climate Change—Special Secretariat for Water. Report on Directive 91/676/EEC—Nitrate Pollution Situation in the Greek Territory (Reference Period: 2020–2023). Available online: https://wfdver.ypeka.gr/wp-content/uploads/2017/04/files/GR08/GR08_P26b_Perilipsi_En.pdf (accessed on 10 October 2024).
7. Skoulou, V.; Mariolis, N.; Zanakis, G.; Zabaniotou, A. Sustainable management of energy crops for integrated biofuels and green energy production in Greece. *Renew. Sustain. Energy Rev.* **2011**, *15*, 1928–1936. [CrossRef]
8. Boukis, I.; Vassilakos, N.; Kontopoulos, G.; Karellas, S. Policy plan for the use of biomass and biofuels in Greece: Part I: Available biomass and methodology. *Renew. Sustain. Energy Rev.* **2009**, *13*, 971–985. [CrossRef]
9. Mitkidis, G.; Magoutas, A.; Kitsios, F. Market and economic feasibility analysis for the implementation of 2nd generation biofuels in Greece. *Energy Strat. Rev.* **2018**, *19*, 85–98. [CrossRef]
10. Giannoulis, K.; Bartzialis, D.; Skoufogianni, E.; Gravalos, I.; Xiradakis, P.; Danalatos, N.G. *Panicum virgatum* L. could be an alternative environmentally friendly feedstock for energy production. *SSRG Int. J. Agric. Environ. Sci.* **2017**, *4*, 14–23. [CrossRef]
11. Chen, Y.; Marek, G.W.; Marek, T.H.; Moorhead, J.E.; Heflin, K.R.; Brauer, D.K.; Gowda, P.H.; Srinivasan, R. Assessment of alternative agricultural land use options for extending the availability of the Ogallala Aquifer in the Northern High Plains of Texas. *Hydrology* **2018**, *5*, 53. [CrossRef]
12. Panagopoulos, Y.; Makropoulos, C.; Gkiokas, A.; Kossida, M.; Evangelou, L.; Lourmas, G.; Michas, S.; Tsadilas, C.; Papageorgiou, S.; Perleros, V.; et al. Assessing the cost-effectiveness of irrigation water management practices in water stressed agricultural catchments: The case of Pinios. *Agric. Water Manag.* **2014**, *139*, 31–42. [CrossRef]
13. First Revision of the Regional River Basin Management Plan. Available online: https://wfdver.ypeka.gr/wp-content/uploads/2017/12/EL08_SDLAP_APPROVED.pdf (accessed on 20 November 2024). (In Greek).

14. Kharel, G.; Zhong, Y.; Will, R.E.; Zhang, T.; Zou, C.B. Potential Hydrological Impacts of Planting Switchgrass on Marginal Rangelands in South Central Great Plains. *Water* **2022**, *14*, 3087. [CrossRef]
15. Kumar, E.; Saraswat, D.; Singh, G. Comparative analysis of bioenergy crop impacts on water quality using static and dynamic land use change modeling approach. *Water* **2020**, *12*, 410. [CrossRef]
16. Feng, Q.; Chaubey, I.; Cibin, R.; Engel, B.; Sudheer, K.P.; Volenec, J.; Omani, N. Perennial biomass production from marginal land in the Upper Mississippi River Basin. *Land Degrad. Dev.* **2018**, *29*, 1748–1755. [CrossRef]
17. Panagopoulos, Y.; Gassman, P.W.; Kling, C.L.; Cibin, R.; Chaubey, I. Water quality assessment of large-scale bioenergy cropping scenarios for the upper Mississippi and Ohio-Tennessee river basins. *JAWRA J. Am. Water Resour. Assoc.* **2017**, *53*, 1355–1367. [CrossRef]
18. Cibin, R.; Trybula, E.; Chaubey, I.; Brouder, S.M.; Volenec, J.J. Watershed-scale impacts of bioenergy crops on hydrology and water quality using improved SWAT model. *GCB Bioenergy* **2016**, *8*, 837–848. [CrossRef]
19. Gassman, P.W.; Valcu-Lisman, A.M.; Kling, C.L.; Mickelson, S.K.; Panagopoulos, Y.; Cibin, R.; Chaubey, I.; Wolter, C.F.; Schilling, K.E. Assessment of bioenergy cropping scenarios for the Boone River watershed in north central Iowa, United States. *JAWRA J. Am. Water Resour. Assoc.* **2017**, *53*, 1336–1354. [CrossRef]
20. Mentzafou, A.; Papadopoulos, A.; Dimitriou, E. Assessing the Impacts of Climatic and Water Management Scenarios in a Small Mountainous Greek River. *Hydrology* **2025**, *12*, 13. [CrossRef]
21. Udias, A.; Pastori, M.; Malago, A.; Vigiak, O.; Nikolaidis, N.P.; Bouraoui, F. Identifying efficient agricultural irrigation strategies in Crete. *Sci. Total. Environ.* **2018**, *633*, 271–284. [CrossRef]
22. Samarinas, N.; Tziolas, N.; Zalidis, G. Improved Estimations of Nitrate and Sediment Concentrations Based on SWAT Simulations and Annual Updated Land Cover Products from a Deep Learning Classification Algorithm. *ISPRS Int. J. Geo-Inf.* **2020**, *9*, 576. [CrossRef]
23. Pisinaras, V.; Herrmann, F.; Panagopoulos, A.; Tziritis, E.; McNamara, I.; Wendland, F. Fully Distributed Water Balance Modelling in Large Agricultural Areas—The Pinios River Basin (Greece) Case Study. *Sustainability* **2023**, *15*, 4343. [CrossRef]
24. Hellenic Statistical Authority, Data on Crop Areas. Available online: https://www.statistics.gr/en/statistics/-/publication/SPG06/- (accessed on 21 December 2024).
25. Arampatzis, G.; Panagopoulos, A.; Pisinaras, V.; Tziritis, E.; Wendland, F. Identifying potential effects of climate change on the development of water resources in Pinios River Basin, Central Greece. *Appl. Water Sci.* **2018**, *8*, 51. [CrossRef]
26. Vasiliades, L.; Loukas, A.; Liberis, N. A water balance derived drought index for Pinios River Basin, Greece. *Water Resour. Manag.* **2010**, *25*, 1087–1101. [CrossRef]
27. Nitrates Directive—Vulnerable Zones Reporting 7—Web Map by Joint Research Centre of European Commission. Available online: https://water.jrc.ec.europa.eu/portal/apps/webappviewer/index.html?id=b33a220c1b284583851e93a245da02ef (accessed on 10 December 2024).
28. Hellenic Ministry for the Environment, Energy and Climate Change-Special Secretariat for Water. Compilation of Management Plan for the River Basins of Thessaly Water District (GR08)-Management Plan; Athens, Greece, 2017. p. 252. Available online: https://wfdver.ypeka.gr/el/project/gr08-26b-approved-management-summaryen-gr/ (accessed on 15 November 2024).
29. Second Revision of the Regional River Basin Management Plan. Available online: https://wfdver.ypeka.gr/wp-content/uploads/2024/07/EL08_2REV_sdlap.pdf (accessed on 25 November 2024). (In Greek).
30. Williams, J.R.; Arnold, J.G.; Kiniry, J.R.; Gassman, P.W.; Green, C.H. History of model development at Temple, Texas. *Hydrol. Sci. J.* **2008**, *53*, 948–960. [CrossRef]
31. Aloui, S.; Mazzoni, A.; Elomri, A.; Aouissi, J.; Boufekane, A.; Zghibi, A. A review of Soil and Water Assessment Tool (SWAT) studies of Mediterranean catchments: Applications, feasibility, and future directions. *J. Environ. Manag.* **2023**, *326*, 116799. [CrossRef]
32. Neitsch, S.L.; Arnold, J.G.; Kiniry, J.R.; Williams, J.R. Soil and Water Assessment Tool (SWAT) Theoretical Documentation. Blackland Research Center, Texas. 2009. Agricultural Experiment Station, Temple, Texas (BRC Report 02–05). Available online: https://swat.tamu.edu/media/69296/swat-io-documentation-2012.pdf (accessed on 2 November 2024).
33. Williams, J.R. Sediment-yield prediction with universal equation using runoff energy factor[1]. In *Present and Prospective Technology for Predicting Sediment Yield and Sources, Proceedings of the Sediment-Yield Workshop, USDA Sedimentation Laboratory, Oxford, MS, USA, 28–30 November 1972. Agricultural Research Service, US Department of Agriculture: 1975; Volume 40, p. 244. Oxford, MS, USA*; Available online: https://upload.wikimedia.org/wikipedia/commons/e/e1/FEDLINK_-_United_States_Federal_Collection_(IA_presentprospecti40sedi).pdf (accessed on 13 December 2024).
34. Wischmeier, W.H.; Smith, D.D. Predicting rainfall-erosion losses from cropland east of the Rocky Mountains. In *Agricultural Handbook*; Agricultural Research Service: Washington, DC, USA, 1965; p. 282.
35. Arabi, M.; Frankenberger, J.R.; Engel, B.A.; Arnold, J.G. Representation of agricultural conservation practices with SWAT. *Hydrol. Process. Int. J.* **2008**, *22*, 3042–3055. [CrossRef]

36. Trybula, E.M.; Cibin, R.; Burks, J.L.; Chaubey, I.; Brouder, S.M.; Volenec, J.J. Perennial rhizomatous grasses as bioenergy feedstock in SWAT: Parameter development and model improvement. *GCB Bioenergy* **2015**, *7*, 1185–1202. [CrossRef]
37. Jiang, R.; Wang, T.T.; Shao, J.; Guo, S.; Zhu, W.; Yu, Y.J.; Chen, S.; Hatano, R. Modeling the biomass of energy crops: Descriptions, strengths and prospective. *J. Integr. Agric.* **2017**, *16*, 1197–1210. [CrossRef]
38. Corine Land Cover 2012—European Environment Agency. Available online: https://www.eea.europa.eu/data-and-maps/data/external/corine-land-cover-2012 (accessed on 15 November 2024).
39. Hiederer, R. Mapping Soil Properties for Europe-Spatial Representation of Soil Database Attributes. EUR26082EN Scientific and Technical Research Series, 2013, Volume 47, pp. 1831–9424. Available online: https://citeseerx.ist.psu.edu/document?repid=rep1&type=pdf&doi=f0031d732307a43e9dca04e46b47dfc6d75a8be9 (accessed on 21 November 2024).
40. Hiederer, R. Mapping Soil Typologies—Spatial Decision Support Applied to European Soil Database. Publications Office of the European Union, 2013, p. 127. Available online: https://esdac.jrc.ec.europa.eu/ESDB_Archive/eusoils_docs/other/EUR25932EN.pdf (accessed on 21 November 2024).
41. Hellenic Statistical Authority, Data on Livestock. Available online: https://www.statistics.gr/en/statistics/-/publication/SPK13/- (accessed on 5 December 2024).
42. General Secretariat of Waste Management Coordination, Wastewater Treatment Control and Planning Department, Web Map of the Urban Wastewater Treatment Plants. Available online: https://astikalimata.ypeka.gr/wtp (accessed on 20 November 2024).
43. First Revision of River Basin Management Plans, Detailed Methodologies on Anthropogenic Pressures. Available online: https://wfdver.ypeka.gr/wp-content/uploads/2017/04/Methodologia_Piesewn_v3.pdf (accessed on 4 November 2024). (In Greek).
44. Inland Water Department of the Hellenic Centre for Marine Research (HCMR), Data on the Automatic Monitoring Network of Hydrological Stations. Available online: https://hydro-stations.hcmr.gr/ (accessed on 5 November 2024).
45. Oo, H.T.; Zin, W.W.; Kyi, C.T. Analysis of streamflow response to changing climate conditions using SWAT model. *Civ. Eng. J.* **2020**, *6*, 194–209. [CrossRef]
46. Panagopoulos, Y.; Dimitriou, E. A Large-Scale Nature-Based Solution in Agriculture for Sustainable Water Management: The Lake Karla Case. *Sustainability* **2020**, *12*, 6761. [CrossRef]
47. Desai, S.; Singh, D.K.; Islam, A.; Sarangi, A. Multi-site calibration of hydrological model and assessment of water balance in a semi-arid river basin of India. *Quat. Int.* **2021**, *571*, 136–149. [CrossRef]
48. Moriasi, D.N.; Arnold, J.G.; Van Liew, M.W.; Bingner, R.L.; Harmel, R.D.; Veith, T.L. Model evaluation guidelines for systematic quantification of accuracy in watershed simulations. *Trans. ASABE* **2007**, *50*, 885–900. [CrossRef]
49. Mapes, K.L.; Pricope, N.G. Evaluating SWAT model performance for runoff, percolation, and sediment loss estimation in low-gradient watersheds of the Atlantic coastal plain. *Hydrology* **2020**, *7*, 21. [CrossRef]
50. Panagopoulos, Y.; Makropoulos, C.; Baltas, E.; Mimikou, M. SWAT parameterization for the identification of critical diffuse pollution source areas under data limitations. *Ecol. Model.* **2011**, *222*, 3500–3512. [CrossRef]
51. WFD Program Conducted by Hellenic Centre for Marine Research HCMR. Monitoring and Recording of the Status (Quality, Quantity, Pressures, and Uses) of Surface Waters of Greece" (MIS 5001676), Under the Operational Program: "Transport Infrastructure, Environment and Sustainable Development" 2014–2020, Financed by the Hellenic Ministry of the Environment and the European Regional Development Fund. Available online: https://wfd.hcmr.gr/ (accessed on 16 November 2024).
52. Skoulikidis, N.T.; Amaxidis, Y.; Bertahas, I.; Laschou, S.; Gritzalis, K. Analysis of factors driving stream water composition and synthesis of management tools—A case study on small/medium Greek catchments. *Sci. Total. Environ.* **2006**, *362*, 205–241. [CrossRef]
53. Giannoulis, K.D.; Vlontzos, G.; Karyotis, T.; Bartzialis, D.; Danalatos, N.G. Assessing the efficiency of switchgrass different cultural practices for pellet production. *Land Use Policy* **2014**, *41*, 506–513. [CrossRef]
54. Giannoulis, K.D.; Karyotis, T.; Sakellariou-Makrantonaki, M.; Bastiaans, L.; Struik, P.C.; Danalatos, N.G. Switchgrass biomass partitioning and growth characteristics under different management practices. *NJAS Wagening. J. Life Sci.* **2016**, *78*, 61–67. [CrossRef]
55. Elbersen, H.W.; Christian, D.G.; El Bassem, N.; Alexopoulou, E.; Pignatelli, V.; Van Den Berg, D. Switchgrass (*Panicum virgatum* L.) as an Alternative Energy Crop in Europe. Initiation of a Productivity Network. Final Report Fair 5-CT97-3701 "Switchgrass", 2001, pp. 1–91. Available online: www.switchgrass.nl (accessed on 9 December 2024).
56. Goldstein, J.; Tarhule, A.; Brauer, D. Simulating the hydrologic response of a semiarid watershed to switchgrass cultivation. *Hydrol. Res.* **2014**, *45*, 99–114. [CrossRef]
57. Schilling, K.E.; Jha, M.K.; Zhang, Y.K.; Gassman, P.W.; Wolter, C.F. Impact of land use and land cover change on the water balance of a large agricultural watershed: Historical effects and future directions. *Water Resour. Res.* **2008**, *44*. [CrossRef]
58. Chen, Y.; Ale, S.; Rajan, N.; Morgan, C.L.S.; Park, J. Hydrological responses of land use change from cotton (*Gossypium hirsutum* L.) to cellulosic bioenergy crops in the Southern High Plains of Texas, USA. *GCB Bioenergy* **2016**, *8*, 981–999. [CrossRef]

59. Chen, Y.; Ale, S.; Rajan, N.; Srinivasan, R. Modeling the effects of land use change from cotton (*Gossypium hirsutum* L.) to perennial bioenergy grasses on watershed hydrology and water quality under changing climate. *Agric. Water Manag.* **2017**, *192*, 198–208. [CrossRef]
60. Femeena, P.V.; Sudheer, K.P.; Cibin, R.; Chaubey, I. Spatial optimization of cropping pattern for sustainable food and biofuel production with minimal downstream pollution. *J. Environ. Manag.* **2018**, *212*, 198–209. [CrossRef] [PubMed]
61. Curk, M.; Glavan, M. Assessing and Mapping the Environmental Impacts of Best Management Practices in Nitrate-Vulnerable Areas. *Water* **2023**, *15*, 2364. [CrossRef]
62. Yang, R.; Liu, W. Nitrate contamination of groundwater in an agroecosystem in Zhangye Oasis, Northwest China. *Environ. Earth Sci.* **2010**, *61*, 123–129. [CrossRef]
63. Giannoulis, K.D.; Danalatos, N.G. Switchgrass (*Panicum virgatum* L.) nutrients use efficiency and uptake characteristics, and biomass yield for solid biofuel production under Mediterranean conditions. *Biomass Bioenergy* **2014**, *68*, 24–31. [CrossRef]

hydrology MDPI

Article

Assessing Hydrological Alterations and Environmental Flow Components in the Beht River Basin, Morocco, Using Integrated SWAT and IHA Models

Fatima Daide [1,*], Thomas Hasiotis [2,*], Soumaya Nabih [1], Soufiane Taia [3], Abderrahim Lahrach [1], Eleni-Ioanna Koutsovili [4] and Ourania Tzoraki [2]

1 Laboratory of Geo-Resources and Environment, Faculty of Sciences and Techniques, University of Sidi Mohamed Ben Abdellah, Fez 30000, Morocco; soumaya.nabih@usmba.ac.ma (S.N.); abderrahim.lahrach@usmba.ac.ma (A.L.)

2 Department of Marine Sciences, School of Environment, University of the Aegean, 81100 Mytilene, Greece; rania.tzoraki@aegean.gr

3 National Thematic Institute for Water Research (INTR-Eau), University Ibn Zohr, Agadir 80000, Morocco; s.taia@uiz.ac.ma

4 Institute of Electronics, Computer and Telecommunication Engineering (IEIIT), Consiglio Nazionale delle Ricerche (CNR), 56122 Pisa, Italy; eleniioannakoutsovili@cnr.it

* Correspondence: fatima.daide@usmba.ac.ma (F.D.); hasiotis@marine.aegean.gr (T.H.)

Abstract: This study presents a comprehensive analysis of hydrological alterations and environmental flow components in the Beht River basin in northwest Morocco, using a coupled approach involving the Soil and Water Assessment Tool (SWAT) for hydrological modeling, the Indicators of Hydrologic Alteration (IHA) for flow regime assessment, and the Standardized Precipitation Index (SPI) for drought characterization. The SWAT model, run on a daily time step, showed satisfactory performance in terms of statistical criteria for both calibration and validation periods, despite encountering limitations, and proved its ability to simulate and reproduce the hydrological behavior of the basin. Using the IHA, we investigated changes in the hydrological regime over two distinct periods, revealing significant hydrological alteration. The SPI analysis supported these findings by highlighting the variable impacts of dry and wet periods on the hydrological regime, thus validating the observed changes in river flow indicators. As a preliminary step toward establishing environmental flows in the Beht River, this study provides foundational insights into the temporal evolution of its hydrology. These findings offer a valuable basis for better water resource management and conservation in the region.

Keywords: hydrological modeling; hydrologic alteration; environmental flow; SWAT; IHA; Beht River basin; SPI

1. Introduction

Climate change has exacerbated extreme weather events in Morocco, with extended droughts followed by strong rainfall episodes resulting in disastrous floods that have devastated infrastructure and agricultural regions [1–4]. Furthermore, as the demand for water resources increases due to population growth, economic development, and climate change, it is apparent that crucial measures must be put in place to ensure rational and integrated water resource management [5–7]. Such management is of paramount importance in the enhancement, mobilization, and protection of water against climatic and anthropogenic hazards [8]. Consequently, these factors significantly modify and alter the natural flow regime of rivers and their ecosystems, leading to serious challenges [9,10].

The comprehension and successful management of water resources in semi-arid watersheds, which are known for their limited water supply, high evapotranspiration rates, and irregular rainfall patterns [11,12], depends mainly on hydrological modeling [13,14]. In particular, it adds critical knowledge to the relationship between climate, land use, and hydrology. Different models can be used in hydrological modeling and take into account a lot of parameters to predict the flow and water quality, requiring a comprehensive set of data, including meteorological, topographical, and hydrological data, which are challenging to obtain particularly in underdeveloped and/or developing countries like Morocco due to the poor quality and scarcity of data [15–17], limited spatiotemporal hydro-meteorological gauging networks, and the complexity of hydrological systems. This understanding helps manage diverse impacts on river system flow regimes, ensuring the continued health and functionality of aquatic ecosystems.

The SWAT (Soil and Water Assessment Tool) model is a widely used hydrological model designed for the comprehensive assessment of water resources, land use, and water quality within a watershed or river basin. The model has demonstrated its resilience and adaptability through extensive implementation in diverse climatic and geographical contexts, including Morocco [13,18–25].

In the context of sustainable freshwater management, it is necessary to assess the degree of alteration and the impact of various changes on the river flow regime and its ecosystem, aiming to ensure adequate environmental flows in rivers [26], which, according to the Brisbane Declaration [27], describe "the quantity, quality and duration of water flows necessary to sustain freshwater and estuarine ecosystems to support subsistence and ecosystem well-being". The assessment of environmental flows requires a detailed study, including a hydrological study to obtain information on the flow regime of the river over the years, i.e., the amount of water flowing in its channels during the different months of the year, the timing of small and large floods, data collection and analysis on fish communities, macro-invertebrates, and riparian vegetation, and an assessment of water quality [28,29]. In addition, the commonly used Indicators of Hydrologic Alteration (IHA) model provides a structure for measuring hydrological changes resulting from human activities and other factors, helping in the assessment of river health and environmental flow needs [28]. To better characterize this climatic variability and support the IHA results, this study also employs the Standardized Precipitation Index (SPI) [30] at multiple timescales.

The aim of the present work is to apply the continuous SWAT model to generate flows that will serve as inputs for evaluating the degree of alteration and the environmental flow components of the Beht River basin using the IHA model. In parallel, SPI analysis is conducted to evaluate the meteorological droughts and wet periods and reinforce the interpretation of long-term hydrological trends. As a preliminary step, which needs to be refined by a comprehensive ecological study, we need to obtain a general idea of the flow conditions on the basis of which ecological flow recommendations should be implemented. This concern has motivated us to investigate and estimate water resources in the Beht River basin, which has a semi-arid climate that is very sensitive to the effects of climate change [31–33], and where few similar studies have been carried out [34–36].

2. Materials and Methods

2.1. Study Area

The basin is situated in the southwest region of the Sebou basin in northwestern Morocco, and it has a surface area of around 4560 km^2 (Figure 1). It is bound to the north by the Gharb plain and the Meknes plateau, to the south by the Oum–Erbia basin, to the west by the Bouregreg basin, and to the east by the Middle Atlas. Its boundaries lie between meridians 5° and 6° west and parallels 33° and 34° north [31].

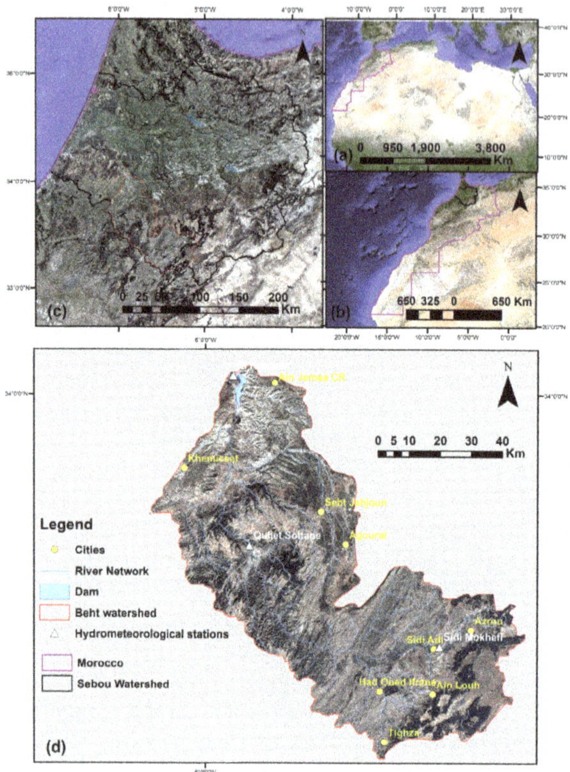

Figure 1. Location of study area: (**a**) North Africa, (**b**) Morocco, (**c**) Sebou watershed, and (**d**) Beht watershed.

It is dominated by primary formations and has a highly varied mountainous appearance. Its altitudes follow a decreasing gradient from upstream to downstream. Its hydrographic network is 451 km long. The main river flows from southeast to northwest, originating at the confluences of the Oueds Tigrigra and Ifrane. Poorly developed soils, brown soils, Vertisols, and similar soils are the main soil types found in the basin. Rangeland predominates the land cover, accounting for 32% of vegetation cover, while forests make up 20% and agricultural land 18%. In total, 23% of the basin's surface is bare land (Appendix A). The region has a Mediterranean climate, with summer droughts and heavy rainfall from violent storms. With 513 mm upstream and 340.7 mm downstream, the average annual rainfall is spatially heterogenous. Maximum monthly mean values are recorded in December for all rainfall stations, while minimum values are recorded in July. Temperatures range from 4.1 °C to 30.4 °C. On the other hand, the hydrological regime is characterized by periods of low water during the dry season and periods of high water, which occur mainly during the wet season [37].

2.2. Overall Methodology

The methodology employed includes initial hydrological modeling using the Soil and Water Assessment Tool (SWAT). Then, the generated flow data serve as a crucial component in the study of hydrological alteration and environmental flow components through the application of the Indicators of Hydrologic Alteration (IHA) model. In addition, a Standardized Precipitation Index (SPI) analysis is conducted to highlight the impact of droughts and wet periods on river flows.

2.3. Model Choices and Description

2.3.1. SWAT Model Setup

The Soil and Water Assessment Tool (SWAT) model is a widely used hydrological model that helps simulate the impacts of land management practices on water resources within a watershed. It is developed and maintained by the United States Department of Agriculture's Agricultural Research Service (USDA) [38,39]. It is primarily used for watershed-scale assessments and provides valuable insights into the quantity and quality of water resources. It takes into account various factors to simulate the movement of water, sediment, nutrients, and agricultural chemicals within the watershed [40].

The SWAT model uses mathematical equations and algorithms to simulate hydrological processes, including precipitation, evaporation, infiltration, surface runoff, groundwater flow, and streamflow [41]. It can estimate the impacts of different land use scenarios, agricultural practices, and climate change on water availability and water quality indicators such as nutrient loading and sediment yield. It uses the following equation for estimating the water balance [41]:

$$SW_t = SW_0 + \sum_{i=1}^{t} (R_{day} - Q_{surf} - E_a - W_{seep} - Q_{gw}) \quad (1)$$

SWt: soil water content [mm];
SW0: initial soil water content [mm];
Rday: daily precipitation [mm];
Qsurf: surface runoff [mm];
Ea: evapotranspiration [mm];
Wseep: infiltration [mm];
Qgw: base flow [mm];
t: time [d].

- Data requirements

The required climate variables for the SWAT model are daily precipitation, daily minimum and maximum temperature, daily mean wind speed, daily relative humidity, and daily solar radiation [39]. For precipitation data, measurements were provided by the Sebou Hydraulic Basin Agency (ABHS), while data for the other variables were sourced from the CFSR (Climate Forecast System Reanalysis) satellite data, which are closest to the observed station, openly available at the SWAT portal "https://globalweather.tamu.edu/ (accessed on 10 March 2023)". CFSR data, produced by the National Centers for Environmental Prediction (NCEP) [42], are widely used to simulate the hydrological system in Moroccan watersheds [18,19,43] and have shown satisfactory results.

- Model Calibration and Validation

The calibration of the SWAT model involved a sensitivity analysis to identify the important SWAT parameters to focus on during the procedure, using the SUFI-2 (Sequential Uncertainty Fitting) program [44,45]. To perform the sensitivity analysis, we first defined the optimized objective function in order to compare the observed and simulated flows. It is based on several statistical indices such as the Nash–Sutcliffe coefficient [46], the coefficient of determination (R^2), and the root mean square error (RSR). Then, the most significant input parameters affecting the flows in the watershed were identified. Finally, a global sensitivity analysis was conducted in order to reduce the number of parameters to be calibrated, excluding those that have a negligible influence on the studied process.

Considering the spatial heterogeneity and the large area of the Beht River basin, multi-site or sequential calibration was implemented, which has been proposed by some researchers [47,48], which consists of calibrating the model at several points that generally represent the gauging stations. Then, for the validation step, the best performing parameter selected during the calibration were used to re-simulate flows on a new dataset.

2.3.2. IHA Model Setup

In order to characterize natural water conditions, assess changes in flow regimes resulting from human activity and hydraulic infrastructure, and determine the degree of hydrological alteration, the US Nature Conservancy developed a technique known as Indicators of Hydrological Alteration, or IHA [49,50]. Almost 2000 hydrologists, researchers, policymakers, and managers of water resources have used the IHA worldwide [51].

The IHA model generates two types of flow statistics: 33 IHA statistics and 34 flow statistics calculated for five different environmental flow components (EFCs).

The 33 IHA parameters are divided into five groups (magnitude, duration, amplitude, frequency, and timing) as part of the annual flow regime. The five main attributes are programmed into the IHA for their influence on aquatic species at different stages of their life cycle. The IHA model enables simple comparison of the hydrological attributes of a site before and after a certain project activity, or of two sites with different types or levels of impact. It provides a statistical measure of change in the central tendency or degree of variation of an attribute of interest [28].

The 34 environmental flow component statistics are a more recent series of hydrological flow parameters and have been developed to identify and calculate statistics on hydrological events such as floods and droughts. These flow components have been determined to be ecologically essential and included in the IHA tool as "environmental flow components" [52]. Each of these flow components has a respective role in the life of an organism. This classification is also based on the fact that each of these groups consists of a flow event that is characteristic of a natural river hydrograph and is ecologically significant.

- Data requirements

In order to define the hydrological regime and assess its alteration, as well as the environmental flow components, it is crucial to conduct a comprehensive analysis of the previous and current states of the hydrological regime, termed pre-impact (1970–1992) and post-impact (1993–2014) periods. The analyses were based on daily flow data obtained by applying the SWAT model, calibrated using measurements of daily precipitation and observed flow records at the El Kansera station from 1970 to 2014. The selection of the pre-impact and post-impact periods was driven by the need to understand the historical hydrological regime and changes that have been caused by human activity and climate change. Notably, these time frames encompass a significant environmental transition, characterized by a prolonged period of drought followed by a rainy period. This succession of contrasting climatic conditions underscores the importance of understanding how such variations influence the hydrological dynamics of the river over time.

This comprehensive approach aims to anticipate long-term trends and potential alterations due to anthropogenic action, coupled with the impacts of climate change on river systems, thereby contributing to informed decision-making and adaptation strategies. All parameters were calculated using non-parametric statistics (median and coefficient of dispersion).

2.3.3. Drought Analysis Using the Standardized Precipitation Index (SPI)

The SPI was calculated at three different timescales, 6, 12, and 36 months, based on monthly precipitation data covering the period from 1970 to 2014 and was computed using the gamma distribution, with distribution parameters estimated through the probability-weighted moments (PWMs) method under the programming language R. In addition, a drought event analysis was carried out to identify drought events and their characteristics in terms of duration, magnitude, intensity, and frequency. These indicators were used to complement the IHA results and to better understand the role of drought dynamics in shaping the hydrological alterations observed in the selected periods.

3. Results

3.1. SWAT Model

Use of the SWAT hydrological model of Beht basin was conducted using model calibration, validation, and sensitivity analysis.

3.1.1. Sensitivity Analysis

In order to optimize the implementation of this procedure for the Beht basin, a sensitivity analysis using the OAT (Once at Time) type was first performed. This sensitivity analysis allowed us to classify the sensitive parameters according to their degree of influence (the significant parameter is the one with the lowest *p*-value and the highest t-stat value). Among 21 parameters, 11 were found to be influential (Table 1).

Table 1. Classification of model sensitive parameters.

Sensitivity Level	Parameters	Optimal Value
1	CN2.mgt	0.14
2	GW_DELAY.gw (days)	31.02
3	ESCO.hru	0.75
4	GW_REVAP.gw	0.2
5	GWQMN.gw (mm H2O)	2190
6	ALPHA_BF.gw (days)	0.8
7	EPCO.hru	0.86
8	SOL_AWC.sol (mm H2O/mm Soil)	0.24
9	REVAPMN.gw (mm H2O)	400.22
10	CH_K2.rte (mm/hr)	83.31
11	CH_N2.rte	0.22

Among the eleven parameters that most influence hydrological modeling (Table 1), we noticed the presence of parameters related to surface flow (Curve Number "CN2"), groundwater flow (aquifer recharge time "GW_DELAY, ALPHA_BF, GWREVAP, GWQMN, REVAPMN"), evaporation (soil evaporation factor as function of depth "ESCO"), soil (soil water content "SOL_AWC"), and basin routing (CH_K2).

The sensitivity of the model to these calibrated parameters confirms the findings of some studies already carried out in areas with similar characteristics in terms of the semi-arid climatic context [19,24,25,53]; so, as can be seen, most of the sensitive parameters are generally groundwater-related.

3.1.2. Model Calibration/Validation

The data presented in Figure 2 correspond to the observed and simulated daily flow calibration and validation for El Kansera gauging station in the outlet of the watershed. This figure shows that the model successfully simulates the flow variation and represents well the different daily peaks, with underestimation of most peak flows and overestimation of base flows. Therefore, the uncertainties in peak flows may be related to the complexity of the watershed characteristics, which may contribute significantly to a relative imperfection in the model.

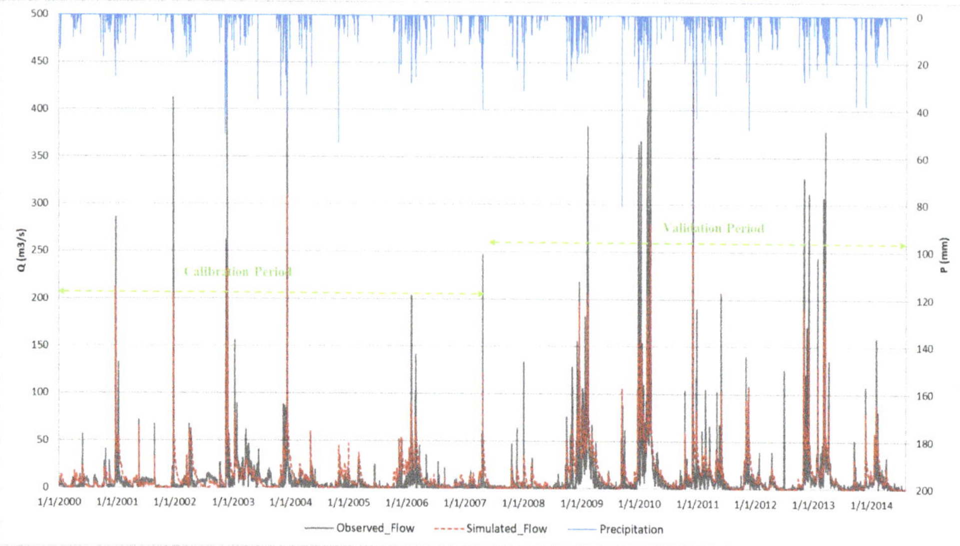

Figure 2. Observed vs. simulated daily flows during calibration/validation.

Previous studies developed for arid and semi-arid basins [24,54–59] have used statistical performance criteria such as R^2, NSE, and RSR. Reference [60] suggested that the NSE value should not exceed 0.5 and the RSR value should not exceed 0.7 to be able to judge the hydrological calibration and validation as satisfactory; [61] proposed that an R^2 value greater than (or equal to) 0.6 is a satisfactory indicator of good SWAT model calibration/validation.

Table 2 lists the calibration evaluation results in daily time step at the El Kansera station. The values indicate good model performance according to the recommended model performance evaluation criteria [60], which resulted in satisfactory simulation of the SWAT model in the studied watershed. Other authors have achieved almost similar results under semi-arid climate conditions [13,19,24,53,62].

Table 2. Performance indicators for calibration/validation.

Indices	Calibration	Validation
NSE	0.68	0.61
R^2	0.7	0.73
RSR	0.4	0.3

3.2. IHA Model

Generally, the IHA model is used to assess the impact of the installation of stations and/or dams on the hydrological regime of a river [63–65] or the impact of climate change [66]. In our case study, the dam was created in 1934, but earlier hydrometeorological data are not available. Nevertheless, a study of the earlier and most recent decades of the hydrological regime, referred to as pre- and post-impact periods, is necessary as an initial step in assessing the impact of changes over the last four decades in the Beht River basin.

The findings from the IHA study clarify alterations in 33 ecological indicators, indicating impacts on the river ecology and water flow caused by human activities, categorized into five groups, along with environmental flow components.

3.2.1. IHA Components

Table 3 provides a comprehensive summary of the IHA results, showing pre- and post-impact values for all five groups.

Table 3. IHA parameter results.

	IHA Factor	Pre-Impact	Post-Impact	Variation (%)
Group 1	October	0.15	0.5	227.45
	November	1.58	2.43	54.38
	December	1.75	3.26	85.92
	January	2.65	3.66	38.04
	February	1.95	3.04	56.26
	March	2.54	2.38	−6.3
	April	1.85	2.28	23.24
	Mai	0.98	0.78	−20.3
	June	0.1	0.28	193.48
	July	0.03	0.05	78.79
	August	0.02	0.02	−2.35
	September	0.03	0.03	7.63
Group 2	1-day minimum	0.012	0.013	2.25
	3-day minimum	0.017	0.015	−9.18
	7-day minimum	0.0185	0.0183	−1.02
	30-day minimum	0.02166	0.02163	−0.13
	90-day minimum	0.027	0.071	155.71
	1-day maximum	61.95	84.58	36.52
	3-day maximum	33.58	51.87	54.46
	7-day maximum	18.83	32.84	74.4
	30-day maximum	8.08	13.86	71.36
	90-day maximum	5.45	6.853	25.53
	Number of zero days	0	0	0
	Base flow index	0.0058	0.0082	40.73
Group 3	Date of minimum flow	231	236	2.16
	Date of maximum flow	317	334	5.36
Group 4	Low pulse count	9	8	−11.11
	Low pulse duration	2	1	−50
	High pulse count	4	5	25
	High pulse duration	13	8.5	−34.61
Group 5	Rise rate	0.0256	0.06288	145.62
	Fall rate	−0.1318	−0.1632	23.82
	Number of reversals	171	183	7.01

1. Group 1: Magnitude of monthly flows

Figure 3 shows the flow duration curves for each month, as well as the annual series. The significance of this graph lies in the ability to assess the probability of exceeding a certain flow magnitude on a given day of a month. Note that the y-axis is represented on a logarithmic scale. The shape of the curve in high and low-flow regimes is particularly important in assessing the characteristics of the river. At high flows, the curve is less abrupt, indicating that high flows occur over several days in both periods. The lower limit, on the other hand, presents a smooth curve, indicating that the capacity of the basin can support low flows during the dry season. These flows are maintained throughout the year due to natural streamflow conditions and/or the high capacity of the groundwater that supports the base flow of the stream. A comparison between the two periods analyzed shows that low and medium flows have increased over the period 1993–2014.

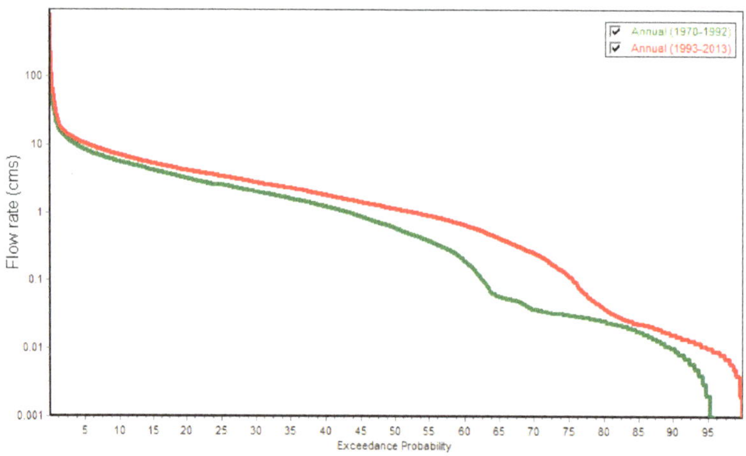

Figure 3. Annual flow duration curve.

Figure 4 shows the variation in monthly flows between the two periods, indicating however an increase in median flows during the humid period from October to February, with an increase recorded in April. These results are reflected in Table 3, which shows an exceptional increase in the months of October, December, and June, indicated by relative variation values of 227%, 85%, and 193%, respectively.

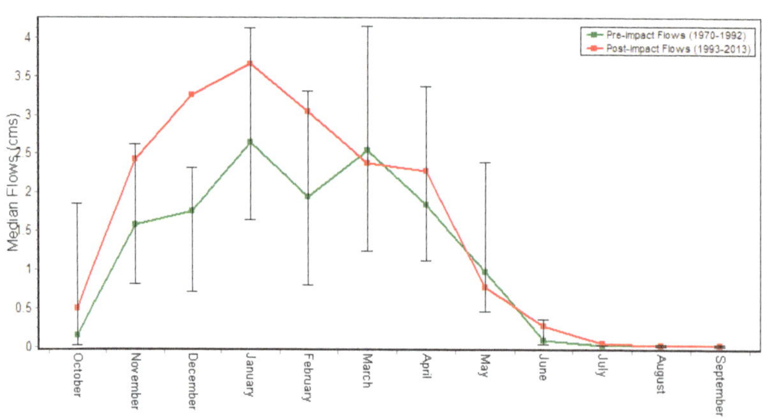

Figure 4. Pre- and post-impact of monthly flows.

Additionally, Figure 5 demonstrates changes in flow distribution for the most wet month (January) and the driest month (July), including the percentiles 25th, 50th, and 75th. These alterations are linked to the different weather conditions observed across the two periods studied: from dry to exceptionally wet, with intense rainfall and flooding events, in the post-period. These findings underline the influence of climate variability on river flow dynamics.

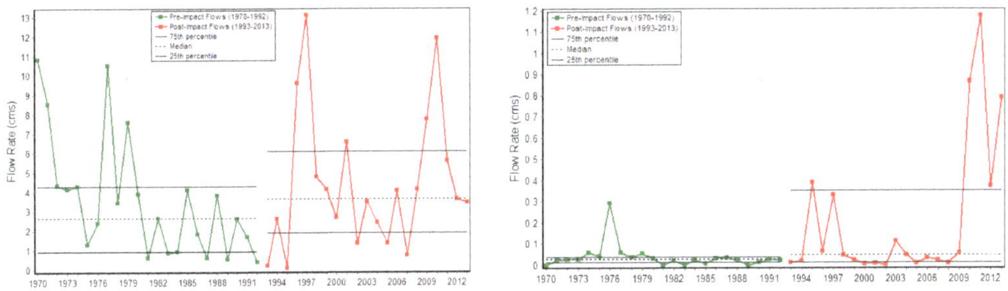

(**a**) (**b**)

Figure 5. Monthly flow for January (**a**) and July (**b**).

2. Group 2: Magnitude and duration of extreme conditions

Figure 6 displays the most significant hydrological changes in the study area, detailing the Range of Variability Approach (RVA) classifications (high, medium, low) for each of the 33 hydrological factors. The analysis of the RVA showed a rise in the occurrence frequency of the majority of indicators, primarily in the high RVA category, identified as a major hydrological change.

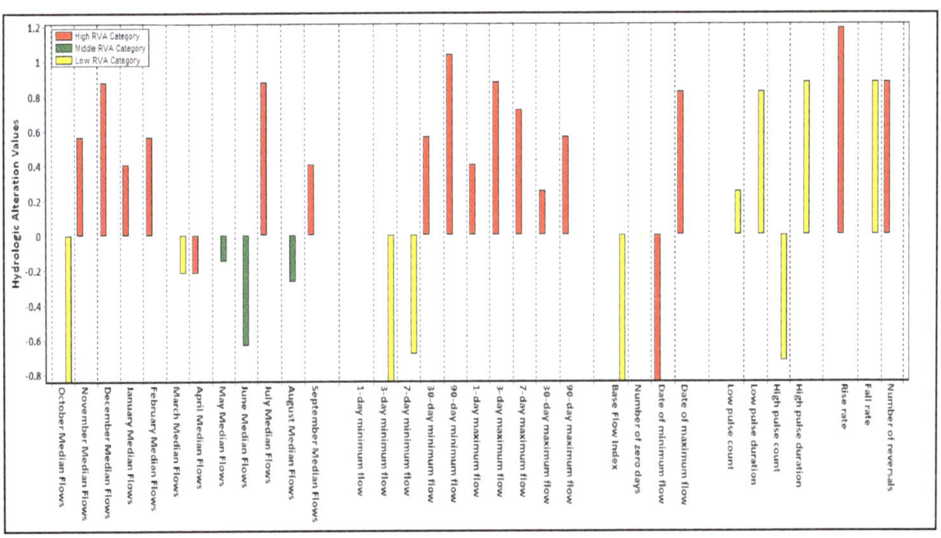

Figure 6. RVA classification of the main hydrological alteration factors.

In particular, there is a marked increase in the frequency of 1-, 3-, 7-, and 90-day maximum flows, indicating higher flood magnitudes and frequency. In terms of minimum flows, the 1- to 7-day minimum flows show negative RVA alterations, indicating lower short duration flows, while 30- and 90-day minimum flows exhibit positive RVA values, suggesting higher baseflow, and this is consistent with the increase in the 90-day minimum flow of 155.71%, in comparison to the 1- to 30-day minimums, which showed lower or even negative flow changes between −9.18% and 2.25%. The significant increase in monthly median flows in October and July supports this trend, with the July increase due to an increase in baseflow, which is confirmed by the baseflow index indicating a negative value. The findings suggest that there has been a notable amount of environmental stress and disruption along the river [67].

3. Group 3: Timing of extreme annual water conditions

Table 3 presents the findings. The post-impact period showed an increase of 2.16% in the Julian date of each 1-day annual minimum, with a negative alteration degree of −0.8. Similarly, the Julian date of each 1-day annual maximum increased by 5% over the last two decades compared to the two decades before, and Julian days have shifted from 317 pre-impact to 334 post-impact, with a positive alteration value of 0.8. Such a shift of almost a month will have an effect on the aquatic ecosystem, such as spawning and migration, leading to a loss of biodiversity.

4. Group 4: High and low-flow pulses

The results of group 4 (Table 3) indicate that there has been no significant change in the frequency of low pulses, with a hydrological alteration value of 0.2. The duration of both low and high pulses has decreased, suggesting shorter periods of drought conditions. In general, an increase in flash drought phenomena can be observed. The decrease is −50% for low pulse duration and −34.6% for high pulse duration. Overall, low and high pulse duration as well as high pulse frequency are all noticeably changed and altered by 0.82, 0.87, and −0.7, respectively. The general health of aquatic ecosystems can be impacted by decreased pulse durations, which may cause problems with biological processes including nutrient cycling and sediment transport.

5. Group 5: Rate and frequency of flow changes

The IHA group 5 parameters show the rate of rise, the rate of fall, and inversions. In general, these factors give details about sudden fluctuations in annual flow. The findings from Table 3 show a 145% increase in the rate of rise, indicating the appearance of more flash flood events, and a 23.8% slight change in the rate of fall, indicating that river biodiversity experiences higher stress and specific measures should be decided for its preservation. More severe drought events cause stress to aquatic habitats and species and may transform some parts of the river network from perennial to intermittent flow. The rate of rise is strongly altered, with a value of 1.19. The rate of fall of the hydrological alteration value is high (0.87). Thus, the number of inversions is strongly altered (0.87).

3.2.2. Environmental Flow Components, Duration and Timing

1. Environmental flow components

Analysis of environmental flow components for the period 1970 to 1992 revealed only one major flood in 1990 (Figure 7). This period also saw small floods, which facilitated rapid growth of aquatic organisms, as flooded shallow areas are often warmer than the main river and rich in nutrients and insects [51]. This phase experienced a long period of drought, manifested by the presence of extremely low flows, which can be stressful for many organisms but can also provide the conditions necessary for certain plant species to regenerate. As for high flow pulses, they occurred with high frequency throughout the period and are important and necessary in low-flow areas, as they provide rich organic matter to sustain aquatic food.

For the second period (1993–2013), the analysis of environmental flow components revealed three major floods, namely in 2002, 2004, and 2010 (Figure 7). Small floods dominated. Consequently, there was movement of organisms upstream–downstream into the floodplain wetlands of the river, which had a positive impact on the migration of aquatic organisms, spawning/reproduction, and the juvenile growth period [68]. These areas, which are often inaccessible, can provide important food resources. On the other hand, low flows showed significant variation and could have an impact on predator–prey relationships, as well as the diversity and number of organisms [69]. High-flow pulses are almost uniform in magnitude, encouraging mobile organisms to move upstream and downstream.

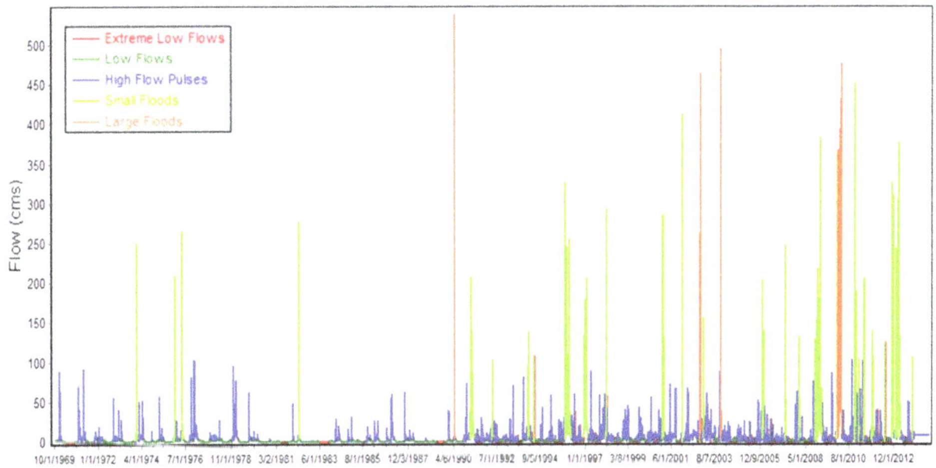

Figure 7. Environmental flow components in Beht River basin (1970–2014).

2. Flow duration and timing of environmental flow components

The duration of all environmental flow components increased, with the exception of extremely low flows (Figure 8). Low flows showed an increasing trend, while extreme low flows otherwise decreased. Based on the analysis of flow component, significant floods occurred following 1989 and 1995, with both large and small floods being the most common. The high flow pulse, which occurred at the beginning of the rainy season, was almost undisturbed, potentially making it easier for organisms to move to upstream locations. This component is inadequate for fully supporting the diverse aquatic organisms found in the river and floodplain wetlands during both small and large floods [69].

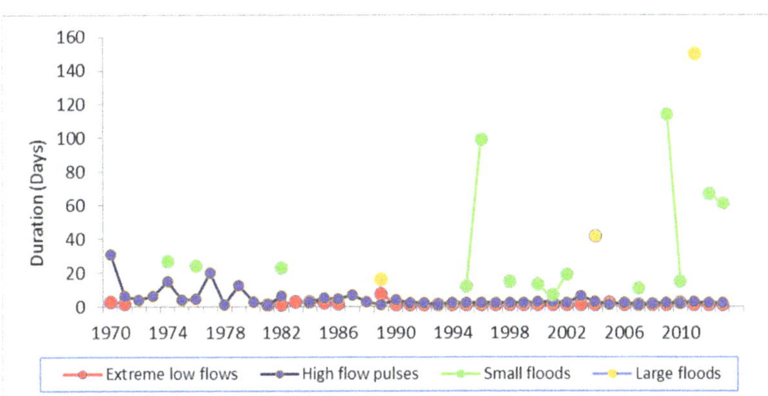

Figure 8. Duration of environmental flow components.

The timing (Julian day) of the occurrence of extreme low flows varies between 140 and 300; this component is steady and represents a median of 216 and a coefficient of dispersion of 0.16. During the first period, these flows occurred in mid-spring (141 Julian day) to autumn (315 Julian day), and for the second period, they generally took place during the summer season and experienced greater variation. Other components of environmental flow are highly variable; high flow pulses experienced greater interannual variation between 1970 and 1992 and occurred during winter and autumn, and between 1993 and 2013 they appeared during all seasons but especially in summer (Figure 9). This can disrupt the reproductive cycle of aquatic organisms living in the river. High-flow pulses indicate to these species to begin moving in rivers toward floodplain spawning grounds. In the case of small floods, they occurred during the winter season and in some years in spring, and were more frequent during the latter period. Extreme floods also occurred during winter, notably between 1993 and 2013, with the exception of the spring flood in 1990.

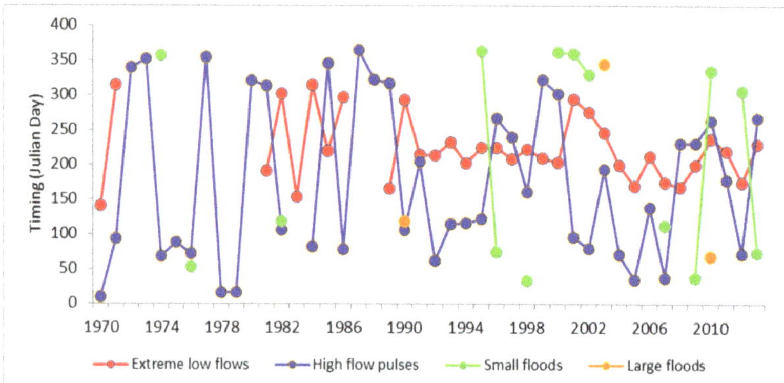

Figure 9. Timing of environmental flow components.

3.3. Drought Analysis Using Standardized Precipitation Index (SPI)

The 6-month SPI (Figure 10a) was calculated to identify and analyze rapid decreases in precipitation over short periods, enabling the detection of sudden transitions between wet and dry conditions.

The most notable decline occurred in the mid-1970s and 1980s—particularly around 1980–1984 and 1988–1990—as well as during the early 2000s. These episodes are marked by sharp transitions from values above +1 to nearly –2 within less than a year.

A total of 12 drought events were identified in the SPI-6 series (Table 4). The durations ranged from 4 to 9 months, and magnitudes varied from –4.55 to –15.0, with drought intensities generally between 0.50 and 0.88. The most intense events occurred between 1980 and 1990 and in 2005. The overall frequency of SPI-6 droughts was estimated as 14.0%, confirming the recurrence of short-term dry spells across the study period.

The 12-month SPI values (Figure 10b) show more persistent drought trends. In the period 1970–1992, the values frequently fall below zero, indicating constant drought and reduced precipitation over shorter periods. Some of the most notable dips occur in the 1980s, with SPI values showing significant negative values (below -1), indicating periods of severe drought.

Positive values predominate during the second period. Peaks even exceed an SPI of 2, which indicates very wet conditions. This suggests that the river basin experienced higher levels of precipitation, particularly in the late 1990s and late 2000s to early 2010s.

In total, nine drought events were detected using SPI-12 (Table 4). Two of the most severe droughts occurred between 1980 and 1982 and between 1988 and 1990, each lasting 16 months with a cumulative magnitude of –29.5. Other moderate droughts occurred in 1995, 2001–2002, and 2005, with durations between 4 and 11 months and intensities ranging from 0.48 to 0.99.

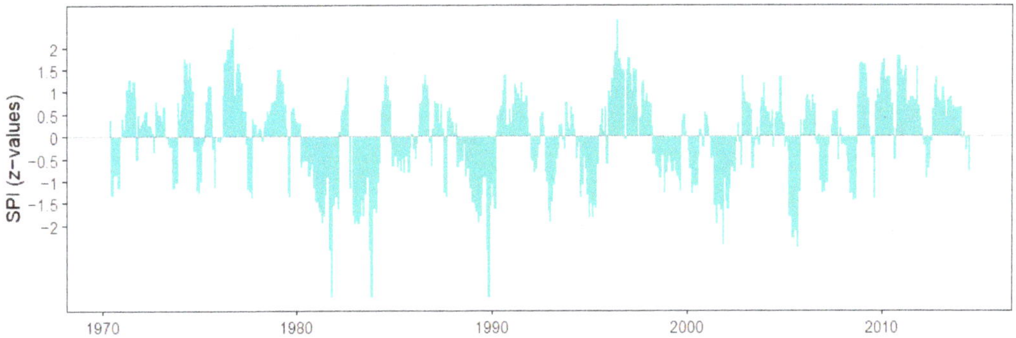

(a)

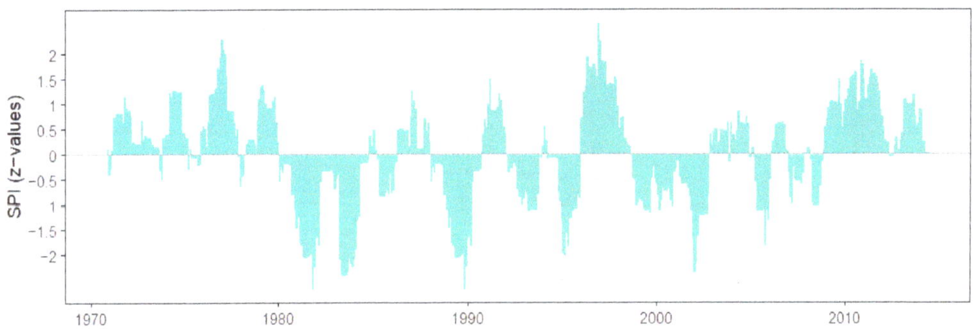

(b)

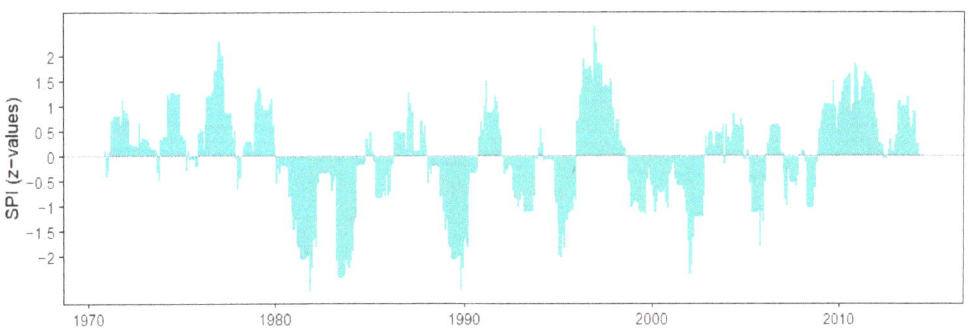

(c)

Figure 10. SPI at 6 months (**a**), 12 months (**b**), and 36 months (**c**).

The 36-month SPI graph (Figure 10c) also indicates a significant negative trend from 1970 to 1992 with long-duration dry spells over several years. This indicates that this area witnessed long droughts with less recovery between dry periods.

In the second period, the SPI shows sustained positive values, including extended wet phases around the late 1990s and in the early 2010s. This indicates that the basin experienced more than just brief wet periods but it was subject to prolonged wet conditions, suggesting a fundamental climatic change.

Five severe long-duration drought events occurred in the SPI-36 record (Table 4). The longest drought occurred from 1981 to 1984, with a duration of 35 months. The other droughts occurred in 1985, from 1989 to 1991, in 1995, and from 2001 to 2002, with durations of 8 to 19 months. The frequency of overall drought was estimated to have been 16.1%, clearly demonstrating that long-duration droughts had significant impacts on water supply and baseflow in the basin.

Table 4. Characteristics of drought events based on SPI of 6, 12, and 36 months.

	Event	Start	End	Duration	Magnitude	Intensity	Frequency
SPI-6	1	12/01/1980	07/01/1981	8	−12.3	0.65	14.00%
	2	10/01/1981	03/01/1982	6	−11.7	0.51	
	3	10/01/1982	07/01/1983	9	−15	0.6	
	4	10/01/1983	03/01/1984	6	−12.1	0.5	
	5	12/01/1988	07/01/1989	8	−12.3	0.65	
	6	10/01/1989	03/01/1990	6	−11.7	0.51	
	7	11/01/1992	03/01/1993	5	−7.12	0.7	
	8	12/01/1994	05/01/1995	6	−9.52	0.63	
	9	04/01/2000	07/01/2000	4	−4.55	0.88	
	10	06/01/2001	11/01/2001	6	−10.6	0.57	
	11	04/01/2005	10/01/2005	7	−13.1	0.53	
	12	05/01/2008	08/01/2008	4	−5.33	0.75	
SPI-12	1	12/01/1980	03/01/1982	16	−29.5	0.54	16.30%
	2	04/01/1983	04/01/1984	13	−26.9	0.48	
	3	12/01/1988	03/01/1990	16	−29.5	0.54	
	4	06/01/1993	09/01/1993	4	−4.41	0.91	
	5	01/01/1995	10/01/1995	10	−14.5	0.69	
	6	05/01/1999	09/01/1999	5	−5.58	0.9	
	7	11/01/2001	09/01/2002	11	−16.1	0.68	
	8	05/01/2005	12/01/2005	8	−9.77	0.82	
	9	05/01/2008	08/01/2008	4	−4.04	0.99	
SPI-36	1	12/01/1981	10/01/1984	35	−59.1	0.59	16.10%
	2	04/01/1985	12/01/1985	9	−12.8	0.7	
	3	11/01/1989	01/01/1991	15	−19.2	0.78	
	4	03/01/1995	10/01/1995	8	−9.56	0.84	
	5	04/01/2001	10/01/2002	19	−24.5	0.78	

4. Discussion

A thorough understanding of hydrological alterations and drought variability is essential for evaluating how climate variability and human activities affect river basins, especially in semi-arid areas like the Beht basin. This study combined hydrological modeling (SWAT), flow regime analysis (IHA), and drought characterization (SPI) to provide a comprehensive understanding of changes over the past four decades.

The SWAT model has been applied in a multitude of research domains, including but not limited to the assessment of soil erosion and sedimentation, as documented in studies by [18,21,70–72]. Furthermore, it has been a significant factor in predicting the environmental impacts of land use changes and climate change, as highlighted in the work of [22]. In the field of hydrology and environmental sciences, the capabilities of the SWAT model extend to the simulation of surface water and groundwater systems, facilitating assessments of water quality and quantity, as illustrated by [23]. Additionally, the model has been a valuable tool in the analysis of environmental risks, as demonstrated by the research conducted by [72]. These diverse applications underscore the model's adaptability and its integral role in advancing our understanding of environmental processes and their interactions within various ecosystems.

The performance of the SWAT model for the Beht River basin was generally satisfactory, especially considering the limited and imprecise data conditions, the spatial distribution of precipitation in the basin which may not be representative, and the heterogeneous characteristics of the basin in terms of land use, soil types, and topography, as well as the flow regime characterized by extremely variable precipitation. Sensitivity analysis revealed that the most sensitive parameter was the Curve Number, due to the presence of the main land cover classes (pasture land, bare land, agricultural areas), which represent more than half of the land uses in the watershed, since runoff is more important in these areas for a rainfall event compared to other land use types. Most of the other sensitive parameters are related to groundwater flow. GW_DELAY and ALPHA_BF regulate water storage in the soil via groundwater to the river. The fact that the predominant soil categories fall under hydrological group B, which is characterized by high drainage capacity, moderate infiltration rates, and water transmission rates, explains the sensitivity to the SOL_AWC and GWQMN parameters. It should be noted that the basin upstream regions are distinguished by the presence of karst geological structures that allow for precipitation infiltration. Following soil saturation, water flows through the deep aquifer and may manifest as base flow after a rainfall event, thereby increasing the total amount of water drained. However, the SWAT model is not able to accurately simulate this aspect. The presence of the sensitive parameters related to evapotranspiration (ESCO, EPCO, REVAPMN) may be due to the significant abstraction of evaporation demands from the lower soil layers.

However, the model's underestimation of extreme peak flows during rainy seasons indicates that the groundwater component of the hydrological system is not accurately simulated. This underestimation can have significant consequences for the accuracy of flood forecasting and management, and it may be connected to the same factors that affect base flows, including the complex distribution of rainfall and potential impacts from small reservoirs. These extreme peak flows are influenced by a combination of factors, including intense rainfall events, rapid runoff due to the karst evolution, and the groundwater component of the hydrological system. The model's inability to accurately capture this groundwater component, especially in the face of heavy rainfall, highlights a significant shortcoming in its representation of the hydrological system's reality. The SWAT model has been shown to perform poorly in simulating rapid hydrological responses in karst areas, largely due to its simplistic representation of groundwater flow [73,74]. Implementing model modifications that account for karst-specific processes, such as those proposed by [75], can improve the simulation of extreme events. The overestimation of base flows can be explained by the poor distribution of rainfall, particularly seepage water in karst areas, and by the potential existence of small reservoirs used to supply water to crops in the catchment, which may affect the flow rate at low levels [76].

Despite these limitations, the SWAT model provides reliable discharge outputs that support the Indicators of Hydrologic Alteration (IHA) analysis. The IHA analysis indicates significant hydrological alteration between the pre-impact (1970–1992) and post-impact (1993–2013) periods. Notably, there was an increase in low and medium monthly flows during the post-impact period, as this was considered a wet period. According to previous studies [77] carried out over the whole of Morocco, the years 1970–1991 were considered a drought period, while the following two decades were wet, which explains this general increase; at the same time, the high flows remained unchanged, which may be linked to climate variability and dam operations [78], since dams are known to reduce the frequency of specified floods [79]. In subsequent years not included in the study, a drought analysis study from 1980 to 2017 showed a drastic increase in drought from 2012 onwards [80]. The variation in monthly flows also showed an exceptional increase in median flows during the wet period from October to February.

The RVA analysis revealed major hydrological changes, with a strong positive alteration mainly in the high RVA category. The increase in maximum flows indicates more frequent and intense flood events driven by short-duration precipitation or human activities like reservoir releases [81]. The increase in minimum flows (30 and 90-day minimum) suggests better maintenance of base flow by improving irrigation practices, monitoring the groundwater level to avoid over pumping, restoring the riparian zone, avoiding high-water-demanding farming practices, and conserving vegetation density. However, short-term minimums remain negatively altered, indicating a temporal divergence in low-flow behavior. This may indicate a seasonal redistribution of flow volumes, where short-term dry events are more intense, but seasonal flow conditions are better maintained, potentially due to human regulation of flow [82].

Alterations were also evident in the timing of extreme annual water conditions, the duration of low and high pulses, the frequency of high pulses, and the rate of rise and fall, thus modifying the flow regime and the aquatic ecosystem. These changes are likely driven by both climatic variability and anthropogenic interventions, including dam operations, groundwater extraction for irrigation, and urban land use expansion.

In addition to climatic variability, anthropogenic pressures, particularly land use and land cover changes, have significantly contributed to the observed hydrological alterations. The Beht basin, as part of the larger Sebou River system, has experienced notable land use transformations over recent decades, including agricultural expansion, urban growth, and deforestation. For example, ref. [83] reported a substantial increase in urbanized areas and cropland in the Sebou Estuary region, changes that likely influenced runoff patterns and the overall river discharge behavior.

Environmental flow components have also been affected, with an increase in the timing and frequency of high-flow pulses during the past-impact period. Major floods occurring in 2002, 2004, and 2010 support these findings. In addition, the decreased duration of extreme low flows suggests more frequent short-term droughts and decreasing baseflow contributions, especially in late summer, which may impact the aquatic habitat dependent on sustained flows.

These hydrological changes, coupled with human impacts, such as the presence of the downstream dam, water pumping for irrigation of cultivated areas, and uncontrolled landfills, have contributed to the degradation of the Beht River basin, which is known for its high diversity of benthic macrofauna [84]. Similar ecological consequences have occurred in other Moroccan and Mediterranean rivers. Reference [85] reported that flow regulation and dam presence brought severe changes in Martil River macroinvertebrate assemblages, with native species decreasing and opportunistic species dominating. Reference [86] showed that reduced flood magnitude in the river affected sediment dynamics and riverbed structure, altering benthic invertebrate habitats. In addition, ref. [87] also demonstrated how minimum flow significantly impacted macroinvertebrates, emphasizing their dependency upon natural low-flow variability. This supports more overall syntheses by [88], which demonstrated that flow timing alterations and reductions in durations of low flow typically destabilize aquatic biodiversity as well as sediment processes in semi-arid regions. These findings underscore the importance of implementing relevant measures aimed at preserving environmental flows. Adaptive management strategies that sustain ecological integrity, while maintaining a balance between ecosystems' water demands and human activity, are needed to deal with these changes.

The SPI analysis further reinforced the hydrological findings. SPI-6 showed that rapid decreases are likely to have significant impacts on monthly flows. For example, the sharp decreases in the late 1980s, 1988–1990, and the early 2000s could explain reduced river flows in those years, reflected in the IHA results for October, June, and other months. Decreases could also correspond to the increase in low-flow counts (1-day, 3-day minimums) as river discharge decreased in response to short-term droughts. In addition, they are associated with higher pulse reductions (Group 4) and an increase in fall rate (Group 5), indicating a rapid decrease in flow in response to reduced precipitation. Similar studies in

Mediterranean catchments confirm that such rapid SPI declines exacerbate flash droughts, altering flow patterns and stressing aquatic ecosystems [89]. Medium-term droughts detected by SPI-12 were also consistent with decreases in minimum flows (Group 2), with negative or stagnant values pre-impact. In contrast, the post-impact period showed more positive SPI values, where monthly flows and maximum flows both showed significant increases (Group 1 and Group 2). As an example, the month of October (+227.45%) showed substantial increases which correspond to the wetter climatic conditions observed in the SPI. Similar trends have been reported in Mediterranean regions, emphasizing susceptibility to alternating periods of severe drought and intense rainfall [90,91]. Finally, the SPI-36 values showed that the basin experienced extended drought conditions, which correspond to low baseflow (as shown by the IHA baseflow index, +40.73%) and minimal changes in river flow regime during the pre-impact period, highlighting the absence of significant hydrological recovery between droughts. In the post-impact period, the trend indicates sustained wet periods, aligned with increases in maximum flows and high-pulse duration (Group 4). The river system has benefited from greater availability of water over longer periods, resulting in more frequent and sustained high flows.

Overall, this combined assessment highlights the importance of maintaining environmental flows and implementing adaptive management strategies to secure the resilience of semi-arid river ecosystems in the face of climate variability and human pressures.

5. Conclusions

This study highlighted significant hydrological changes in the Beht River basin over the past four decades. The IHA analysis showed a general increase in low and medium monthly flows during the post-impact period (1993–2013), which was most likely caused by a combination of dam regulation and wetter climatic conditions. While monthly and minimum flows showed positive trends, the frequency and duration of extreme flow events changed as well, with shorter extreme low-flow durations indicating more frequent short-term drought.

The SPI analysis at multiple timescales supported these findings by demonstrating the high variability between dry and wet periods. These climatic fluctuations, combined with human activities such as dam operations and land use changes, have altered the flow regime and stressed the aquatic ecosystem.

Overall, the findings show ongoing hydrological alterations in a semi-arid Mediterranean basin, emphasizing the critical need to set adequate environmental flow requirements to protect riverine ecosystems. As a first step, this study provides important hydrological insights that can inform future ecological studies and guide sustainable water resource management in the Beht River basin.

Author Contributions: Conceptualization, F.D., O.T. and A.L.; methodology, F.D., S.N. and O.T.; software, S.N. and F.D.; validation, S.N.; investigation, F.D. and O.T.; resources, S.T.; writing—original draft preparation, F.D.; writing—review and editing, F.D., S.N., O.T., T.H. and E.-I.K.; supervision, O.T. and A.L. All authors have read and agreed to the published version of the manuscript.

Funding: This research received no external funding.

Data Availability Statement: The data are contained within the article.

Acknowledgments: O.T. received ERASMUS+ funding for practice mobility in the University of Sidi Mohamed Ben Abdellah, Fez, Morocco, and is grateful to the ERASMUS office support of the University of the Aegean, particularly to G. Koufos, E. Spathopoulou, and Ch. Varkaraki. The authors also gratefully acknowledge the valuable assistance of Vaggelis Livadiotis during the development of the SPI analysis.

Conflicts of Interest: The authors declare no conflicts of interest.

Appendix A

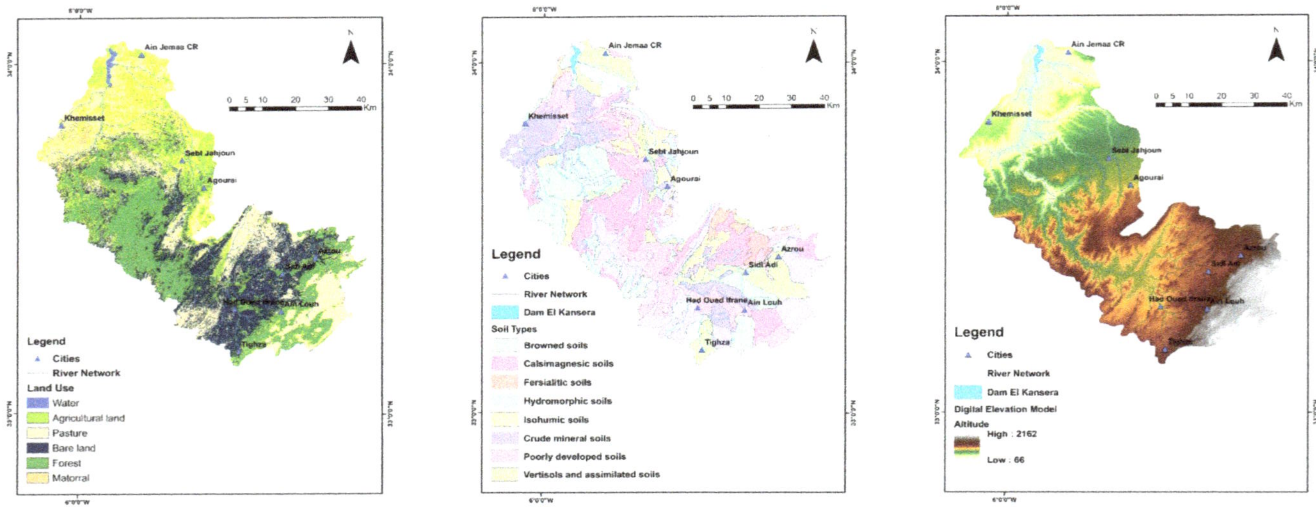

(a) (b) (c)

Figure A1. Spatial datasets used in SWAT modeling: (**a**) land use/land cover; (**b**) soil types; (**c**) digital elevation model (DEM).

References

1. IPCC. Summary for policymakers. In *Climate Change 2021: The Physical Science Basis. Contribution of Working Group I to the Sixth Assessment Report of the Intergovernmental Panel on Climate Change*; Masson-Delmotte, V., Zhai, P., Pirani, A., Connors, S.L., Péan, C., Berger, S., Caud, N., Chen, Y., Goldfarb, L., Gomis, M.I., et al., Eds.; Cambridge University Press: Cambridge, UK; New York, NY, USA, 2021; pp. 3–32. [CrossRef]
2. Kantoush, S.A.; Saber, M.; Abdel-Fattah, M.; Sumi, T. Integrated Strategies for the Management of Wadi Flash Floods in the Middle East and North Africa (MENA) Arid Zones: The ISFF Project. In *Wadi Flash Floods: Challenges and Advanced Approaches for Disaster Risk Reduction*; Springer: Singapore, 2022; pp. 3–34. [CrossRef]
3. Loudyi, D.; Hasnaoui, M.D.; Fekri, A. Flood risk management practices in Morocco: Facts and challenges. In *Wadi Flash Floods. Natural Disaster Science and Mitigation*; Sumi, T., Kantoush, S.A., Saber, M., Eds.; Springer: Singapore, 2022; pp. 35–94. [CrossRef]
4. World Bank Group. Climate Risk Profile: Morocco. 2021. Available online: https://www.worldbank.org (accessed on 21 September 2022).
5. Taleb, H. Water management in Morocco. In *Management of Intentional and Accidental Water Pollution*; Springer: Dordrecht, The Netherlands, 2006; pp. 177–180. [CrossRef]
6. Ait Kadi, M.; Ziyad, A. Integrated water resources management in Morocco. In *Global Water Security*; Water Resources Development and Management; Springer: Singapore, 2018; pp. 143–163. [CrossRef]
7. World Bank Group. Supporting Morocco's Journey to Disaster and Climate Resilience. Morocco Integrated Disaster Risk Management and Resilience. 2022. Available online: https://www.worldbank.org/en/results/2022/08/11/supporting-morocco-s-journey-to-disaster-and-climate-resilience (accessed on 21 September 2023).
8. Han, J.; Lee, D.; Lee, S.; Chung, S.W.; Kim, S.J.; Park, M.; Lim, K.J.; Kim, J. Evaluation of the effect of channel geometry on streamflow and water quality modeling and modification of channel geometry module in SWAT: A case study of the Andong Dam Watershed. *Water* **2019**, *11*, 718. [CrossRef]
9. McAllister, D.E.; Craig, J.F.; Davidson, N.; Delany, S.; Seddon, M. *Biodiversity Impacts of Large Dams*; Background Paper No.1; IUCN/UNEP/WCD: Gland, Switzerland, 2001.
10. Kuriqi, A.; Ardiçlioglu, M.; Muceku, Y. Investigation of seepage effect on river dike's stability under steady state and transient conditions. *Pollack Period.* **2016**, *11*, 87–104. [CrossRef]
11. de Andrade, E.M.; Sena, M.G.T.; da Silva, A.G.R.; Pereira, F.J.S.; Lopes, F.B. Uncertainties of the rainfall regime in a tropical semi-arid region: The case of the State of Ceará. *Rev. Agro@Mbiente On-Line* **2016**, *10*, 88–95. [CrossRef]
12. Zhang, L.; Xiao, J.; Zhou, Y.; Zheng, Y.; Li, J.; Xiao, H. Drought events and their effects on vegetation productivity in China. *Ecosphere* **2016**, *7*, e01591. [CrossRef]
13. Bouadila, A.; Benaabidate, L.; Bouizrou, I.; Aqnouy, M. Implementation of distributed hydrological modeling in a semi-arid Mediterranean Catchment Azzaba; Morocco. *J. Ecol. Eng.* **2019**, *20*, 236–254. [CrossRef]
14. Pulighe, G.; Bonati, G.; Colangeli, M.; Traverso, L.; Lupia, F.; Altobelli, F.; Dalla Marta, A.; Napoli, M. Predicting streamflow and nutrient loadings in a semi-arid Mediterranean watershed with ephemeral streams using the SWAT model. *Agronomy* **2019**, *10*, 2. [CrossRef]
15. Näschen, K.; Diekkrüger, B.; Leemhuis, C.; Steinbach, S.; Seregina, L.S.; Thonfeld, F.; Van der Linden, R. Hydrological modeling in data-scarce catchments: The Kilombero floodplain in Tanzania. *Water* **2018**, *10*, 599. [CrossRef]
16. Beck, H.E.; Vergopolan, N.; Pan, M.; Levizzani, V.; van Dijk, A.I.; Weedon, G.P.; Brocca, L.; Pappenberger, F.; Huffman, G.J.; Wood, E.F. Global-scale evaluation of 22 precipitation datasets using gauge observations and hydrological modeling. In *Satellite Precipitation Measurement. Advances in Global Change Research*; Levizzani, V., Kidd, C., Kirschbaum, D., Kummerow, C., Nakamura, K., Turk, F., Eds.; Springer: Cham, Switzerland, 2020; Volume 69, pp. 625–653.
17. Wanzala, M.A.; Stephens, E.M.; Cloke, H.L.; Ficchi, A. Hydrological model preselection with a filter sequence for the national flood forecasting system in Kenya. *J. Flood Risk Manag.* **2025**, *18*, e12846. [CrossRef]
18. Boufala, M.; El Hmaidi, A.; Chadli, K.; Essahlaoui, A.; El Ouali, A.; Taia, S. Hydrological modeling of water and soil resources in the basin upstream of the Allal El Fassi dam (Upper Sebou watershed; Morocco). *Model. Earth Syst. Environ.* **2019**, *5*, 1163–1177. [CrossRef]
19. Bouslihim, Y.; Kacimi, I.; Brirhet, H.; Khatati, M.; Rochdi, A.; Pazza, N.E.; Miftah, A.; Yaslo, Z. Hydrologic Modeling Using SWAT and GIS; Application to Subwatershed Bab-Merzouka (Sebou; Morocco). *J. Geogr. Inf. Syst.* **2016**, *8*, 20–27. [CrossRef]
20. Briak, H.; Mrabet, R.; Moussadek, R.; Aboumaria, K. Use of a calibrated SWAT model to evaluate the effects of agricultural BMPs on sediments of the Kalaya river basin (North of Morocco). *Int. Soil Water Conserv. Res.* **2019**, *7*, 176–183. [CrossRef]
21. Markhi, A.; Laftouhi, N.; Grusson, Y.; Soulaimani, A. Assessment of potential soil erosion and sediment yield in the semi-arid N′fis basin (High Atlas; Morocco) using the SWAT model. *Acta Geophys.* **2019**, *67*, 263–272. [CrossRef]
22. Choukri, F.; Raclot, D.; Naimi, M.; Chikhaoui, M.; Nunes, J.P.; Huard, F.; Hérivaux, C.; Sabir, M.; Pépin, Y. Distinct and combined impacts of climate and land use scenarios on water availability and sediment loads for a water supply reservoir in northern Morocco. *Int. Soil Water Conserv. Res.* **2020**, *8*, 141–153. [CrossRef]

23. Zettam, A.; Taleb, A.; Sauvage, S.; Boithias, L.; Belaidi, N.; Sanchez-Perez, J.M. Applications of a SWAT model to evaluate the contribution of the Tafna catchment (north-west Africa) to the nitrate load entering the Mediterranean Sea. *Environ. Monit. Assess.* **2020**, *192*, 510. [CrossRef]
24. Brouziyne, Y.; Abouabdillah, A.; Bouabid, R.; Benaabidate, L.; Oueslati, O. SWAT manual calibration and parameters sensitivity analysis in a semi-arid watershed in North-western Morocco. *Arab. J. Geosci.* **2017**, *10*, 427. [CrossRef]
25. Moumen, Z.; Nabih, S.; Elhassnaoui, I.; Lahrach, A. Hydrologic Modeling Using SWAT: Test the Capacity of SWAT Model to Simulate the Hydrological Behavior of Watershed in Semi-Arid Climate. In *Decision Support Methods for Assessing Flood Risk and Vulnerability*; IGI Global: Hershey, PA, USA, 2020; pp. 162–198. [CrossRef]
26. Hayes, D.S.; Brändle, J.M.; Seliger, C.; Zeiringer, B.; Ferreira, T.; Schmutz, S. Advancing towards functional environmental flows for temperate floodplain rivers. *Sci. Total Environ.* **2018**, *633*, 1089–1104. [CrossRef]
27. Declaration, B. The Brisbane Declaration: Environmental flows are essential for freshwater ecosystem health and human well-being. In Proceedings of the 10th International River Symposium, Brisbane, Australia, 3–6 September 2007.
28. Gopal, B. Methodologies for the assessment of environmental flows. In *Environmental Flows: An Introduction for Water Resources Managers*; National Institute of Ecology: New Delhi, India, 2013; pp. 129–182.
29. Dutta, V.; Sharma, U.; Kumar, R. Assessment of river ecosystems and environmental flows: Role of flow regimes and physical habitat variables. *Clim. Change Environ. Sustain.* **2017**, *5*, 20–34. [CrossRef]
30. McKee, T.B.; Doesken, N.J.; Kleist, J. The relationship of drought frequency and duration to time scales. In Proceedings of the 8th Conference on Applied Climatology, Boston, MA, USA, 17–22 January 1993; Volume 17, pp. 179–183.
31. Abdallaoui, A. Contribution à l'étude Du Phosphore et Des Métaux Lourds Contenus Dans Les Sédiments et de Leur Influence Sur Les Phénomènes d'eutrophisation et de La Pollution. Cas Du Bassin Versant de l'oued Beht et de La Retenue de Barrage El Kansera. Ph.D. Thesis, Faculté des Sciences, Université Moulay Ismail, Meknès, Morocco, 1998.
32. Lakhili, F.; Mohammed, B.; Nouzha, B.; Hamid, L.; Abrerrahim, L. Etude de la qualité physicochimique et de la contamination métallique des eaux de surface du bassin versant de Beht (Maroc). *Eur. Sci. J.* **2015**, *11*, 132–147.
33. El Gaatib, R. Cartographie Décisionnelle de l'aménagement Intégré Du Bassin Versant de l'Oued Beht: Analyse Spatiale de l'aléa Érosif Des Sols Pour La Gestion Des Risques Naturels d'envasement de La Retenue Du Barrage El Kansera (Maroc). Ph.D. thesis, Université Mohammed V—Rabat, Ecole Mohammadia d'Ingénieurs, Rabat, Morocco, 2017.
34. El Gaatib, R.; Larabi, A.; Faouzi, M. Integrated elaboration of priority planning of vulnerable areas to soil erosion hazard using Remote Sensing and GIS techniques: A pilot case of the Oued Beht Watershed (Morocco). *J. Mater. Environ. Sci.* **2015**, *6*, 3110–3126.
35. El Gaatib, R.; Larabi, A. Integrated evaluation of soil erosion hazard and risk management in the Oued Beht watershed using remote sensing and GIS techniques: Impacts on El Kansra Dam Siltation (Morocco). *J. Geogr. Inf. Syst.* **2014**, *6*, 677–689. [CrossRef]
36. Yacine, E.A.; Essahlaoui, A.; Oudija, F.; Mimich, K.; Nassiri, L. Assessment of soil erosion by (RUSLE) using remote sensing and GIS case of watershed of Beht in upstream of Ouljat Sultan dam (Morocco). *ARPN J. Eng. Appl. Sci.* **2019**, *14*, 1765–1776.
37. Daide, F.; Afgane, R.; Lahrach, A.; Chaouni, A.A. Beht watershed (Morocco) rainfall-runoff simulation with the HEC-HMS hydrological model. *Ecol. Eng. Environ. Technol.* **2022**, *23*, 142–155. [CrossRef] [PubMed]
38. Arnold, J.G.; Srinivasan, R.; Muttiah, R.S.; Williams, J.R. Large area hydrologic modeling and assessment part I: Model development. *J. Am. Water Resour. Assoc.* **1998**, *34*, 73–89. [CrossRef]
39. Neitsch, S.L.; Arnold, J.G.; Kiniry, J.R.; Williams, J.R. *Soil and Water Assessment Tool Theoretical Documentation, Version 2005*; USDA-ARS Grassland Soil and Water Research Laboratory, and Texas A&M University: Temple, TX, USA; Blackland Research and Extension Center: Temple, TX, USA, 2005.
40. Gassman, P.W.; Reyes, M.R.; Green, C.H.; Arnold, J.G. The Soil and Water Assessment Tool: Historical Development, Applications, and Future Research Directions. *Trans. ASABE* **2007**, *50*, 1211–1250. [CrossRef]
41. Neitsch, S.L.; Arnold, J.G.; Kiniry, J.R.; Williams, J.R. *Soil and Water Assessment Tool. Theoretical Documentation. Version 2009*; Technical Report No.406; Texas Water Resources Institute: Forney, TX, USA, 2011. [CrossRef]
42. Saha, S.; Moorthi, S.; Pan, H.L.; Wu, X.; Wang, J.; Nadiga, S.; Tripp, P.; Kistler, R.; Woollen, J.; Behringer, D.; et al. The NCEP climate forecast system reanalysis. *Bull. Am. Meteorol. Soc.* **2010**, *91*, 1015–1057. [CrossRef]
43. Brirhet, H. Intégration Des Modèles Globaux et Distribués Dans La Modélisation Hydrologique Du Bassin Versant d'Issan, Région de Souss. Ph.D. Thesis, Université Sidi Mohamed Ben Abdellah, Fez, Morocco, 2016.
44. Abbaspour, K.C.; Johnson, C.A.; Van Genuchten, M.T. Estimating uncertain flow and transport parameters using a sequential uncertainty fitting procedure. *Vadose Zone J.* **2004**, *3*, 1340–1352. [CrossRef]
45. Abbaspour, K.C.; Yang, J.; Maximov, I.; Siber, R.; Bogner, K.; Mieleitner, J.; Zobrist, J.; Srinivasan, R. Modelling hydrology and water quality in the pre-alpine/alpine Thur watershed using SWAT. *J. Hydrol.* **2007**, *333*, 413–430. [CrossRef]
46. Nash, J.E.; Sutcliffe, J.V. River flow forecasting through conceptual models part I—A discussion of principles. *J. Hydrol.* **1970**, *10*, 282–290. [CrossRef]
47. Cao, W.; Bowden, W.B.; Davie, T.; Fenemor, A. Multi-variable and multi-site calibration and validation of SWAT in a large mountainous catchment with high spatial variability. *Hydrol. Process.* **2006**, *20*, 1057–1073. [CrossRef]

48. Van Liew, M.W.; Veith, T.L. Guidelines for Using the Sensitivity Analysis and Auto-Calibration Tools for Multi-Gage or Multi-Step Calibration in SWAT. 2010, pp. 1–30. Available online: https://www.researchgate.net/publication/268344153_Guidelines_for_Using_the_Sensitivity_Analysis_and_Auto-calibration_Tools_for_Multi-gage_or_Multi-step_Calibration_in_SWAT (accessed on 2 June 2024).
49. Richter, B.D.; Baumgartner, J.V.; Powell, J.; Braun, D.P. A method for assessing hydrologic alteration within ecosystems. *Conserv. Biol.* **1996**, *10*, 1163–1174. [CrossRef]
50. Richter, B.D.; Baumgartner, J.V.; Wigington, R.; Braun, D. How much water does a river need? *Freshw. Biol.* **1997**, *37*, 231–249. [CrossRef]
51. The Nature Conservancy. *Indicators of Hydrologic Alteration Version 7.1 User's Manual*; The Nature Conservancy: Arlington, VA, USA, 2009. Available online: https://rdrr.io/rforge/IHA/ (accessed on 28 December 2023).
52. Mathews, R.; Richter, B.D. Application of the Indicators of hydrologic alteration software in environmental flow setting. *JAWRA J. Am. Water Resour. Assoc.* **2007**, *43*, 1400–1413. [CrossRef]
53. Fadil, A.; Rhinane, H.; Kaoukaya, A.; Kharchaf, Y.; Bachir, O.A. Hydrologic modeling of the Bouregreg watershed (Morocco) using GIS and SWAT model. *J. Geogr. Inf. Syst.* **2011**, *3*, 279. [CrossRef]
54. Molina-Navarro, E.; Hallack-Alegría, M.; Martínez-Pérez, S.; Ramírez-Hernández, J.; Mungaray-Moctezuma, A.; Sastre-Merlín, A. Hydrological modeling and climate change impacts in an agricultural semiarid region. Case study: Guadalupe River basin, Mexico. *Agric. Water Manag.* **2016**, *175*, 29–42. [CrossRef]
55. De Almeida Bressiani, D.; Srinivasan, R.; Jones, C.A.; Mendiondo, E.M. Effects of different spatial and temporal weather data resolutions on the stream flow modeling of a semi-arid basin, Northeast Brazil. *Int. J. Agric. Biol. Eng.* **2015**, *8*, 1–16. [CrossRef]
56. Qi, Z.; Kang, G.; Chu, C.; Qiu, Y.; Xu, Z.; Wang, Y. Comparison of SWAT and GWLF model simulation performance in humid south and semi-arid north of China. *Water* **2017**, *9*, 567. [CrossRef]
57. Bouadila, A.; Bouizrou, I.; Aqnouy, M.; En-Nagre, K.; El Yousfi, Y.; Khafouri, A.; Hilal, I.; Abdelrahman, K.; Benaabidate, L.; Abu-Alam, T.; et al. Streamflow Simulation in Semiarid Data-Scarce Regions: A Comparative Study of Distributed and Lumped Models at Aguenza Watershed (Morocco). *Water* **2023**, *15*, 1602. [CrossRef]
58. Aqnouy, M.; El Messari, J.E.S.; Ismail, H.; Bouadila, A.; Moreno Navarro, J.G.; Loubna, B.; Mansour, M.R.A. Assessment of the SWAT model and the parameters affecting the flow simulation in the watershed of Oued Laou (Northern Morocco). *J. Ecol. Eng.* **2019**, *20*, 104–113. [CrossRef]
59. M'barek, S.A.; Rochdi, A.; Bouslihim, Y.; Miftah, A. Multi-site calibration and validation of swat model for hydrologic modeling and soil erosion estimation: A case study in el grou watershed; Morocco. *Ecol. Eng. Environ. Technol.* **2021**, *22*, 45–52. [CrossRef]
60. Moriasi, D.N.; Arnold, J.G.; Van Liew, M.W.; Bingner, R.L.; Harmel, R.D.; Veith, T.L. Model evaluation guidelines for systematic quantification of accuracy in watershed simulations. *Trans. ASABE* **2007**, *50*, 885–900. [CrossRef]
61. Arnold, J.G.; Moriasi, D.N.; Gassman, P.W.; Abbaspour, K.C.; White, M.J.; Srinivasan, R.; Santhi, C.; Harmel, R.D.; Van Griensven, A.; Van Liew, M.W.; et al. SWAT: Model use; calibration; and validation. *Trans. ASABE* **2012**, *55*, 1491–1508. [CrossRef]
62. Abouabdillah, A.; White, M.; Arnold, J.G.; De Girolamo, A.M.; Oueslati, O.; Maataoui, A.; Lo Porto, A. Evaluation of soil and water conservation measures in a semi-arid river basin in Tunisia using SWAT. *Soil Use Manag.* **2014**, *30*, 539–549. [CrossRef]
63. Zhang, Z.; Huang, Y.; Huang, J. Hydrologic alteration associated with dam construction in a medium-sized coastal watershed of southeast China. *Water* **2016**, *8*, 317. [CrossRef]
64. Šarauskienė, D.; Adžgauskas, G.; Kriaučiūnienė, J.; Jakimavičius, D. Analysis of hydrologic regime changes caused by small hydropower plants in lowland rivers. *Water* **2021**, *13*, 1961. [CrossRef]
65. Ali, R.; Kuriqi, A.; Abubaker, S.; Kisi, O. Hydrologic alteration at the upper and middle part of the yangtze river, China: Towards sustainable water resource management under increasing water exploitation. *Sustainability* **2019**, *11*, 5176. [CrossRef]
66. Brouziyne, Y.; Belaqziz, S.; Benaabidate, L.; Aboubdillah, A.; El Bilali, A.; Elbeltagi, A.; Tzoraki, O.; Chehbouni, A. Modeling long term response of environmental flow attributes to future climate change in a North African watershed (Bouregreg watershed, Morocco). *Ecohydrol. Hydrobiol.* **2022**, *22*, 155–167. [CrossRef]
67. Uday Kumar, A.K.V.J. Assessment of flow regime alteration in the Krishna River basin. *ISH J. Hydraul. Eng.* **2022**, *28*, 212–218. [CrossRef]
68. Spurgeon, J.J.; Pegg, M.A. Juvenile growth of a macrohabitat generalist tied to hydrological character of large-river system. *Freshw. Biol.* **2017**, *62*, 291–302. [CrossRef]
69. Abebe, W.B.; Tilahun, S.A.; Moges, M.M.; Wondie, A.; Derseh, M.G.; Nigatu, T.A.; Mhiret, D.A.; Steenhuis, T.S.; Camp, M.V.; Walraevens, K.; et al. Hydrological foundation as a basis for a holistic environmental flow assessment of tropical highland rivers in ethiopia. *Water* **2020**, *12*, 547. [CrossRef]
70. Bouslihim, Y.; Rochdi, A.; El Amrani Paaza, N.; Liuzzo, L. Understanding the effects of soil data quality on SWAT model performance and hydrological processes in Tamedroust watershed (Morocco). *J. Afr. Earth Sci.* **2019**, *160*, 103616. [CrossRef]
71. Briak, H.; Moussadek, R.; Aboumaria, K.; Mrabet, R. Assessing sediment yield in Kalaya gauged watershed (Northern Morocco) using GIS and SWAT model. *Int. Soil Water Conserv. Res.* **2016**, *4*, 177–185. [CrossRef]

72. Brouziyne, Y.; Abouabdillah, A.; Chehbouni, A.; Hanich, L.; Bergaoui, K.; McDonnell, R.; Benaabidate, L. Assessing hydrological vulnerability to future droughts in a Mediterranean watershed: Combined indices-based and distributed modeling approaches. *Water* **2020**, *12*, 2333. [CrossRef]
73. Yactayo, G.A. Modification of the SWAT Model to Simulate Hydrologic Processes in a Karst-Influenced Watershed. Doctoral Dissertation, Faculty of Virginia Polytechnic Institute and State University, Blacksburg, VA, USA, 2009.
74. Al Khoury, I.; Boithias, L.; Labat, D. A Review of the Application of the Soil and Water Assessment Tool (SWAT) in Karst Watersheds. *Water* **2023**, *15*, 954. [CrossRef]
75. Dai, J.; Pan, L.; Deng, Y.; Wan, Z.; Xia, R. Modified SWAT Model for Agricultural Watershed in Karst Area of Southwest China. *Agriculture* **2025**, *15*, 192. [CrossRef]
76. Toledo, C.E.; Alcantara, N.R. Sensitivity of hydrological connectivity in a semiarid basin with a high-density reservoir network. *Rev. Ambient. Água.* **2019**, *14*, e2367. [CrossRef]
77. Chbouki, N.; Stockton, C.W.; Myers, D.E. Spatio-temporal patterns of drought in Morocco. *Int. J. Climatol.* **1995**, *15*, 187–205. [CrossRef]
78. Hrour, Y.; Thomas, Z.; Fovet, O.; Sebari, K.; Rousseau-Gueutin, P. Changes in precipitation and discharge in a Mediterranean catchment as a response to climate change and human activities. *J Water Clim Change* **2022**, *13*, 3253–3273. [CrossRef]
79. Fox, H.R.; Wilby, R.L.; Moore, H.M. The impact of river regulation and climate change on the barred estuary of the Oued Massa, southern Morocco. *River Res. Appl.* **2001**, *17*, 235–250. [CrossRef]
80. Hakam, O.; Baali, A.; El Kamel, T.; Ahouach, Y.; Azennoud, K. Comparative evaluation of various drought indices (DIs) to monitor drought status: A case study of Moroccan Lower Sebou basin. *Kuwait J. Sci.* **2022**, *49*, 1–25. [CrossRef]
81. Ghalley, W.; Niemann, J.D.; Shrestha, S. Hydrologic alteration by hydropower dams and climate change in the Sesan Basin. *J. Appl. Water Eng. Res.* **2024**, *12*, 388–406. [CrossRef]
82. Zuo, Q.; Liang, S. Effects of dams on river flow regime based on IHA/RVA. *Proc. Int. Assoc. Hydrol. Sci.* **2015**, *368*, 275–280. [CrossRef]
83. El Adnani, A.; Habib, A.; Zourarah, B. Spatio-temporal dynamics and evolution of land use land cover using remote sensing and GIS in Sebou Estuary, Morocco. *J. Geogr. Inf. Syst.* **2019**, *11*, 551–566. [CrossRef]
84. Lamhasni, N.; Lahcen, C.; Timallouka, M. Bio-évaluation de la qualité des eaux de surface d'oued Beht (Maroc) Indice Biologique Global des réseaux de contrôle et de surveillance (IBG-RCS). *Eur. Sci. J.* **2017**, *13*, 94–111.
85. Guellaf, A.; Kassout, J.; Boselli, V.A.; Bennas, N.; El Alami, M.; Errochdi, S.; Kettani, K. Short-term responses of aquatic ecosystem and macroinvertebrate assemblages to rehabilitation actions in martil river (North-Western Morocco). *Hydrobiology* **2023**, *2*, 446–462. [CrossRef]
86. Lobera Galán, G.; Muñoz Gràcia, I.; López Tarazón, J.A.; Vericat Querol, D.; Batalla, R.J. Effects of flow regulation on river bed dynamics and invertebrate communities in a Mediterranean river. *Hydrobiologia* **2017**, *784*, 283–304. [CrossRef]
87. Belmar, O.; Velasco, J.; Gutiérrez-Cánovas, C.; Mellado-Díaz, A.; Millán, A.; Wood, P.J. The influence of natural flow regimes on macroinvertebrate assemblages in a semiarid Mediterranean basin. *Ecohydrology* **2013**, *6*, 363–379. [CrossRef]
88. Poff, N.L.; Zimmerman, J.K.H. Ecological responses to altered flow regimes: A literature review to inform the science and management of environmental flows. *Freshw. Biol.* **2010**, *55*, 194–205. [CrossRef]
89. Vicente-Serrano, S.M.; López-Moreno, J.I. Hydrological response to different time scales of climatological drought: An evaluation of the Standardized Precipitation Index in a mountainous Mediterranean basin. *Hydrol. Earth Syst. Sci.* **2005**, *9*, 523–533. [CrossRef]
90. Vélez-nicolás, M.; García-lópez, S.; Ruiz-ortiz, V.; Zazo, S.; Molina, J.L. Precipitation Variability and Drought Assessment Using the SPI: Application to Long-Term Series in the Strait of Gibraltar Area. *Water* **2022**, *14*, 884. [CrossRef]
91. Lopez-Bustins, J.A.; Pascual, D.; Pla, E.; Retana, J. Future variability of droughts in three Mediterranean catchments. *Nat. Hazards* **2013**, *69*, 1405–1421. [CrossRef]

hydrology

MDPI

Article

An Application of the Ecosystem Services Assessment Approach to the Provision of Groundwater for Human Supply and Aquifer Management Support

Malgorzata Borowiecka, Mar Alcaraz * and Marisol Manzano

Departamento de Ingeniería Minera y Civil, Universidad Politécnica de Cartagena, Paseo Alfonso X III, 48, 30203 Cartagena, Spain; malgorzata.borowiecka@edu.upct.es (M.B.); marisol.manzano@upct.es (M.M.)
* Correspondence: mar.alcaraz@upct.es

Abstract: Increasing pressures on groundwater in the last decades have led to a deterioration in the quality of groundwater for human consumption around the world. Beyond the essential evaluation of groundwater dynamics and quality, analyzing the situation from the perspective of the Ecosystem Services Assessment (ESA) approach can be useful to support aquifer management plans aiming to recover aquifers' capacity to provide good quality water. This work illustrates how to implement the ESA using groundwater flow and nitrate transport modelling for evaluating future trends of the provisioning service Groundwater of Good Quality for Human Supply. It has been applied to the Medina del Campo Groundwater Body (Spain), where the intensification of agricultural activities and groundwater exploitation since the 1970s caused severe nitrate pollution. Nitrate status and future trends under different fertilizer and aquifer exploitation scenarios were modelled with MT3DMS coupled to a MODFLOW model calibrated with piezometric time series. Historical land use and fertilizer data were compiled to assess nitrogen loadings. Besides the uncertainties of the model, the results clearly show that: (i) managing fertilizer loads is more effective than managing aquifer exploitation; and (ii) only the cessation of nitrogen application by the year 2030 would improve the evaluated provisioning service in the long term. The study illustrates how the ESA can be incorporated to evaluate the expected relative impact of different management actions aimed at improving significant groundwater services to humans.

Keywords: groundwater ecosystem services assessment; provision of good quality groundwater; nitrate transport modelling; Medina del Campo Groundwater Body

1. Introduction

Increased industrialization, agricultural intensification, excessive use of synthetic chemicals, and unsustainable groundwater abstraction have led to a significant deterioration in groundwater quality and quantity worldwide. One of the most pressing environmental concerns in recent decades has been the contamination of groundwater by nitrates, primarily due to agricultural activities [1–3]. The widespread use of nitrogen-based fertilizers, livestock manure, and irrigation return flows has resulted in elevated nitrate concentrations, causing serious environmental and health dangers. Many studies indicate that nitrogen leaching into aquifers has more than doubled since pre-industrial times due to human-induced pressures. Nitrate contamination not only affects groundwater quality but also disrupts the natural nitrogen cycle, leading to broader ecological consequences [4,5].

Groundwater quality degradation directly impacts essential benefits to both humans and natural ecosystems, such as the provision of good-quality drinking water and water for agricultural irrigation, or supporting groundwater-associated surface aquatic ecosystems such as rivers, wetlands, and springs [6]. As groundwater quality declines, its ability to sustain these critical services has led to economic, social, and environmental losses [7,8]. Groundwater quality deterioration affects not only water availability [9] but also the broader ecological balance, disrupting biodiversity, reducing agricultural productivity, and increasing water treatment costs [8–14].

The Ecosystem Services Assessment (ESA) approach is a useful tool to support natural resources management decisions, as is the case with groundwater and its associated ecosystems. The ESA has been successfully applied to the qualitative and quantitative bio-physical evaluation of diverse ecosystem services, e.g., those related to forests, grasslands [15,16], wetlands, lakes, surface hydrology [17–19], land planning [20,21], and groundwater-dependent ecosystems [22].

The incorporation of the ESA approach into groundwater management has emerged as a valuable assessment tool for balancing resource use with conservation objectives [23,24]. The ESA approach allows technicians and managers to assess the present state and future evolution of groundwater ecosystem services (GWESs). This approach considers the diverse benefits that groundwater provides and helps delineate strategies for its preservation and sustainable use, facilitating informed decision-making for sustainable groundwater management [25,26]. Spanish legislation has integrated this methodology through the River Basin Hydrological Plans, as outlined in the Royal Decree 1159/2021 [27], emphasizing the importance of ecosystem-based management strategies.

Although the application of the ESA approach to groundwater has begun recently, studies remain scarce [6–30]. More recently, the ESA approach has been applied to evaluate two groundwater ecosystem services in the Medina del Campo Groundwater Body (MCGB) [9]: the provision of groundwater for irrigation, and the regulation of surface ecosystems dependent on groundwater (rivers, wetlands, and riparian forests).

This study was part of the European Project NAIAD (H2020-SC5), in which the authors focused on the physical assessment of selected GWESs that play a key role in mitigating drought hazards and supporting surface ecosystems in the MCGB [31]. The MCGB, located in the Duero River Basin in northwestern Spain, serves as a representative case study for assessing the impact of agricultural practices on GWESs. The region relies on groundwater for irrigation, urban water supply, and the maintenance of surface aquatic ecosystems. Over the past 40 years, intensive agricultural activities, excessive groundwater extraction, and increased fertilizer use have caused significant hydrogeological changes, including declining groundwater levels and nitrate contamination. These disruptions to GWESs have led to reduced groundwater availability and quality, with both economic and environmental consequences [32,33].

The primary aim of this research is to illustrate the usefulness of integrating the ESA approach into groundwater evaluation by assessing the current state and potential future evolution of GWESs in the MCGB with numerical modeling of groundwater flow and nitrate transport. The assessment focuses on a key GWES in the study area, the provision of Groundwater of Good Quality for Human Supply. This service represents the capacity of aquifers to provide fresh water for human consumption, a function increasingly threatened by nitrate contamination and groundwater depletion.

2. Study Area

2.1. Location and General Characteristics

The MCGB is situated in the Duero River Basin, in NW Spain (Figure 1), and has a surface area of around 3670 km^2. The MCGB is bordered by the Duero River to the north, the Central System mountain range to the south, the Adaja River to the east, and the Guareña River to the west [32–34]. The average altitude is 812 m a.s.l.; the lowest zone (645 m a.s.l.) is located on the northern edge of the study area, and the highest (1252 m a.s.l.) is located on the southern edge. There is a general decreasing slope from S to N.

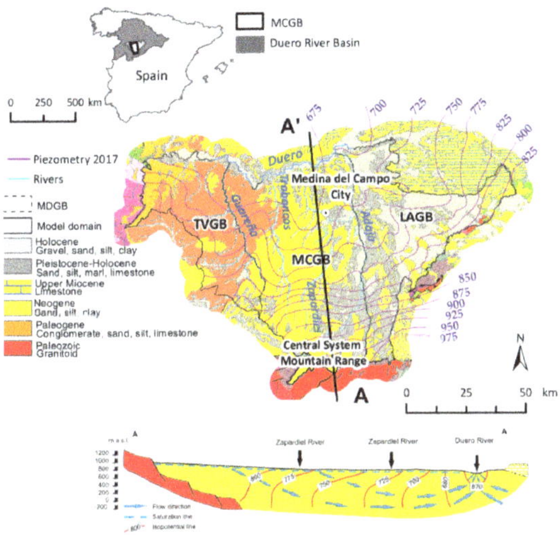

Figure 1. Location, hydrographic network, regional piezometry of the modelled area in 2017, and geological map [34]. MCGB: Medina del Campo Groundwater Body; LAGB: Los Arenales Groundwater Body; TVGB: Tierra del Vino Groundwater Body.

The hydrographic network includes the main river courses Adaja, Zapardiel, Trabancos, and Guareña, all flowing to the Duero River in the north (Figure 1). A few decades ago, there were many small groundwater-associated wetlands, but almost all of them are now dry.

Although the NAIAD project focused on the MCGB, for a better understanding and definition of the boundary conditions two more groundwater bodies have been included in the study: Los Arenales (LAGB) and Tierra del Vino TVGB), placed to the E and to the W of the Medina del Campo, respectively (Figure 1). Some results will refer to the three groundwater bodies, but most will refer just to the MCGB.

The area has a Mediterranean climate characterized by hot summers and cold winters, with a mean annual temperature of 12 °C (between the years 1940 and 2005) [35]. Winter values are around 4 °C, and summer values are around 22 °C. Average annual precipitation over the basin is 404 mm, most of it taking place in autumn. Average annual potential evapotranspiration is 750 mm [34–38]. The area is prone to climatic and geomorphological hazards such as floods, droughts, and soil erosion.

2.2. Water and Land Use

The MCGB spreads across five different provinces, with 130 municipalities and 70,000 inhabitants. The economy is mainly agricultural, with about 3200 km^2 of cultivated area and only 39 km^2 of urban surface. Medina del Campo city is the only town with a significant industrial and services sector. The rest of the surface is covered by forests and scrub vegetation, mainly with species from the pine family (*Pinus pinea*, *Pinus pinaster*) or poplars (*Populus nigra*, *Populus alba*) [34]. Currently, riparian forests are located only in the Duero and Adaja rivers (in the latter, artificially reintroduced). In the other rivers, riparian forests have disappeared, mainly due to water-table drawdown because of aquifer overexploitation [39]. Rain-fed crops, primarily consisting of cereals, legumes, and sunflowers, dominate in the MCGB; irrigated crops, including beet, potato, corn, and rapeseed, represent only 10% of the total cultivated area [40].

The Autonomous Region of Castilla y León, where the MCGB is located, is Spain's largest fertilizer consumer, with a total of 441,378 tons used in 2020. This represents 22.69% of the total national consumption, and has increased 19.58% since 2010 [41].

The MCGB provides diverse ecosystem services to humans and the environment, such as water supply for small villages and agricultural irrigation [42], and support for aquatic surface ecosystems (wetlands, riverine forests, and streams). However, agricultural intensification, overuse of fertilizers since the late 1970s, and intensive groundwater exploitation from the mid-1970s to the early 2000s have severely reduced and degraded groundwater quality and quantity.

2.3. Geology and Hydrogeology

The MCGB is part of the large Duero River sedimentary basin. With a tectonic origin, the basin is filled with eroded materials from several mountain ranges that isolated the basin for hundreds of years. The geological basement consists of Paleozoic igneous and metamorphic rocks [43]. The filling is composed of Tertiary sediments, mainly fine alluvium, eolian deposits, and a silty-sandy-marly matrix of lacustrine origin with gypsum layers in the northeast, deposited during the Paleogene and Neogene [44], though most sediments date to the Upper-Middle Miocene. Plio–Quaternary deposits include alluvial terraces and extensive eolian sandy layers [45]. The total sediment thickness increases from 500 m in the south to 2000 m in the north, influenced by deep regional faults that affect the basement and part of the sedimentary filling [34–47].

Despite high textural heterogeneity, the sedimentary fill behaves as a well-connected multilayer aquifer system. The upper part forms a heterogeneous (sometimes discontinuous) water-table aquifer, while deeper layers act as a semi-confined to confined system, with silty layers acting as aquitards. According to this, the conceptual hydrogeological model is a system formed by three more or less continuous layers: (1) a semiconfined lower aquifer, composed of Paleogene sands, silts, and gravels reaching depths of up to 2000 m near the Duero River to the north; (2) an intermediate aquitard, consisting of Neogene fine-grained sediments with a thickness ranging from 35 to 50 m; and (3) an upper aquifer, formed by Plio–Quaternary gravel, sand, and conglomerates with up to 15 m thickness [38–48]. The hydraulic permeability is high in the Quaternary alluvial and eolian sands ($>10^2$ m/day), and low in the Tertiary sediments (between 10^4 and 10^2 m/day) [48].

Natural recharge occurs through rainfall infiltration, with slow groundwater transfer through the aquitard from the upper to the lower aquifer. The regional horizontal groundwater flow is from the south to the north, according to [34] (Figure 1). From the groundwater resources accounting standpoint, irrigation does not contribute additional recharge, as it relies on local groundwater. However, from the management and environmental angle, groundwater pollution due to excess irrigation water recharge is a very relevant issue [48].

Until the mid-20th century, natural groundwater discharge in the MCGB upper aquifer occurred through evapotranspiration and to the surface drainage network. Since the mid-1970s, natural discharge has been progressively substituted by groundwater abstraction for irrigation and for the supply of small towns. Pumping intensification during the 1980s and 1990s led to lowered groundwater levels by up to 30 m between 1972 and 2012 [49], and disconnected rivers and wetlands from the saturated zone. Today, only the Duero and the Adaja rivers maintain a permanent flow (the last one due to river regulation by dams), while the other rivers in the zone only have flow after heavy rainfall. Natural discharge from the deeper, semiconfined layer occurs toward the Duero River and through groundwater abstraction.

Groundwater chemical composition in the MCGB is mainly of the calcium–magnesium bicarbonate, calcium–sodium bicarbonate, and sulfate–calcium facies. The electrical conductivity ranges from 150 to 1500 μS/cm, with average values around 500 μS/cm. Nitrate concentrations vary between 0 and 100 mg/L in the whole aquifer system (DRBA database for the period 1977–2019) [42].

3. Methodology

The study methodology consisted of selecting a specific GWES adequately representing groundwater quality problems in the MCGB. The actual situation and possible future evolution of this GWES were assessed under different groundwater and fertilizer management scenarios with a nitrate transport numerical model. In this section, we describe the numerical model first and provide details on the methodology used to assess the selected GWES later.

3.1. *Identification of the Groundwater Ecosystem Service to Be Evaluated*

According to DRBA [32–51], one of the most severe problems in the MCGB is the risk of not complying with the environmental quantity and quality objectives of the European Union for the year 2027. Increased agricultural production, intensive use of fertilizers, and excessive groundwater exploitation for agricultural irrigation have caused severe problems. Groundwater contamination in the MCGB by nitrate reached concentrations of 190 mg/L. Thus, the provision of groundwater with good quality for human consumption is not suitable in many parts of the area.

Due to the relevance of this problem, we selected a specific GWES related to water quality to evaluate its current status and the probable future evolution under different scenarios, the Abiotic Provisioning Service Groundwater of Good Quality for Human Supply (APS-GWGQHS), as classified under the European Union's Common International Classification of Ecosystem Services (CICES) [52]. In the MCGB, this service reflects the aquifer system's ability to provide water suitable for human use. Improving the provision of this service requires reducing nitrate concentrations in groundwater.

3.2. *Nitrate Transport Numerical Model*

A nitrate transport numerical model was developed to assess the current state and future evolution of the selected GWES. The numerical model is based on an existing groundwater flow numerical model covering the period from 1950 to 2018, which was calibrated to present-day conditions. This flow model is based on the three-dimensional MODFLOW 2005 [53] code, which solves groundwater flow equations in porous media, using the finite difference method. MODFLOW [53,54] can simulate both steady and transient state flow across multiple layers and different types of aquifers. Specific details about this numerical model can be found in [9]; only the main aspects are presented here.

The modelled area covers a surface of 8728.75 km^2, with the MCGB occupying 3702.75 km^2 in the central part of the zone (Figure 1). The whole model domain is discretized into 34,915 active quadrilateral cells, each measuring 500 $\times$ 500 m. It has three layers: (1) the upper layer represents the unconfined aquifer, extending down to a depth of 15 m; (2) the intermediate layer represents the aquitard, from 15 to 50 m, which does not contribute to the evaluated ecosystem service; and (3) the lower layer represents the semiconfined aquifer, which ranges from 50 m to the geological basement. The simulation period extends from 1950 to 2018, with 1950 representing nearly undisturbed conditions. The simulation time step is one year.

Key parameters such as hydraulic conductivity (K), storage coefficient (S), and pumping rates were manually calibrated using piezometric data from 1950 to 2018, and historical records of groundwater extractions obtained from the databases and other documents of the DRBA and the Geological and Mining Institute of Spain (see data sources and additional details in [9]). The lower layer was divided into 17 hydraulic conductivity zones based on hydrogeological maps covering the study area [48]. The flow model was calibrated using this zoning and the piezometric time series of 102 observation points scattered across the full modeled domain. Despite the relatively coarse grid resolution and annual time steps, the model achieved an average absolute error of 12 m for the full modelled area and 11 m for the MCGB subarea, with a regression coefficient of 0.962 (Figure S1). This high level of accuracy makes the model reliable for groundwater forecasting at a regional scale.

For the forecasting simulations of the flow model, two types of scenarios were considered: groundwater exploitation management and climate change projections. The groundwater exploitation scenarios were:

(1) Maintaining current abstraction rates (Business as Usual, BAU).
(2) Reducing abstractions to an Exploitation Index (EI) = 0.8 by the year 2050, in alignment with the settled objectives of the Duero River Basin Authority. The EI is the mean annual total withdrawal of groundwater relative to the average annual groundwater renewable resources.

Climate change scenarios were based on Representative Concentration Pathways (RCPs) commonly used in regional hydrological assessments:

(A) RCP 4.5, corresponding to a moderate stabilization scenario. It was associated with a projected 3% increase in annual precipitation in the study area, based on climate change analyses of a climate-change specialist NAIAD partner.
(B) RCP 8.5, a high-emission scenario. It was associated with an 8% decrease in precipitation in the area, based on national assessments (CEDEX, 2017).
(C) A No-Change scenario was also included as a baseline.

Climate change simulations were implemented as changes in groundwater recharge. To simulate future recharge, the model assumed that the current recharge coefficient (ratio of recharge to precipitation) remained constant throughout the simulation period. This ratio was applied to the projected precipitation time series for each climate scenario to generate extended recharge inputs.

The results of the prospective flow model indicated that precipitation changes had a relatively minor effect on groundwater levels when compared to the impact of controlling groundwater abstraction [9]. Following this result, in the nitrate transport model we retained only the two groundwater management scenarios (BAU and EI = 0.8) and assumed stable climatic conditions, focusing on the most influential and policy-relevant drivers of aquifer response.

The code MT3DMS [55] was coupled with MODFLOW 2005 [53] to simulate nitrate transport. This code simulates how groundwater contaminants, like nitrate (NO_3^-), change over time due to advection, dispersion, and chemical reactions, and considers different sources and sinks for contaminants.

Nitrate concentrations in groundwater were simulated in a three-step approach: (1) in order to quantify nitrogen (N) input to the unsaturated zone, two models based on empirical data were developed to obtain information on the spatial and temporal evolution of N supply to the soil in the study area between years 1950 and 2018; (2) to generate a N input time function to groundwater for the numerical model, several models were used to estimate N input to the aquifer saturated zone between years 1950 and 2018. The results were converted to equivalent nitrate (NO_3^-) concentrations and compared to measured NO_3^- values in the MCGB, and the model best matching the observations was used to assess the present state and the future evolution of the evaluated groundwater ecosystem service; and (3) the NO_3^- input model selected in step 2 was used to assess the possible future evolution of the ecosystem service.

3.2.1. Nitrate Observations

Nitrate historical data in groundwater were obtained from the DRBA database of the groundwater quality observation network. Chemical analyses with ionic balance errors greater than 10% were discarded. As a result, 352 data records from 172 observation points were used for the study of NO_3^- spatial and temporal evolution.

A major portion of the data was obtained from observation boreholes deeper than 50 m, which tap the lower aquifer layer. For many of them, there is neither geological nor technical information, and they probably mix water from all layers. Due to the lack of information, data were grouped and analyzed in two depth ranges: $\leq$50 m, representing the Upper Aquifer and Aquitard layers together, and >50 m, representing the Lower Aquifer layer. However, due to the mentioned potential mixing of water from different flow lines at observation points with long screens, there is notable uncertainty about which depth represents the data.

3.2.2. Temporal and Spatial Evolution of Nitrogen Supply to the Soil

The following data were integrated to estimate nitrogen (N) supply to the soil in the study area: the official 2018 land-use map [40]; N doses recommended for the crops in the area [56]; temporal evolution of surfaces cultivated with different crops [57–59]; and fertilizers sales history, both in Spain [60] and in the Autonomous Community of Castilla and León [61].

The official crops map of the year 2018 includes 42 crop classes, natural surfaces, and urban areas, grouped into various categories with a total surface of 8728.75 km^2. To ease the next steps, this map was simplified by grouping the original categories into five broader types: Group 1 includes fallow land, forest and park areas, and urban zones such as roads and water bodies; Group 2 corresponds to cereals; Group 3 comprises sunflower, herbaceous, and leguminous crops; Group 4 includes olive groves and fruit trees; and Group 5 consists of maize, garden crops, potatoes, and sugar beet. This reclassification reduced the complexity of the original map while preserving essential agricultural characteristics for the modeling process. The original map and the one resulting from this simplification are shown in Figure 2.

After map simplification, the surface area requiring N fertilization (mainly dominated by cereals) was 5620.5 km^2, and the surface area not requiring N fertilization (forest vegetation, fallows, natural, urban, and other surfaces) was 3108.25 km^2.

Given the uncertainties associated with the type of data and the spatial and temporal scales, two methods were used to better account for the N input to the soil: Method 1 is based on cultivated surface data and recommended N doses for the five crop groups in Figure 2, while Method 2 is based on available N fertilizer sales data and their temporal evolution.

Method 1. Nitrogen input to the soil based on the modified land-use map, recommended doses, and the temporal evolution of the cultivated surface.

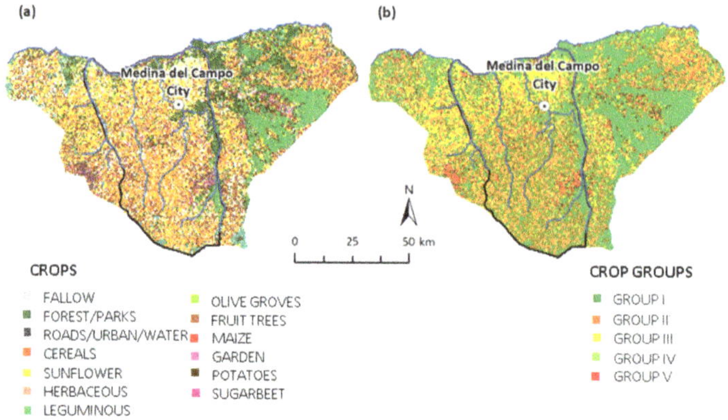

Figure 2. (**a**) Official land-use map of year 2018 [40]; (**b**) simplified crops map after merging crops into five major groups with similar nitrogen needs.

To estimate N supply to the soil using the simplified crops map, an average value of N doses for each crop group was calculated using the recommended N doses for each individual crop within the group, weighted by its area.

The temporal evolution of the cultivated area in the study zone between the years 1952 and 2018 was analyzed to build an N input time function for the numerical model. Raw data were obtained from the Spanish National Statistics Institute, the Autonomous Community of Castilla and León, and the Spanish Ministry of Agriculture, Fisheries and Food [57–59], and was elaborated as shown in Figure 3. The figure shows that the cultivated surface has been decreasing since the early 1990s. According to Seoane [62], this could be due to several different causes: crop abandonment as a result of new agricultural terms and regulations promoted by the European Union since the late 1980s; emigration from rural to urban areas; population aging [63]; and decreased precipitation (for rain-fed crops) [64], among others.

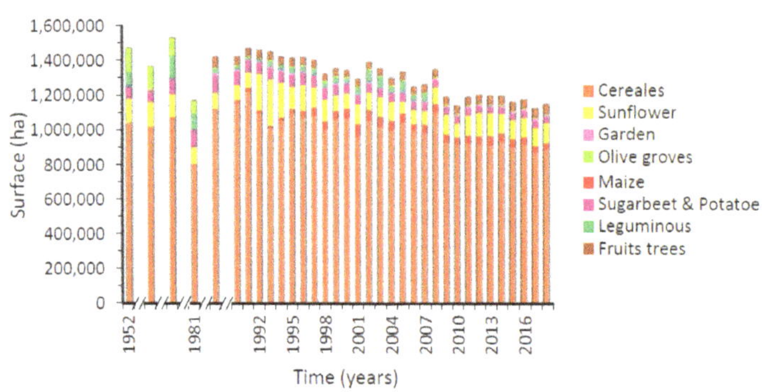

Figure 3. Temporal evolution of surface area (ha) dedicated to the main crops in the study area during the period 1952–2018. Own elaboration based on data obtained from [57–61].

Method 2. Nitrogen input to the soil according to fertilizer sales.

The recorded N fertilizer consumption was analyzed at the regional [59] and national [60] scales. Data were collected for the entire country for the period 1950–2005, and for the Autonomous Community of Castilla and León from 2005 onwards. The available data were extrapolated to the total modelled surface. Figure 4 shows the evolution of fertilizer sales extrapolated to the modelled area from 1950 to 2018.

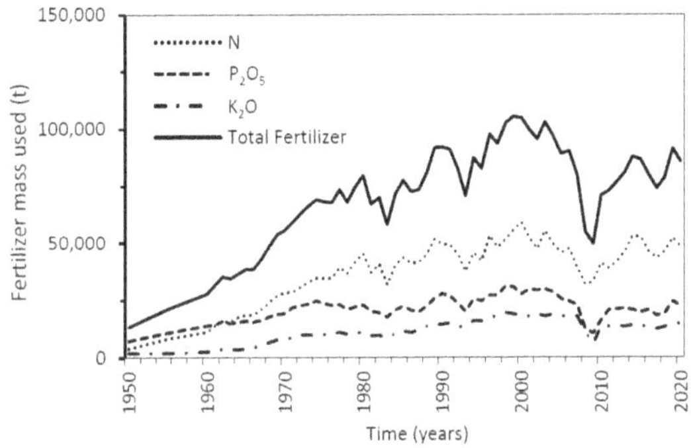

Figure 4. Temporal evolution of fertilizer sales (t) estimated for the whole modelled surface. Own elaboration with data from the Autonomous Community of Castilla and León [59] and the Spanish Ministry of Agriculture, Fisheries and Food [60].

3.2.3. Nitrogen Leaching to Groundwater

To generate representative data of NO_3^- input to groundwater, two methods were used to account for nitrogen leaching from the soil to the saturated zone of the aquifer: Method A used the Pratt equation [65], and Method B used the Nitrogen Index (NI) model [66].

Method A. Nitrogen leaching to the saturated zone according to the Pratt equation.

Fertilizers are essential for profitable agricultural production [67], but intensive farming has led to increased use in recent decades. Fertilizer overdoses are not absorbed by crops, leading to N losses to the atmosphere and runoff, or to N moving below the root zone [68], eventually leaching into the saturated zone and thus polluting groundwater [69].

The reviewed studies show wide ranges of N leaching to the saturated zone with respect to the total N applied: 14% to 38% [68–70], 10% to 30% [71], or 25% [5]. Ramos et al. [69] estimated losses for crops like onion, potatoes, and citrus using the empirical equation of Pratt (Equation (1)), which considers that 20% of N applied leaches to the soil [65]:

$$N_L = 0.20 \times (N_i \times D)^{0.712} \tag{1}$$

where:

N_L is N leaching to the saturated zone (kg/ha/year),

N_i is the maximum recommended N dose (kg/ha/year),

D is recharge to the water table in cm/year.

Method B. Nitrogen leaching to the saturated zone according to Nitrogen Index.

This method used the Nitrogen Index (NI) software (Version 1.0) [66] to calculate leaching percentages by estimating the annual balance of N inputs and outputs in the soil and its availability for subsequent crops. The model included different crop combinations and different N doses, both under rain-fed and irrigated conditions.

A total of 14 simulations were conducted, applying the maximum recommended N doses (kg/ha/yr). The range of leaching percentage calculated by NI goes from 5 to 35.6%, with an average value of 10.25%. Table 1 shows the percentage of N leaching for the considered crop groups.

These methods were combined with the two methods used to estimate N input to the soil, resulting in four models to reproduce N input to the saturated zone (Table 2). A numerical model was implemented and run for each of the four models, and the one whose results were most coherent with the observed values of NO_3^- was selected and subsequently calibrated to evaluate the future management scenarios considered by the

DRBA. Model 1.B Doses + NI was selected as the best representation of the conditions in the studied zone (see Section 4.1).

Table 1. Nitrogen leaching for the considered crop groups, calculated with Nitrogen Index.

Crop Groups	N applied to Soil (kg/ha/yr)	N leaching to the Saturated Zone (%)	N Leaching to the Saturated Zone (kg/ha/yr)
Group I	0	0	0
Group II	40	5	2
Group III	50	5	2.5
Group IV	110	16.6	18.26
Group V	250	35.6	89

Table 2. Models initially used to estimate nitrogen input to groundwater.

		Nitrogen (N) Application	
		Method 1. Recommended N Doses and Temporal Evolution of Cultivated Surface Per Crop Groups	**Method 2. Temporal Evolution of N Fertilizer Sales**
Nitrogen leaching to groundwater	**Method A. Pratt equation**	1.A. Doses + Pratt	2.A. Sales + Pratt
	Method B. Nitrogen Index model	1.B. Doses + NI	2.B. Sales + NI

3.2.4. Sensitivity Analysis of Transport Parameters

Given the uncertainties associated with the nitrate observations and the regional scope of the model, full calibration was not feasible. To enhance the utility of the model, a sensitivity analysis was conducted on key parameters that influence NO_3^- concentrations, such as the hydraulic conductivity, porosity, storage, and the dispersion coefficient. A relative sensitivity criterion was applied as defined by Equation (2):

$$\frac{CR}{CP} = \frac{(R_a - R_b)/R_b}{(P_a - P_b)/P_b} \quad (2)$$

where:

CR is the relative change of NO_3^- concentration.

CP is the relative change in each parameter or variable with respect to a reference value (typically, the calibrated values in the flow model). The magnitude of the perturbation was set between −50% and +50% [72,73] relative to the reference model.

P_a y P_b are the values used in the sensitivity analysis of the parameter in question (perturbed value and reference value),

R_a y R_b are the perturbed and reference NO_3^- concentration values.

The parameters of the reference model (the one considering Model 1.B Doses + NI for nitrate input to groundwater) to be perturbed were: horizontal hydraulic conductivity Kh = 0.008 to 20 m/day; vertical hydraulic conductivity Kv = 0.0001 to 0.1 m/day; and specific storage S = 0.0001 to 0.1. A dispersion value of 1000 m was assigned.

The sensibility analysis performed indicated that porosity is the most influential parameter on NO_3^- concentrations in a similar magnitude (Figure 5). Based on the results of this analysis, the efforts have focused on finding the most representative values for porosity to best reflect the observed conditions.

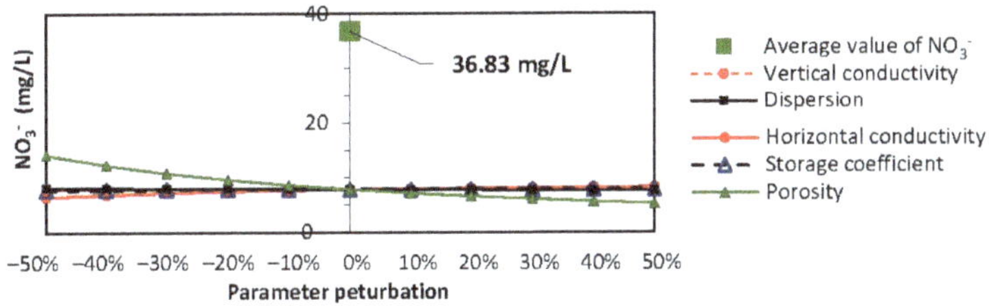

Figure 5. Effect of parameter perturbation in the sensitivity analysis on NO_3^- concentration entering the saturated zone. The mean observed concentration in the Upper Aquifer is indicated (36.83 mg/L).

3.2.5. Porosity Calibration

The porosity of the reference model was calibrated for the zones established in the groundwater flow model (see Figure S1). Initial porosity values were derived from the literature for sediments similar to those in the aquifer system [74–76]. For layer 1, porosity values were derived from the storage coefficient values used in the flow model, ranging from 0.1 to 0.001, so they were not calibrated.

The initial porosity of layers 2 and 3 was set at 0.25. Those values were manually calibrated and validated (Table S1 and Figure S2) considering the range of porosity values for similar sediment textures reported in the literature [74–76], weighted by their thicknesses obtained from lithological columns in the DRBA database [77,78].

Manual calibration of porosity allowed obtaining average NO_3^- concentrations in groundwater closer to the observed values at the regional level. Due to the characteristics of the available data (see Section 3.2.1) and the scope of the model, extreme values and local variations of NO_3^- concentrations could not be adjusted. The Supplementary Material (Figures S3 and S4) includes temporal evolution graphs of observed and calculated nitrate concentrations for the 75 observation points used.

During porosity calibration, particular emphasis was placed on the public water supply wells, as data from these points are generally more reliable: their technical characteristics—such as screen depth and length—are well known, and the chemical analyses are considered more representative, since these wells are in continuous use and water is renewed. The overall root mean square error (RMSE) between the observed and simulated nitrate concentrations across all observation points was 36.2 mg/L, whereas the RMSE for the subset of supply wells was significantly lower, at 12.2 mg/L. This improved performance is illustrated in Figure 6, which shows observed versus simulated nitrate concentrations for these supply points.

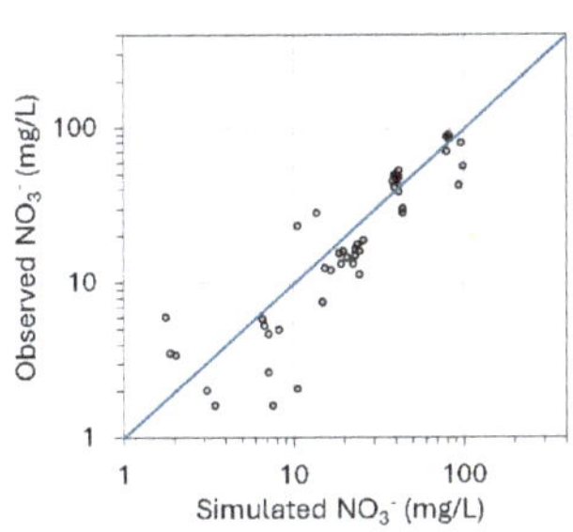

Figure 6. Observed versus simulated nitrate concentrations (mg/L) for the public water supply wells used in the calibration of porosity. These points were prioritized due to their higher reliability and representativeness.

Maximum and average NO_3^- values, and number of observation points with concentrations above 37.5 mg/L before and after porosity calibration were compared (Table 3). Modelled values obtained after porosity calibration approached the observations significantly. Then, given the regional scope of the nitrate model, the calibrated model was used to perform nitrate simulations. The final porosity values (ranging from 0.001 to 0.25) and their spatial distribution are available in Table S1 and Figure S2.

Table 3. Comparison of observed and simulated (with the selected model 1.B) NO_3^- concentrations (mg/L) for the period 2000–2018 at two depth ranges, ≤50 m (Upper Aquifer plus Aquitard layers, UAq-Aqt) and >50 m (Lower Aquifer, LAq), before and after porosity calibration.

Observed and Simulated NO_3^- (mg/L) in Observation Points		Observed	Model 1.B Doses + NI	Model 1.B Doses + NI After Porosity Calibration
Average	UAq-Aqt	54.0	15.5	41.0
	LAq	27.0	0.8	19.6
Maximum	UAq-Aqt	321.5	92.0	166.4
	LAq	220.0	8.0	101.0
Nº of points with NO_3^- > 37.5 mg/L	UAq-Aqt	78	21	81
	LAq	72	0	23

3.3. *Assessment of the Current Status of the Groundwater Ecosystem Service Evaluated*

The current status of the Abiotic Provisioning Service Groundwater of Good Quality for Human Supply was evaluated by modelling NO_3^- concentrations in groundwater by the year 2018. The initial (year 1950) NO_3^- values were set to zero, representing natural conditions. As stated, NO_3^- inputs to groundwater from fertilizers were modelled using the results of Model 1.B Doses + NI (the N results were converted to equivalent NO_3^- concentrations). Nitrate input was introduced as an annual recharge through a mass flow (kg/ha/year). It was assumed that rivers and wetlands do not contribute to NO_3^- concentrations and instead act as output zones for NO_3^-.

The spatial and temporal evolutions of NO_3^- concentrations measured in the DRBA observation network were compared to the NO_3^- concentrations calculated by the calibrated numerical model. The assessment has been performed using the proportion of aquifer surface area with NO_3^- concentrations between 10 and 37.5 mg/L as an indicator. The different number and spatial distribution of NO_3^- measured and modelled data make it impossible to extract rigorous information on temporal and spatial changes. Consequently, the transport model was developed using average NO_3^- concentration values and could not be optimally calibrated for individual observation points. Nevertheless, the model remains a valuable tool for large-scale basin assessments, considering its limitations. Thus, the model providing the best consistency between modelled and measured NO_3^- values in the MCGB was used to assess the present state and the future evolution of the selected ecosystem service.

3.4. *Assessment of the Possible Future Evolution of the Groundwater Ecosystem Service Evaluated*

The potential future evolution of the selected GWES service was evaluated for two groundwater management scenarios and two fertilizer use scenarios, covering the period from 2018 to 2100. The groundwater management scenarios were based on the Exploitation Index (EI). The two scenarios considered are:

(1) Business as Usual (BAU): No changes in the EI with respect to the current value in the MCGB (EI = 2.0, according to [9]).
(2) EI0.8: Reduction in EI from 2.0 to 0.8 by the year 2050 and beyond (this is the target of the DRBA).

The two scenarios related to N fertilizer use are:

(A) N20%: A linear reduction of 20% in the application of N fertilizers from current numbers (in 2018) by 2030 and beyond.
(B) ZeroN: A linear reduction in N fertilizer application from current levels (year 2018) to zero by year 2030 and beyond.

Scenario A is the objective set by the Duero River Basin Hydrological Plan for the period 2022–2027, adapted from the European Strategy of the European Green Deal project established at the end of 2019 [51] to reduce nitrate concentrations below 37.5 mg/L by the year 2030 [79]. Although scenario B is improbable, in this work it was considered a theoretical extreme to explore the long-term legacy effects of nitrate contamination.

The evaluation involved simulating NO_3^- concentrations in both aquifers and identifying concentration trends along the study period for the four models resulting from combining all the scenarios: 1.A, 1.B, 2.A, and 2.B (Table 4).

Table 4. Models considered for evaluating the hypothetical future evolution of nitrate concentrations in groundwater.

N Fertilizer Use Scenarios	Groundwater Exploitation Scenarios	
	1. IE = BAU	**2. IE Reduction to 0.8 by Year 2050 and Beyond**
A. 20% reduction by 2030 and beyond	1.A BAU-N20%	2.A EI0.8-N20%
B. Reduction to zero by 2030 and beyond	1.B BAU-ZeroN	2.B EI0.8-ZeroN

The relative impact of the different scenarios on the selected GWES was assessed by comparing the aquifer surface areas with different NO_3^- concentrations resulting from the four models in Table 4 relative to 2018. Three concentration ranges were considered to perform the comparison: NO_3^- < 10 mg/L (unpolluted); NO_3^- between 10 and 37.5 mg/L (polluted but under the legal limit); and NO_3^- > 37.5 mg/L (polluted beyond the legal limit). The analysis focused on the short-term (by 2050, when both the application of fertilizers and the EI reduction would stabilize) and long-term (by 2100). In addition, the modeled period was extended to evaluate the timeframe over which nitrate concentrations are expected to decrease significantly because of the proposed management measures. Due to the uncertainties associated with projections beyond 2100, all input parameters related to boundary conditions were held constant during this extended simulation period.

4. Results and Discussion

4.1. Nitrate Input to Groundwater

Nitrogen leaching estimations were calculated considering nitrogen (N) masses and concentrations. In order to compare with nitrate (NO_3^-) observations in the aquifer, the calculated N values were converted to equivalent NO_3^- values. Figure 7 shows a summary of the temporal evolution of NO_3^- inputs to groundwater in the whole modelled aquifer surface, as simulated by the four leaching models considered.

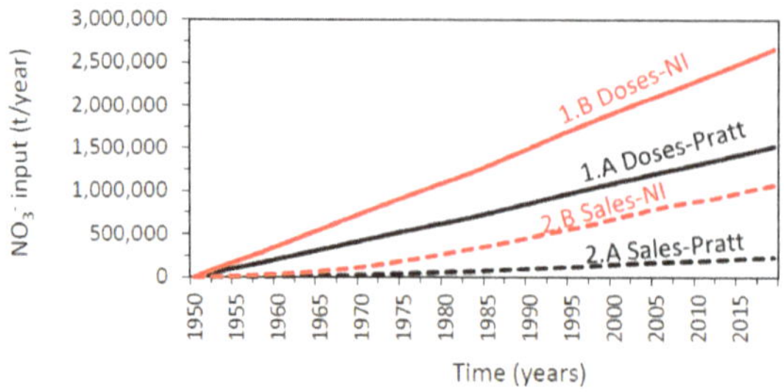

Figure 7. Temporal evolution of NO_3^- inputs to groundwater (t/year in the modelled surface) between 1950 and 2018 according to the four leaching models considered.

A clear discrepancy is observed when comparing the nitrogen inputs estimated using recommended agronomic doses with those estimated from fertilizer sales statistics, with sales-based estimates yielding consistently higher values. This difference is primarily attributed to the limitations of the spatial allocation method used to downsize fertilizer sales data to the study area. For data before 2005, national-level fertilizer sales were downscaled to the modelled area using proxies such as agricultural production. After 2005, data from the regional statistics office of the Autonomous Community of Castilla and León were employed. In both cases, data downscaling to the Medina del Campo Groundwater Body involved interpolation based on the extent of cultivated land. This coarse approximation, combined with the lack of detailed farm-level data, tends to underestimate the actual nitrogen applied to crops. These limitations support the selection of the "recommended dose" approach as more consistent with the agronomic reality of the area.

NO_3^- concentrations simulated with the four leaching models were compared to observed data from 2000 to 2018, obtained from the groundwater quality observation network of the DRBA. Maximum and average NO_3^- concentrations, as well as the number of observation points with concentrations exceeding 37.5 mg/L, were compared for the Upper Aquifer–Aquitard layers together, and for the Lower Aquifer (Table 5). Model 1.B Doses-NI was selected as the one best representing the observed data.

Table 5. Comparison of observed and simulated NO_3^- concentrations (mg/L) with the four leaching models during the 2000–2018 period at two depth ranges, ≤50 m (Upper Aquifer and Aquitard, UAq-Aqt) and >50 m (Lower Aquifer, LAq).

Observed and Simulated NO_3^- (mg/L)		Observed	Model 1.A Doses + Pratt	Model 1.B Doses + NI	Model 2.A Sales + Pratt	Model 2.B Sales + NI
Average	Uaq-Aqt	54.0	9.7	15.5	1.9	8.7
	Laq	27.0	0.4	0.8	0.07	0.3
Maximum	UAq-Aqt	321.5	29.0	92.0	8.7	41.8
	LAq	220.0	2.5	8.0	0.3	1.4
Nº of points with NO_3^- > 37.5 mg/L	UAq-Aqt	78	0	21	0	0
	LAq	72	0	0	0	0

4.2. *Assessment of the Current Status of the Abiotic Provisioning Service Groundwater of Good Quality for Human Supply*

The assessment of the APS-GGQHS service was based on observing the aquifer surface with nitrate concentrations under 37.5 mg/L. The spatial distribution of NO_3^- in groundwater was analyzed by creating maps for three different time periods: 1975–1985 (117 observation points with 224 data records available), 2001–2002 (130 observation points with 213 data records available), and 2018 (325 observation points with 772 data records available). To generate a map for a specific period, enough observed data with accepted analytical quality was required. If the data for a given year were insufficient to support reliable interpolation, a broader period was selected instead. In such cases, concentration averages over the selected period were used to ensure adequate spatial data coverage and to enable the creation of an interpolation map. Many of the observations correspond to the Lower Aquifer, and some correspond to the Aquitard. There were fewer observation points in the Upper Aquifer (4), so it was not possible to create maps for the Upper Aquifer due to data scarcity.

As stated previously, the different number and spatial distribution of observed NO_3^- concentration data made it impossible to extract rigorous information on temporal and spatial changes. Moreover, the very different values in nearby boreholes suggest that nitrate values are influenced by local conditions that cannot be adequately represented by a regional model. Thus, a preliminary assessment of the measured data did not reveal clear temporal trends, but significant local changes in NO_3^- concentrations (Figure 8).

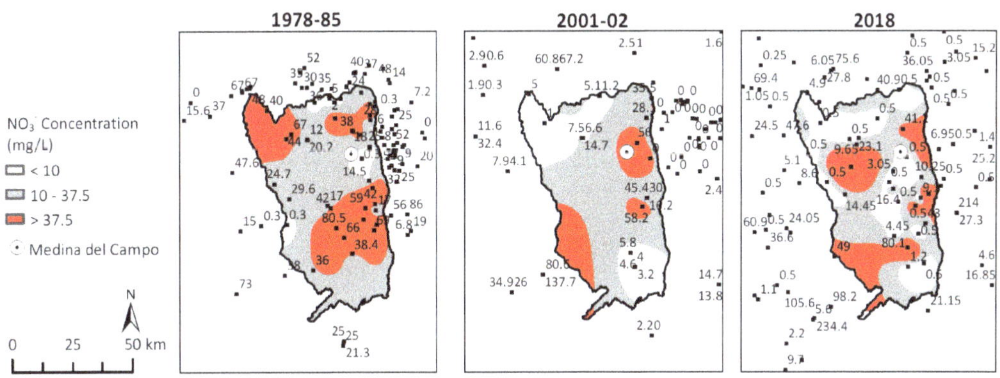

Figure 8. Distribution maps of NO_3^- (mg/L) measured in the groundwater observation network of the MCGB and its surroundings for three different periods: 1978–1985, 2001–2002, and 2018. Data represent the Lower Aquifer. Images created using the Multi-spline interpolation method.

Despite the impossibility of deducing spatial and temporal nitrate trends at the regional scale from the observations, nitrate values simulated with the porosity-calibrated Model 1.B Doses-NI are realistic on a broad scale (see Table 3). They show that almost 70% of the modelled area in layer one (Upper Aquifer) and 13% in layer three (Lower Aquifer) has NO_3^- concentrations higher than 37.5 mg/L (Figure 9). This result suggests that, at present, the status of the Abiotic Provisioning Service Groundwater of Good Quality for Human Supply is of poor quality, and its flow to society is probably nonexistent or close to zero. This is consistent with the fact that in many areas of the MCGB, groundwater cannot be used for human supply.

Figure 10 shows the spatial distribution of observed and simulated NO_3^- concentrations in 2018, just in the Lower Aquifer layer of the MCGB. The model cannot reproduce the high NO_3^- concentrations observed in the southern part of the MCGB, but it does reasonably reproduce the distribution in the central part. This better model performance for the central area can be explained by the higher data density of the data and the prioritization of calibration water supply wells near Medina del Campo city, while the southern region—characterized by limited data availability and low agricultural pressure—shows greater discrepancies. This model was used to simulate the future evolution of the selected GWES.

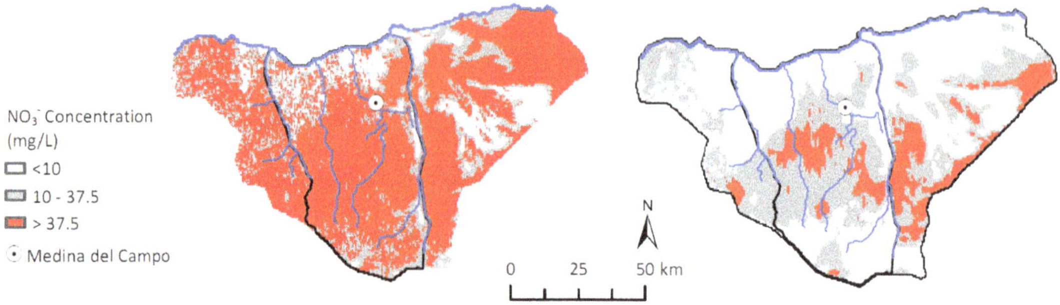

Figure 9. Spatial distribution of NO_3^- concentrations in the whole modelled area in 2018, expressed as surface area with different concentration ranges. (**Left**): layer 1 (Upper Aquifer). (**Right**): layer 3 (Lower Aquifer).

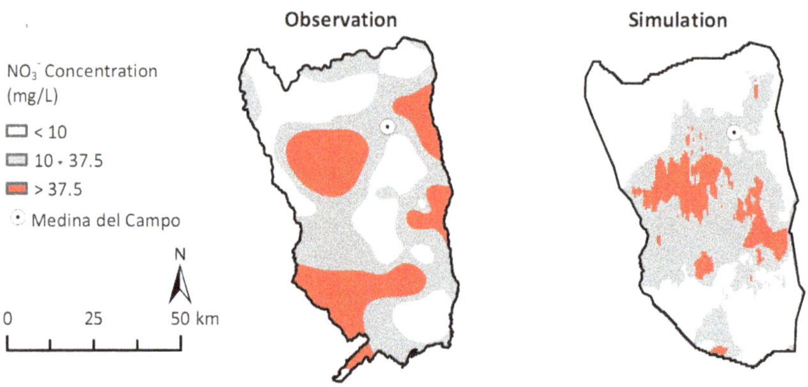

Figure 10. Spatial distribution of observed and simulated NO_3^- concentrations in 2018 in the Lower Aquifer of the MCGB. The small differences in the groundwater body contours are due to the fact that the model contour was simplified, especially in the SW, where the aquifer thickness is reduced and there are no observation points.

4.3. Assessment of the Possible Future Evolution of the Abiotic Provisioning Service Groundwater of Good Quality for Human Supply

The forecasted temporal evolution of the total mass of NO_3^- stored in the aquifer system, simulated with the four models in Table 4, is shown in Figure 11. The spatial distribution of the aquifer surfaces with different NO_3^- concentrations resulting from the four models considered is available in the Supplementary Material (Figures S5–S8).

Changes in aquifer surface area with $NO_3^- > 37.5$ mg/L in the years 2050 and 2100 with respect to those in 2018 are summarized in Table 6, showing percentage increases or decreases for each scenario. The results show that a 20% reduction in fertilizer application by 2030 and beyond leads to continued NO_3^- accumulation and a gradual increase in concentrations over time. In models 1.A BAU-N20% and 2.A EI0.8-N20%, the aquifer surface area with NO_3^- concentrations above 37.5 mg/L will continue increasing until the year 2100. Only the complete cessation of nitrogen fertilizer use significantly reduces nitrate inputs. According to the extended-time simulation, this scenario would result in a 50% decrease in the surface area affected by nitrate contamination by 2350.

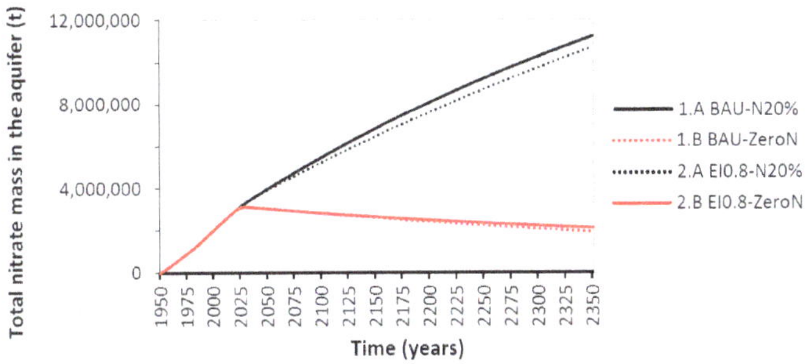

Figure 11. Temporal evolution of the total mass of nitrate (t) stored in the aquifer system simulated with the four models in Table 4.

Table 6. Changes of aquifer surface area with NO_3^- > 37.5 mg/L in years 2050 and 2100 relative to year 2018, in layer 1 (Upper Aquifer) and layer 3 (Lower Aquifer), and for the four combinations of future scenarios considered. Changes are expressed as percentages; positive values indicate an increase and negative values indicate a decrease.

N Fertilizer Use Scenarios			**Groundwater Exploitation Scenarios**	
			1. IE = BAU	**2. IE Reduction to 0.8 by Year 2050 and Beyond**
A. 20% reduction by 2030 and beyond	Layer 1	In 2050	+4.4	+4.8
		In 2100	+7.9	+8.5
	Layer 3	In 2050	+9.4	+8.7
		In 2100	+20.1	+18.2
B. Reduction to zero by 2030 and beyond	Layer 1	In 2050	−49.4	−49.1
		In 2100	−65.9	−65.8
	Layer 3	In 2050	+2.7	+2.5
		In 2100	−4.9	−4.3

When comparing groundwater management scenarios to nitrogen fertilizer reduction scenarios, the trends in NO_3^- concentrations at three observation points in the MCGB (Figure 12) show clearly that reducing fertilizer use has a greater impact on lowering nitrate concentrations than reducing groundwater extractions. These points correspond to public water supply wells, which are of particular relevance for groundwater management, and are spatially well distributed across the groundwater body. Moreover, combining fertilizer cessation and reduced groundwater pumping provides only a marginal additional benefit. Even under the most favorable scenario, the total nitrate mass is expected to be reduced by only about half by 2350, indicating that full recovery to natural conditions would require several centuries.

These results show that neither reducing groundwater abstraction nor reducing fertilizer application—or even stopping it—will significantly improve the conditions in the aquifer for a significant improvement of the Abiotic Provisioning Service Groundwater of Good Quality for Human Supply in the short (by 2030) and medium (by 2050) term. Such significant enhancement of the service would only be possible in the long term (by 2350), and would be achieved mostly by control of fertilizer doses.

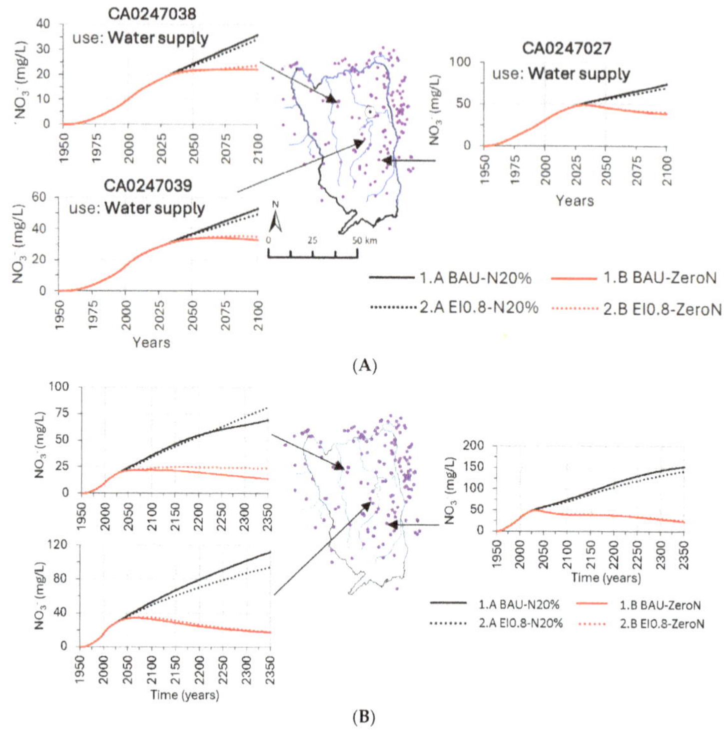

Figure 12. Simulated temporal evolution of nitrate concentrations at three water-supply observation points in the MCGB for two groundwater exploitation scenarios: 1. IE = BAU and 2. IE reduction to 0.8 by 2050 combined with the two fertilizer use scenarios: (**A**) 20% reduction by 2030 and beyond, and (**B**) reduction to 0 by 2030.

4.4. Uncertainty of the Numerical Modelling

The results presented in this study illustrate the utility of numerical modelling to support the assessment of groundwater ecosystem services, with particular focus on the provision of good-quality groundwater for human supply. In particular, the model has proven valuable for estimating the surface area of the aquifer affected by excessive nitrate concentrations under various fertilizer application strategies and groundwater exploitation scenarios. This spatial indicator, aligned with environmental quality objectives, provides a more robust and policy-relevant metric than isolated point concentrations—especially when dealing with regional-scale systems and fragmented monitoring data.

The regional scale of the model (with a horizontal resolution of 500 × 500 m) does not allow for the representation of local heterogeneity or small-scale pollution sources, such as manure hotspots or faulty septic systems. While this resolution is appropriate for simulating broader hydrogeological trends and assessing spatial patterns of contamination, it cannot capture sub-grid variations that may significantly affect nitrate concentrations at individual points.

Several limitations must be acknowledged. First, the observed nitrate concentration data used for calibration and validation suffer from considerable spatial and temporal inconsistencies. The groundwater quality monitoring network lacks systematic coverage and continuity; most wells are fully screened and therefore mix water from multiple depths, which masks vertical gradients and limits interpretability. In many locations, long-term time series are unavailable, and only sporadic measurements exist, particularly in the earlier decades of the simulation period. These limitations hinder the accurate reproduction of point-scale trends in nitrate contamination.

These constraints guided our decision to focus the model evaluation on the consistency of nitrate concentration orders of magnitude and spatial patterns, rather than on matching observed values at specific wells. The model was not intended as a predictive tool for local concentration values, but rather as a means to validate the conceptual model and assess the evolution of the aquifer's functional status under different management scenarios.

Despite these limitations, the model offers useful insights into the potential for long-term aquifer recovery, the relative importance of land-use versus pumping interventions, and the inertia of nitrate pollution in groundwater systems. The scenarios explored—while not all realistic—serve as boundary conditions that help to delineate the temporal dynamics of ecosystem service provision. In particular, the observed delay between management action and water quality improvement reinforces the importance of early intervention and sustained monitoring.

Finally, the modelling framework developed here supports the ecosystem services assessment by linking physical processes with human benefits in a spatially explicit and policy-relevant way. It enables the identification of vulnerable areas and critical time horizons for action, offering valuable input for planning and decision-making at the aquifer level.

5. Conclusions

This study illustrates how to incorporate the Ecosystem Services Assessment (ESA) approach to evaluate the expected relative impact of different management actions aiming to improve the flow of significant groundwater services to humans, such as the provision of good quality groundwater for human supply. The provisioning service assessed here, the Abiotic Provisioning Service Groundwater of Good Quality for Human Supply, was used as a case study in the Medina del Campo Groundwater Body (Spain), where groundwater quality and use have been severely affected by nitrate pollution due to intensive agricultural practices.

The study results confirm the usefulness of coupling groundwater flow and nitrate transport numerical modelling with the ESA framework to assess both the current status and the potential long-term evolution of the service under various feasible groundwater exploitation and fertilizer use management scenarios. Even though the representativity of the observed data is uncertain and modelling results cannot match them with fidelity, simulated nitrate concentrations reasonably reproduce the order of magnitude and the spatial distribution at a regional scale. This provides confidence in the results and validates the use of groundwater modelling to assess prospective analyses of groundwater ecosystem services.

The analysis of available nitrate concentration data revealed widespread contamination, with concentrations exceeding the regulatory limit of 37.5 mg/L for human consumption in several areas of the studied groundwater body, reaching up to 212 mg/L in some zones. After refining the dataset to include only observation points with acceptable ionic balance, the assessment showed that nitrate contamination has substantially reduced the capacity of the aquifer to provide good quality water, and also that this degradation will continue over time.

Model simulations indicate that a 20% reduction in nitrogen fertilizer use, as proposed in current EU strategies, will be insufficient to reverse nitrate contamination even in the long term (up to the year 2100). Only a complete cessation of fertilizer application would lead to significant improvements, with a projected 50% reduction in nitrate concentrations in the very long term (by 2350), provided no major climate changes occur. These results underscore the urgent need to implement stricter nitrogen management policies and to consider multi-decadal recovery horizons when designing effective groundwater protection strategies.

Overall, the findings highlight that without ambitious and sustained reductions in fertilizer inputs, the availability of the Abiotic Provisioning Service Groundwater of Good Quality for Human Supply will continue to decline well beyond the year 2050.

Supplementary Materials: The following supporting information can be downloaded at: https://www.mdpi.com/article/10.3390/hydrology12060137/s1, Table S1. Final porosity values after their adjustment for the 3 layers of the model and the 17 zones shown in Figure S1. Figure S1. Groundwater flow model configuration and calibration. (a) Perspective view of the model layers, showing the location of observation points assigned to each layer. (b) Hydraulic conductivity (HK) zones used for the calibration of flow and transport parameters. (c) Calibration plot of observed versus simulated hydraulic heads for the flow model. Figure S2. Spatial distribution of observation points colored by the model layer to which they are assigned. The total number of points and concentration data available per layer are also indicated. Observation points corresponding to public water supply wells are highlighted in bold, as they were prioritized during calibration due to their greater reliability. Figure S3. Map of calibrated porosity used in the transport model, based on the zonification of the hydraulic conductivity zones (HK zones). Figure S4. (a) Temporal evolution of simulated and observed NO_3^- concentration for public water supply wells within the study area. These observation points were prioritized during the calibration process due to their higher data reliability and relevance for groundwater management; (b) Temporal evolution of simulated and observed NO_3^- concentration of remaining observation points used for porosity calibration, grouped according to the zonification scheme (HK zones) applied during model calibration. Figure S5. Modelled surface areas in layer 1 (upper aquifer) showing nitrate concentrations in years 2018, 2050, and 2100 for groundwater management scenario (1) (no change), and fertilizer use scenarios (a) (20% reduction by 2030 and beyond) and (b) (cessation by 2030 and beyond). Figure S6. Modelled surface areas in layer 3 (lower aquifer) showing nitrate concentrations in years 2018, 2050, and 2100 for groundwater management scenario (1) (no change), and fertilizer use scenarios (a) (20% reduction by 2030 and beyond) and (b) (cessation by 2030 and beyond). Figure S7. Modelled surface areas in layer 1 (Upper Aquifer) showing nitrate concentrations in years 2018, 2050, and 2100 for groundwater management scenario (2) (EI = 0.8), and fertilizer use scenarios (a) (20% reduction by 2030 and beyond) and (b) (cessation by 2030 and beyond). Figure S8. Modelled surface areas in layer 1 (Upper Aquifer) showing nitrate concentrations in 2018, 2050, and 2100 for groundwater management scenario (2) (EI = 0.8), and fertilizer use scenarios (a) (20% reduction by 2030 and beyond) and (b) (cessation by 2030 and beyond).

Author Contributions: Conceptualization, M.M.; Data curation, M.B.; Formal analysis, M.B. and M.A.; Funding acquisition, M.M.; Investigation, M.B. and M.M.; Methodology, M.A. and M.M.; Project administration, M.M.; Resources, M.B.; Software, M.B.; Supervision, M.M.; Validation, M.A. and M.M.; Visualization, M.B.; Writing—original draft, M.B.; Writing—review and editing, M.A. and M.M. All authors have read and agreed to the published version of the manuscript.

Funding: This paper is based on research performed in the framework of the project Nature Insurance Value: Assessment and Demonstrations (NAIAD, grant agreement number: 730497) funded by the European Commission under the Horizon 2020 program. The authors express their gratitude to the two anonymous reviewers whose comments and suggestions notably improved this manuscript.

Data Availability Statement: The datasets generated during the current study are openly available in Zenodo at https://doi.org/10.5281/zenodo.15323735. All other relevant data are included in the article and its Supplementary Materials. Further inquiries can be directed to the corresponding author.

Acknowledgments: The authors thank the Duero River Basin Authority for the comments and information provided during the meetings of the NAIAD project, and to all the partners of NAIAD involved in the Medina del Campo Demo for helping to delineate the hydrogeological conceptual model and the future scenarios. This study reflects only the author's interpretation of the work performed.

Conflicts of Interest: The authors declare no conflicts of interest. The funders had no role in the design of the study; in the collection, analyses, or interpretation of data; in the writing of the manuscript; or in the decision to publish the results.

Abbreviations

The following abbreviations are used in this manuscript:

MCGB	Medina del Campo Groundwater Body
GWES	Groundwater ecosystem service
APS-GWGQHS	Abiotic Provisioning Service—Groundwater of Good Quality for Human Supply
ESA	Ecosystem Service Assessment

References

1. Harrison, S.; McAree, C.; Mulville, W.; Sullivan, T. The problem of agricultural 'diffuse' pollution: Getting to the point. *Sci. Total Environ.* **2019**, *677*, 700–717. [CrossRef] [PubMed]
2. Vinod, P.N.; Chandramouli, P.N.; Koch, M. Estimation of Nitrate Leaching in Groundwater in an Agriculturally Used Area in the State Karnataka, India, Using Existing Model and GIS. *Aquat. Procedia* **2015**, *4*, 1047–1053. [CrossRef]
3. Gleeson, T.; Wada, Y.; Bierkens, M.F.P.; van Beek, L.P.H. Water balance of global aquifers revealed by groundwater footprint. *Nature* **2012**, *488*, 197–200. [CrossRef] [PubMed]
4. Worrall, F.; Davies, H.; Burt, T.; Howden, N.J.K.; Whelan, M.J.; Bhogal, A.; Lilly, A. The flux of dissolved nitrogen from the UK—Evaluating the role of soils and land use. *Sci. Total Environ.* **2012**, *434*, 90–100. [CrossRef] [PubMed]
5. Galloway, J.N.; Dentener, F.J.; Capone, D.G.; Boyer, E.W.; Howarth, R.W.; Seitzinger, S.P.; Asner, G.P.; Cleveland, C.C.; Green, P.A.; Holland, E.A.; et al. Nitrogen Cycles: Past, Present, and Future. *Biogeochemistry* **2004**, *70*, 153–226. [CrossRef]
6. Griebler, C.; Avramov, M. Groundwater ecosystem services: A review. *Freshw. Sci.* **2015**, *24*, 355–367. [CrossRef]
7. Abbasi, M.R.; Sepaskhah, A.R. Nitrogen leaching and groundwater N contamination risk in saffron/wheat intercropping under different irrigation and soil fertilizers regimes. *Sci. Rep.* **2023**, *13*, 6587. [CrossRef]
8. Aziz, T.; Frey, S.K.; Lapen, D.R.; Preston, S.; Russell, H.A.J.; Khader, O.; Erler, A.R.; Sudicky, E.A. Economic valuation of subsurface water contributions to watershed ecosystem services using a fully integrated groundwater–surface-water model. *Hydrol. Earth Syst. Sci.* **2025**, *29*, 1549–1568. [CrossRef]
9. Borowiecka, M.; Alcaraz, M.; Manzano, M. Assessment of Ecosystem Services with numerical modelling to support groundwater dependent ecosystems and aquifer management: A demo study in the Medina del Campo Groundwater Body, Spain. *Case Stud. Chem. Environ. Eng.* **2024**, *10*, 100914. [CrossRef]
10. Saccò, M.; Mammola, S.; Altermatt, F.; Alther, R.; Bolpagni, R.; Brancelj, A.; Brankovits, D.; Fišer, C.; Gerovasileiou, V.; Griebler, C.; et al. Groundwater is a hidden global keystone ecosystem. *Glob. Change Biol.* **2024**, *30*, 17066. [CrossRef]
11. Nsoh, W. Achieving Groundwater Governance: Ostrom's Design Principles and Payments for Ecosystem Services Approaches. *Transnatl. Environ. Law.* **2022**, *11*, 381–406. [CrossRef]
12. Griebler, C.; Malard, F.; Lefébure, T. Current developments in groundwater ecology—From biodiversity to ecosystem function and services. *Curr. Opin. Biotechnol.* **2014**, *27*, 159–167. [CrossRef] [PubMed]
13. Bastani, M. Source area management practices as remediation tool to address groundwater nitrate pollution in drinking supply wells. *J. Contam. Hydrol.* **2019**, *226*, 103521. [CrossRef] [PubMed]
14. Yang, L.; Zheng, C.; Andrews, C.B.; Wang, C. Applying a Regional Transport Modeling Framework to Manage Nitrate Contamination of Groundwater. *Groundwater* **2021**, *59*, 292–307. [CrossRef]
15. Whiteis, A.M.; Zou, C.B.; Joshi, O.; Ferguson, B.; Roberts, S. Quantitative assessment of ecosystem services in diverse land uses within the forest-grassland transition zone of southern Great Plains, USA. *Ecosyst. Serv.* **2025**, *71*, 101697. [CrossRef]
16. Lovrić, M.; Torralba, M.; Orsi, F.; Pettenella, D.; Mann, C.; Geneletti, D.; Plieninger, T.; Primmer, E.; Hernandez-Morcillo, M.; Thorsen, B.J.; et al. Mind the income gap: Income from wood production exceed income from providing diverse ecosystem services from Europe's forests. *Ecosyst. Serv.* **2025**, *71*, 101689. [CrossRef]
17. Kok, S.; Le Clec'h, S.; Penning, W.E.; Buijse, A.D.; Hein, L. Trade-offs in ecosystem services under various river management strategies of the Rhine Branches. *Ecosyst. Serv.* **2025**, *72*, 101692. [CrossRef]

18. Sheng, J. Collaborative models and uncertain water quality in payments for watershed services: China's Jiuzhou River eco-compensation. *Ecosyst. Serv.* **2024**, *70*, 101671. [CrossRef]
19. Younesi, M.; Saadatpour, M.; Afshar, A. Integration of the system of environmental economic accounting-ecosystem accounting (SEEA-EA) framework with a semi-distributed hydrological and water quality simulation model. *Ecosyst. Serv.* **2024**, *70*, 101672. [CrossRef]
20. Peña, L.; de Manuel, B.F.; Méndez-Fernández, L.; Viota, M.; Ametzaga-Arregi, I.; Onaindia, M. Co-Creation of Knowledge for Ecosystem Services Approach to Spatial Planning in the Basque Country. *Sustainability* **2020**, *12*, 5287. [CrossRef]
21. Peña, L.; Onaindia, M.; de Manuel, B.F.; Ametzaga-Arregi, I.; Casado-Arzuaga, I. Analysing the Synergies and Trade-Offs between Ecosystem Services to Reorient Land Use Planning in Metropolitan Bilbao (Northern Spain). *Sustainability* **2018**, *10*, 4376. [CrossRef]
22. Murray, B.R.; Hose, G.C.; Eamus, D.; Licari, D. Valuation of groundwater-dependent ecosystems: A functional methodology incorporating ecosystem services. *Aust. J. Bot.* **2006**, *54*, 221. [CrossRef]
23. Danielopol, D.L.; Gibert, J.; Griebler, C.; Gunatilaka, A.; Hahn, H.J.; Messana, G.; Notenboom, J.; Sket, B. Incorporating ecological perspectives in European groundwater management policy. *Environ. Conserv.* **2004**, *31*, 185–189. [CrossRef]
24. Quevauviller, P. Groundwater monitoring in the context of EU legislation: Reality and integration needs. *J. Environ. Monit.* **2005**, *7*, 89. [CrossRef]
25. Anzaldua, G.; Gerner, N.V.; Lago, M.; Abhold, K.; Hinzmann, M.; Beyer, S.; Winking, C.; Riegels, N.; Jensen, J.K.; Termes, M.; et al. Getting into the water with the Ecosystem Services Approach: The DESSIN ESS evaluation framework. *Ecosyst. Serv.* **2018**, *30*, 318–326. [CrossRef]
26. EU-GWD. *Directive 2006/118/EC of the European Parliament and of the Council of 12 December 2006 on the Protection of Groundwater Against Pollution and Deterioration. Official Journal of the European Communities*; European Parliament and of the Council: Brussels, Belgium, 2006. Available online: https://eur-lex.europa.eu/legal-content/EN/TXT/PDF/?uri=CELEX:32006L0118 (accessed on 1 May 2025).
27. BOE. Real Decreto 1159/2021, de 28 de Diciembre, por el Que se Modifica el Real Decreto 907/2007, de 6 de Julio, por el que se Aprueba el Reglamento de la Planificación Hidrológica. 2021. Available online: https://www.boe.es/eli/es/rd/2021/12/28/1159 (accessed on 5 March 2025).
28. Manzano, M.; Lambán, L.J. Una aproximación a la evaluación de los servicios de las aguas subterráneas al ser humano en España. *Ambient. Rev. Del Minist. De Medio Ambiente* **2012**, *1*, 32–41. Available online: https://doi.org/10261/277184 (accessed on 5 March 2025).
29. Guswa, A.J.; Brauman, K.A.; Brown, C.; Hamel, P.; Keeler, B.L.; Sayre, S.S. Ecosystem services: Challenges and opportunities for hydrologic modeling to support decision making. *Water Resour. Res.* **2014**, *50*, 4535–4544. [CrossRef]
30. Iliopoulos, V.G.; Damigos, D. Groundwater Ecosystem Services: Redefining and Operationalizing the Concept. *Resources* **2024**, *13*, 13. [CrossRef]
31. Mayor, B.; Gunn, E.L.; Marcos, C.; Vay, L. *PROYECTO NAIAD: Evaluación del Estado Y Simulación Hidrogeológica Del Comportamiento Del Acuífero de Medina del Campo*; Ministerio para la Transición Ecológica y el reto Demográfico. Confederación Hidrográfica del Duero: Valladolid, Spain, 2021. Available online: https://www.chduero.es/documents/20126/427605/NAIAD+Evaluaci%C3%B3n+del+estado+y+simulaci%C3%B3n+hidrogeol%C3%B3gica+del+comportamiento+del+acu%C3%ADfero+de+Medina+del+Campo.pdf/e5e07db6-e542-614a-317f-4548bad6b254?t=1648728521852 (accessed on 1 May 2025).
32. DRBA. Plan Hidrológico de la Parte Española de la Demarcación Hidrográfica del Duero 2009-15. 2012; Volume 776. Available online: https://www.chduero.es/documents/20126/104329/Memoria_20131023_AGS_v12_Negro.pdf (accessed on 1 May 2025).
33. Mayor, B.; de la Hera-Portillo, M.; Llorente, Á.; Heredia, J.; Calatrava, J.; Martínez, D.; Manzano, M.; Robles-Arenas, V.; Borowiecka, G.; Mediavilla, R.; et al. *Greening Water Risks*; Springer International Publishing: Cham, Switzerland, 2023. [CrossRef]
34. IGME. Mapa Geológico de España 1: 50000, hoja 427 Medina del Campo. 2007, Volume 75. Available online: http://info.igme.es/cartografiadigital/geologica/Magna50Hoja.aspx?Id=427&language=es (accessed on 14 September 2021).
35. MIRAME—IDEDuero. Distribución Espacial de la Precipitación en el Periodo 1940–2005. 2020. Available online: https://mirame.chduero.es/chduero/viewer (accessed on 1 May 2025).
36. Llorente, M.; Bejarano, M.D.; De la Hera, A.; Aguilera, H. Aguilera, Precipitation trends in the Medina del Campo groundwater body region (Spain): Towards implementing nature-based solutions for droughts and floods events. In Proceedings of the European Geosciences Union (EGU 2018), Vienna, Austria, 8–13 April 2018. [CrossRef]
37. Roca, N.L.; Lozano, P.J.; Cadiñamos, J.A.; Latasa, I.; Longares, L.A.; Meaza, G. Los Mantos Eólicos del Sector Sudoccidental de la Provincia de Valladolid. Una investigación geomorfológica y Edafológica. 2015, pp. 1699–1708. Available online: https://www.researchgate.net/publication/283485210_Los_mantos_eolicos_del_sector_sudoccidental_de_la_provincia_de_Valladolid_Una_investigacion_geomorfologica_y_edafologica (accessed on 14 September 2021).

38. IGME. Apoyo a la caracterización adicional de las masas de agua subterránea en riesgo de no cumplir los objetivos medioambientales en 2015. Demarcación Hidrográfica del Duero. Masa de Agua Subterránea: 47 Medina del Campo. Encomienda de Gestión Para la Realización de Trabajos Científico-Técnicos de Apoyo a la Sostenibilidad y Protección de las Aguas Subterráneas. (Actividad 2 2009); pp. 1–42. Available online: http://info.igme.es/SidPDF/139000/899/139899_0000019.pdf (accessed on 14 September 2021).
39. Porée, L. Evolución de las Zonas Riparias de Los Ríos Trabancos, Zapardiel, Adaja y Guareña (Cuenca del Duero) Entre 1956 y la Actualidad. 2019, Volume 37. Available online: https://repositorio.upct.es/server/api/core/bitstreams/1b8d4923-59da-47eb-9e70-49ad99de1237/content (accessed on 27 September 2021).
40. ITACyL. Mapa de Cultivos y Superficies Naturales de 2018. Instituto Tecnológico Agrario de Castilla y León. 2023. Available online: https://mcsncyl.itacyl.es/ (accessed on 22 June 2024).
41. del Campo, S.G. El uso de Fertilizantes, Bajo la Lupa. Diario de Castilla y León, 2023. Available online: https://www.diariodecastillayleon.es/mundo-agrario/230410/27956/fertilizantes-lupa.html (accessed on 21 May 2024).
42. MIRAME-DRBA. Ficha Técnica de la Masa de Agua Subterránea de Medina del Campo—400047. 2024. Available online: https://mirame.chduero.es/chduero/public/groundWaterBody/gwb/search/general/400047 (accessed on 21 June 2024).
43. Sánchez-Moya, Y.; Sopeña, A. El rift mesozoico ibérico. In *Proc. 1a Jorn. Geol. Médica España, Salamanca, Vera, A., Ed.*; Geología de España, Sociedad Geológica de España-Instituto Geológico de España: Salamanca, España, 2004; pp. 484–522.
44. Alonso-Gavilán, G.; Armenteros, I.; Carballeira, A.; Corrochano, A.; Huerta, P.; Rodríguez, J.M. Cuencas Cenozoicas. In *Geol. España*; Vera, J.D., Civis, J., Eds.; Sociedad Geológica de España e Instituto Geológico y Minero de España: Salamanca, España, 2004; p. 890. Available online: https://oa.upm.es/4007/2/TORRES_CL_2004_01.pdf (accessed on 5 March 2025).
45. Cunha, P.P.; de Vicente, G.; Martín-González, F. *Cenozoic Sedimentation Along the Piedmonts of Thrust Related Basement Ranges and Strike-Slip Deformation Belts of the Iberian Variscan Massif*; Springer: Cham, Switzerland, 2019; pp. 131–165. [CrossRef]
46. Marin, C.; Constaán, A.R.; López-Gutiérez, J.C.; Rubio-Sánchez, F.M.; De la Hera, A. Modelo Geológico 3D del Acuífero de Medina del Campo (SE Cuenca del Duero)—Dialnet. 2021. Available online: https://dialnet.unirioja.es/servlet/articulo?codigo=8474291 (accessed on 26 May 2024).
47. Rey-Moral, C.; Gómez-Ortiz, D.; Giménez, E.; López, M.T. Control morfoestructural de la distribución de arsénico en el sur de la Cuenca del Duero. In *Proc. 1a Jorn. Geol. Médica España, Salamanca 2016*; Giménez-Forcada, E., Ed.; Instituto Geológico y Minero de España: Madrid, Spain, 2016; pp. 83–88. Available online: https://www.igme.es/Publicaciones/publiFree/Libro%20en%20flash/mobile/index.html (accessed on 1 May 2025).
48. IGME. Plan Hidrológico de la Cuenca del Duero. Proyecto de Investigación Hidrogeológica de la Cuenca del Duero, Sistemas 8 y 12. Instituto Geológico y Minero de España. 75 p + gráficos y mapas. 1980. Available online: http://info.igme.es/SidPDF/018000/318/Tomo%201/18318_0001.pdf (accessed on 1 May 2025).
49. MIRAME-DRBA. Presión Difusa Sobre la Masa Medina del Campo—30100045. DRBA. 2024. Available online: https://mirame.chduero.es/chduero/public/pressures/groundContamination/search/technical/30100045 (accessed on 22 June 2024).
50. DRBA. La Masa de Agua Subterránea Medina del Campo. Meeting on the Status of Groundwater Bodies in Duero River Basin. 2014. Available online: https://www.chduero.es/documents/20126/77728/24140618_LA_MASA_DE_AGUA_MEDINA_DEL_CAMPO.pdf (accessed on 17 September 2021).
51. DRBA. Plan Hidrológico de la Parte Española de la Demarcación Hidrográfica del Duero Revisión de Tercer Ciclo (2022–2027). 2022, Volume 5, p. 998. Available online: https://www.chduero.es/web/guest/propuesta-de-proyecto-de-plan-hidrol%C3%B3gico (accessed on 26 February 2024).
52. Haines-Young, R.; Potschin, M. Common International Classification of Ecosystem Services (CICES V5.1). One Ecosystem. 2018; Volume 3. Available online: https://www.zemeunvalsts.lv/documents/view/8b6dd7db9af49e67306feb59a8bdc52c/Common%20International%20Classification%20of%20Ecosystem%20Services%20Guidance-V51-01012018.pdf (accessed on 1 May 2025).
53. Harbaugh, A.W. *MODFLOW-2005: The U.S. Geological Survey Modular Ground-Water Model-the Ground-Water Flow Process*; US Department of the Interior, US Geological Survey: Reston, VA, USA, 2005. [CrossRef]
54. Harbaugh, A.W.; Banta, E.R.; Hill, M.C.; McDonald, M.G. *MODFLOW-2000, The U.S. Geological Survey Modular Ground-Water Model: User Guide to Modularization Concepts and the Ground-Water Flow Process*; University of Alabama Tuscaloosa: Tuscaloosa, AL, USA, 2000. [CrossRef]
55. Zheng, C. *MT3DMS: A Modular Three-Dimensional Multispecies Transport Model for Simulation of Advection, Dispersion, and Chemical Reactions of Contaminants in Groundwater Systems*; Documentation and User's Guide; US Army Corps of Engineers: Washington, DC, USA, 1999; Volume 220.
56. BOCyL. Código de Buenas Prácticas Agrarias. 2020, pp. 22346–22399. Available online: https://agriculturaganaderia.jcyl.es/web/es/produccion-agricola/codigo-buenas-practicas-agrarias.html (accessed on 1 May 2025).
57. INE. Fondo Documental—Anuarios Historicos de España. *Instituto Nacional de Estadística.* 2024. Available online: https://www.ine.es/inebaseweb/libros.do?tntp=25687 (accessed on 22 June 2024).

58. MAPA. Anuario de Estadística. Ministerio de Agricultura, Pesca y Alimentación. 2024. Available online: https://www.mapa.gob.es/es/estadistica/temas/publicaciones/anuario-de-estadistica/ (accessed on 22 June 2024).
59. JCyL. Anuario de estadística agraria de Castilla y León. Junta de Castilla y León. 2024. Available online: http://www.jcyl.es/web/jcyl/AgriculturaGanaderia/es/Plantilla100/1284228463984/_/_/_ (accessed on 22 June 2024).
60. MAPA. Estadística de Consumo de Fertilizantes en la Agricultura. Ministerio de Agricultura, Pesca y Alimentación. 2024. Available online: https://www.mapa.gob.es/es/estadistica/temas/estadisticas-agrarias/agricultura/estadisticas-medios-produccion/fertilizantes.aspx (accessed on 26 June 2024).
61. JCyL. Agricultura y Ganadería. Junta de Castilla y León. 2024. Available online: https://agriculturaganaderia.jcyl.es/web/es/produccion-agricola/situacion-castilla-leon.html (accessed on 18 May 2024).
62. Seoane, S.S. Efectos Ecológicos del Abandono de Tierras de Cultivo en la Provincia de León (Municipio de Chozas de Abajo). 1998, p. 265. Available online: https://buleria.unileon.es/bitstream/handle/10612/16851/Efectos_Ecol%C3%B3gicos_Derivados_Abandono_Tierras.PDF?sequence=1 (accessed on 1 May 2025).
63. Alonso, F.; Martínez-Hernández, C.; Serrato, F.B.; Angel, M.; Carrillo, F. Principales Causas Del Abandono de Cultivos en la Región de Murcia. Abandono de cultivos en la Región de Murcia. Consecuencias Ecogeomorfológicas, 2016; pp. 203–226. Available online: https://docta.ucm.es/entities/publication/a049ceac-26dc-4eec-9126-167e569a23af (accessed on 1 May 2025).
64. Benito, C.; Tobar, S. De Andalucía a Castilla y León: La Sequía Amenaza al Cultivo de los Cereales, Frutos Secos, Aceite de Oliva y Arroz. Prensa El Español, 2023. Available online: https://www.elespanol.com/invertia/empresas/distribucion/20230416/andalucia-castilla-leon-sequia-amenaza-cultivo-cereales/755924856_0.html (accessed on 1 May 2025).
65. Pratt, P.F. *Nitrogen Use and Nitrate Leaching in Irrigated Agriculture*; American Society of Agronomy: Madison, WI, USA, 1984. [CrossRef]
66. Delgado, J.; Gagliardi, P.M.; Rau, E.J.; Fry, R.; Figueroa, U.; Gross, C.; Cueto-Wong, J.; Shaffer, M.J.; Kowalski, K.; Neer, D.; et al. Nitrogen Index 4.4 User Manual. *Data Base* **2012**, *3304*, 1–148. Available online: https://vtechworks.lib.vt.edu/server/api/core/bitstreams/26bc1f82-85bd-479d-8f0c-59731e108548/content (accessed on 1 May 2025).
67. Vramontes, U.F.; Delgado, J.A.; Wong, J.A.C. *Manual del Usuario—Índice de Nitrógeno Ver. 4.4*; Instituto Nacional de Investigaciones Forestales Agrícolas y Pecuarias: Mexico City, Mexico, 2016.
68. Meisinger, J.; Jorge, D.; Alva, A. Nitrogen Leaching Management. *Encycl. Soils Environ.* **2006**, *2*, 1122–1124. Available online: https://www.ars.usda.gov/research/publications/publication/?seqNo115=157476 (accessed on 1 May 2025).
69. Ramos, C.; Agut, A.; Lidón, A.L. Nitrate leaching in important crops of the Valencian Community region (Spain). *Environ. Pollut.* **2002**, *118*, 215–223. [CrossRef] [PubMed]
70. Hussain, M.Z.; Robertson, G.P.; Basso, B.; Hamilton, S.K. Leaching losses of dissolved organic carbon and nitrogen from agricultural soils in the upper US Midwest. *Sci. Total Environ.* **2020**, *734*, 139379. [CrossRef] [PubMed]
71. Craswell, E.; Craswell, E. Fertilizers and nitrate pollution of surface and ground water: An increasingly pervasive global problem. *SN Appl. Sci.* **2021**, *3*, 518. [CrossRef]
72. Jiménez-Martínez, J.; Candela, L.; Molinero, J.; Tamoh, K. Groundwater recharge in irrigated semi-arid areas: Quantitative hydrological modelling and sensitivity analysis. *Hydrogeol. J.* **2010**, *18*, 1811–1824. [CrossRef]
73. Hornero, J.; Manzano, M.; Ortega, L.; Custodio, E. Integrating soil water and tracer balances, numerical modelling and GIS tools to estimate regional groundwater recharge: Application to the Alcadozo Aquifer System (SE Spain). *Sci. Total Environ.* **2016**, *568*, 415–432. [CrossRef]
74. Hiscock, K.M.; Bense, V.F. *Hydrogeology Principles and Practice*, 2nd ed.; John Wiley & Sons Ltd.: Oxford, UK, 2014. Available online: https://books.google.com/books/about/Hydrogeology.html?hl=es&id=j9njAgAAQBAJ (accessed on 1 May 2025).
75. Custodio, E.; Llamas, M.R. *Hidrología Subterránea*, 2nd ed.; Ediciones Omega: Barcelona, Spain, 1983.
76. IGME. Conceptos básicos de la hidrogeología. *Angew. Chem. Int. Ed.* **1967**, *6*, 951–952.
77. CHD. Red de Control del Nivel—CHDuero. 2019. Available online: https://www.chduero.es/web/guest/red-de-control-del-nivel (accessed on 26 May 2019).
78. CHD. Red de Control del Estado Químico—CHDuero. 2019. Available online: https://www.chduero.es/web/guest/red-control-estado-quimico (accessed on 20 June 2019).
79. Parlamento Europeo. Reglamento UE 2019/1009 del Parlamento Europeo por el Que se Establecen Disposiciones Relativas a la Puesta en Disoiosición en el Mercado de los Productos Fertilizantes y se Modifican los Reglamentos (CE) no 1069/2009 y (CE) no 1107/2009 y se deroga el Re. 2019. Available online: https://www.boe.es/doue/2019/170/L00001-00114.pdf (accessed on 1 May 2025).

hydrology

MDPI

Article

Application of the Groundwater Data Mapper Tool to Assess Storage Changes in a Groundwater-Driven Basin in the Klamath Watershed, Oregon, USA

Daniel Shepard [1], Norman L. Jones [2,*] and Gustavious P. Williams [2]

1 Klamath Falls Field Office, US Fish and Wildlife Service, 1936 California Avenue, Klamath Falls, OR 97601, USA; daniel_shepard@fws.gov

2 Department of Civil and Construction Engineering, Brigham Young University, Provo, UT 84602, USA; gus.p.williams@byu.edu

* Correspondence: njones@byu.edu; Tel.: +1-801-422-7569

Abstract: Streamflow in the Upper Williamson Basin of the Klamath Watershed is groundwater dominated with year-to-year fluctuations in both volume and duration, including multi-year periods with no streamflow. The relationship between precipitation, groundwater, and streamflow is difficult to characterize because of the limited number of monitoring wells, large data gaps, and a unique geologic structure that controls flow. To understand why surface flow has ceased entirely, we use the Groundwater Data Mapper Tool to impute gaps in the well data using machine learning and open-source Earth observation data and then compute changes in groundwater storage over time. Our research confirms that groundwater storage is correlated to streamflow and finds that there is a control groundwater storage below which flow does not occur. Furthermore, we find that groundwater storage is correlated to rainfall with a three- to four-year delay. This lag and the geologic structural control mean that even with several years of above-average precipitation, live flow may take years to resume. This insight allows water managers to understand and adjust for this highly irregular streamflow for better management decisions.

Keywords: drought; climate change; groundwater; remote sensing; machine learning

1. Introduction

The interaction between precipitation, groundwater storage, and streamflow can be complex [1]. In most areas, streamflow is predominantly governed by precipitation with baseflow more coupled to groundwater levels. In some cases, especially in mountain basins controlled by snowpack, streamflow can be dependent on groundwater storage with a time lag, sometimes on the order of a few years from precipitation events that provide recharge [2]. These interactions can be byzantine, especially when geologic structures are involved that can control or influence interactions between recharge, groundwater storage, and streamflow [3,4].

The Williamson River Basin is located on the northern (upper) end of the Klamath River Watershed near the border of Oregon and California, USA (Figure 1). It is located in a high desert environment on the eastern slope of the Cascade Range, and most of the precipitation in the basin is from snowfall. Because of its unique geology, the streamflow in the basin is heavily influenced by groundwater and can take years to reflect trends in precipitation. Beginning in 2001, there has been an overall decreasing trend in the volume of water leaving the basin as surface flow. Prior to 2000, streamflow leaving the basin, as measured at a United States Geological Survey (USGS) gage, could be roughly correlated with the mountain snowpack from the previous 2 to 3 years [3–5]. Since the turn of the millennia, climate change, drought, and supplemental groundwater use have had as-yet unquantified impacts on the aquifer, potentially disrupting a previously balanced hydrologic system [6]. Notably, in 2015 and from 2021 to the present, no live flow has been measured leaving the basin. The downward trend of streamflow volume resulting in a complete absence of flow from the headwaters of this basin is concerning and has an impact on irrigators, federal agencies, NGOs, and water managers.

Figure 1. Location of the Williamson Basin as it relates to Upper Klamath Lake and the Klamath River. Klamath River Watershed in Northern California and Southern Oregon.

The Williamson Basin has been studied by various researchers since the 1950s, but the precise relationship between precipitation, groundwater, and streamflow has been poorly understood and not well characterized. Streamflow leaving the basin is dominated by fluctuations in groundwater levels. The groundwater is correlated to precipitation but exhibits a significant time lag because of the underlying geology of the basin. The lag occurs because of the time required between the infiltration of precipitation into the groundwater system, aquifer recharge, and emergence of the Williamson River at the basin mouth. The atypical hydrology of this watershed means that traditional rainfall-runoff models, such as PRMS, are not reliable, and hydrologic forecasts can be inaccurate. A better understanding of the precipitation–groundwater–streamflow dynamics is needed to improve forecasting and predict when live streamflow could be expected to return to this watershed.

Figure 2 illustrates the topographic profile of the Williamson Basin, a unique geomorphological structure characterized as a "buried bowl" encircled by mountainous terrain.

The basin's distinctive configuration consists of sediment infill on top of a porous volcanic aquifer with a lip at its mouth. This impermeable "lip" forces groundwater to the surface in order to leave the watershed. Recharge occurs through precipitation, primarily in the form of snow at higher elevations. The upper Williamson River originates in the southeastern mountainous regions of the basin (Figure 2, area A) and is fed by seeps and springs within the Yamsay Volcanic Complex as it flows north. The river then flows into the sediment-filled basin with a 200 km^2 (124 mi^2) farmed wetland complex and National Wildlife Refuge (N.W.R) at the center. Here, the river sheet flows across the marsh and is dissipated by high levels of evapotranspiration and groundwater infiltration [2,3,5,7,8] (Figure 2, area B). Groundwater, originating from the Cascades on the west side, flows sub-surface under the basin floor to re-fill the porous sediments and aquifer under the N.W.R. and then re-emerges through the various springs and seeps at the southwest portion of the marsh (Figure 2, area B). There is no defined channel as water flows from the marsh, but the flow gradually coalesces into a single stream as it flows over a topographic high created by a basaltic intrusion at the mouth of the basin, locally named "Kirk Reef" (Figure 2 area C).

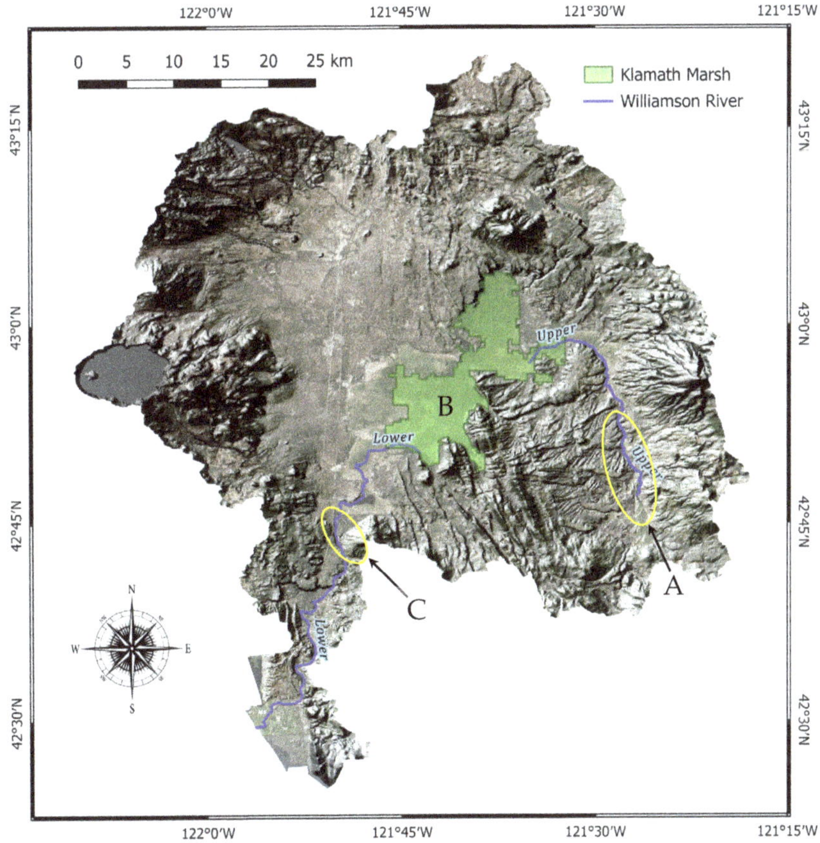

Figure 2. Unique geology of the Williamson Basin impacting the interactions between groundwater and streamflow. (A) Upper reaches of the Williamson River fed by springs and seeps. (B) Sediment-filled wetland complex (National Wildlife Refuge) where the river disappears. (C) Southern region where the river is measured using a USGS gage (Oregon Digital Elevation Model DEM, https://geohub.oregon.gov, accessed on 14 November 2024).

Using a bathtub as an analogy, (ground)water in this basin must re-fill the aquifer up to a certain volume before it can overflow and leave the basin (as Williamson River). Because of this, there can be a considerable lag between precipitation and streamflow. This study delves into stream flow from the Williamson Basin, the lag phenomenon, and the correlation with precipitation patterns. Comprehensive groundwater characterization in the Williamson Basin remains complex due to the limited number of monitoring wells, predominantly situated in the sedimentary-capped lower basin areas, and gaps in the historical groundwater level records from these wells.

Background

Prior research on the Williamson Basin showed that the river receives as much as 99% of its flow from groundwater discharge during periods of low precipitation [2]. The USGS found a substantial groundwater flow system consisting of eight aquifers in the Upper Klamath Basin, mainly composed of late Tertiary to Quaternary volcanic rocks with high transmissivity and primarily recharged by snowmelt [9]. The USGS and the United States Bureau of Reclamation developed a model using the Precipitation–Runoff Modeling System (PRMS) to forecast streamflow in the Upper Klamath Basin [10]. PRMS is a physics-based deterministic model that simulates processes such as streamflow, evapotranspiration, snow, and groundwater processes [11]. While the model has been calibrated to historical data and gives reasonably good results, it may become less reliable as the region experiences unprecedented droughts [10]. Chang and Jones [12] predict that climate change is expected to have a dramatic impact on water resources in the 21st century in Oregon, particularly in basins that are highly dependent on groundwater. Carroll et al. [13] show that declining groundwater storage will reduce streamflow in a warming climate. Mayer and Naman [14] examined 25 basins in the region and showed that warming temperatures and snowpack reductions have altered seasonal streamflow characteristics and have resulted in sometimes dramatic reductions in baseflow, especially in groundwater-dependent basins like the Williamson Basin where long-term groundwater depletion could result in periods where even higher than normal precipitation does not result in flow. Hess and Stonewall [15] examined the drought of 2013 that caused a near-complete cessation of streamflow in the basin and found that changes in regulated diversions mitigated the impact of the drought. Tague et al. [16] examined streamflow responses to climate change in the Oregon Cascades and found reductions in summer streamflow with increasing temperatures.

The interplay between groundwater storage and baseflow in streams has long been a focus of research [17–22]. Most streamflow forecasting is focused on short-term (7–28 day) forecasts where the emphasis is typically on predicting peak flows in response to precipitation events [23–29]. For longer-term processes, groundwater models can simulate groundwater–surface water interaction in aquifers where both groundwater and surface water are well characterized [30–32]. Relationships between streamflow at specific gages and groundwater levels at nearby monitoring wells can be insightful, but a more holistic approach is generally required for basin-scale analysis, which is driven by total aquifer storage changes rather than local water levels. Some researchers have examined the correlation between long-term streamflow and overall groundwater storage in a basin [13,33,34], and while this process is promising, it has been hindered by scarcity of groundwater data and large gaps in the time history of the wells.

One approach for assessing groundwater storage in a basin is to use a water balance model such as the PRMS model described above. However, PRMS uses the following water balance equation:

$$P = ET + Q + \Delta S$$

where P = precipitation, Q = runoff, and ΔS = change in storage (soil moisture + groundwater). One can solve this equation for groundwater storage, but that number is simply the groundwater recharge potential. To find the actual groundwater storage change, one would have to know how much groundwater is being withdrawn due to pumping, runoff, and ET volumes. These data are not available for this basin. While runoff and ET could potentially be estimated, there is limited to no data on groundwater withdrawals available. Thus, a PRMS analysis of groundwater with no accounting for withdrawals would overestimate the groundwater storage in the basin.

Another approach for characterizing groundwater storage history in a basin, that does not require well data, is using Earth observations such as data from the NASA Gravity Recovery and Climate Experiment (GRACE) mission [35–37]. GRACE data can be combined with simulated terrestrial water datasets from the Global Land Data Assimilation System (GLDAS) to derive groundwater storage change estimates [38–47]. While this method does not require in situ data, it only works well for large basins, as the resulting groundwater storage anomaly grid resolution is at a scale of 1×1 degree making it too coarse for use with smaller basins such as the Williamson Basin.

To avoid the issues listed above, researchers at Brigham Young University have developed computational tools such as the Groundwater Data Mapper (GWDM) for assessing groundwater storage in difficult watersheds like the Williamson, which has limited in situ measurements. The GWDM tools consist of a web application and a set of Python scripts that combine in situ water level data with GLDAS soil moisture data using a machine learning algorithm to impute missing well data [48,49]. The imputed well data are then subsequently used in a spatio-temporal interpolation process to generate time-varying water level rasters representing the groundwater surface elevation at selected timesteps and ultimately a time series of estimated groundwater storage change for a basin or aquifer [50,51]. Because of the size of this watershed, the scarce well data, and a broken rainfall–runoff relationship, we selected the GWDM to characterize the groundwater storage in the Williamson Basin.

The GWDM tools are unique since they use machine learning algorithms in concert with GLDAS soil moisture anomalies to impute missing well data and then use the composite well data to generate an aquifer storage curve. By comparison, current applications of machine learning to solve groundwater issues are still somewhat limited in their temporal scale and are generally aimed at short-term predictive modeling, rather than generating accurate time histories of storage change over long periods. Tao et al. [52] noted that less than 1% of the reviewed research has applied groundwater data on a yearly timestep, critical for prediction and management purposes, with most work focusing on predicting groundwater levels on a daily to monthly timestep. Our aim is to address the research gap noted by Tao et al. [52]. In the Williamson Basin, in spite of sparse well locations in gaps in historical measurements, we can calculate aquifer storage as far back as the late 1960s. Looking back in time, we can better understand the current state of groundwater in this high desert basin and its relationship to precipitation and streamflow.

The GWDM tools do not use machine learning to develop a hydrologic model of the flow dynamics in the basin. Rather, this tool is used to impute gaps in water level records for each well to enable a more accurate groundwater storage change estimate. This algorithm was recently used in the Central Valley in California to assess long-term groundwater storage change [53]. This Central Valley study imputed water level observations for hundreds of wells distributed throughout the valley going back to the 1940s and generated groundwater storage change over time, depicting groundwater depletion in the valley. This depletion curve matched the shape of the depletion curve found using groundwater storage change estimates from the GRACE mission, but it does not suffer from the "leakage effect" that is common with GRACE data due to the coarse grid spacing of the GRACE data [39,54–56].

Our research objective is to employ the GWDM tools described above to analyze available data and characterize the precipitation–groundwater–streamflow process and interactions. This technique combines in situ water level measurements with Earth observations to impute gaps in the water level data using a machine learning algorithm. Using these complete well records, we then interpolated these data over the basin at selected timesteps. We used the difference in these surfaces to compute a volumetric groundwater storage time history for the basin. This storage history allows us to analyze relationships between precipitation, groundwater, and streamflow, a capability previously not possible in the basin as the required data were not available. We use this analysis to more precisely understand the conditions leading to the cessation of live streamflow and predict when streamflow could be expected to return based on forecast precipitation patterns. This detailed understanding of the interrelated physical processes provides resource managers with a better characterization of the basin hydrology basin and provides new insights and tools to support decision-making. The larger objective of this study is to demonstrate, through this case study, the use of new techniques and methods that can provide resource managers and researchers with the tools to better understand processes and issues in areas where streamflow is governed by groundwater interactions.

2. Details of the Field Site

2.1. Lithology of the Watershed

The Williamson Basin is approximately 3720 km^2 (1436 mi^2) in size and is located on the eastern edge of the Cascade volcanic range. The northern, eastern, and southern boundaries are enclosed by tertiary volcanics and the fault-block geology of the basin and range province. Mount Mazama is the high point of the basin at 2486 m (8157 ft), which then slopes downward to an elevation of 1245 m (4090 ft) at the Klamath Marsh. A simplified geology of the basin subdivides the lithology into three major geological units: (1) the surficial pumice blanket; (2) intertwined beds of sand, clay, gravel, pyroclastics, and basalt (Yonna formation); and (3) the underlying andesitic/basaltic bedrock, which is the main water-bearing unit [2,4,5,7,8,57]. On the surface, pyroclastic fall pumice, ash, and other volcaniclastics from Mt. Mazama blanket the basin from the Crater Lake down to the Klamath Marsh. Newcomb and Hart [57] labeled this the "Mazama Pumice Plain" and documented its highly porous nature with infiltration rates on the order of 10^{-4} m/s (3.28×10^{-4} ft/s) [7]. Internally, the bedrock aquifer has interbedded vesicular scoriaceous tuff, is highly fractured, and displays columnal jointing. Aquifer pumping tests have a high transmissivity with average values of approximately 15,800 m^2/day (170,000 ft^2/day) [6].

Historically, the groundwater levels have fluctuated in response to climatic cycles, ranging from years to decades, with the groundwater levels and springs responding to wet and dry periods. Areas with groundwater level changes of 1.5 m (5 ft) have been noted by researchers as common, with fluctuations up to 3.6 m (12 ft) observed in regions closer to the Cascades [9].

2.2. Basin Precipitation

The Cascade Mountains (including Mount Mazama) receive upward of 150 cm (60 in) of precipitation per year, mostly as snow. The central and eastern portions of the range, which lie in the rain shadow, receive as little as 45 cm (18 in) of precipitation per year [2,5]. Temperatures range from highs of 40° C to lows of −31° C. The revised Köppen–Geiger classification identifies the Williamson Basin as warm, dry summer continental, and the high elevation rim is classified as being dry, sub-arctic [58].

Snowfall and snowmelt recharge the groundwater system through the highly porous desert soils and alluvium along the high elevations of the Cascade Range. Previous research estimates that this spongelike geology can absorb and move upwards of 3.7×10^7 m^3 (30,000 acre-ft) of water into the aquifer [4,57]. Because of this, many named and unnamed tributaries start in the high mountains and disappear into the sub-surface, leaving the area absent of typical drainage morphology. Maps show that the Sand, Scott, Miller, and Cottonwood Creek all disappear into the alluvium before reaching confluence with the Williamson River.

Because the Klamath Marsh acts as both a terminus for surface flow and acts as the groundwater-driven headwaters of the Williamson River, rainfall/runoff models of this watershed are typically inaccurate, as streamflow is not directly driven by precipitation. We performed an analysis of the 12-month cumulative precipitation volume to gage the flows and confirmed that a correlation exists between precipitation and flow on a year-to-year basis but is weak (Figure 3). Streamflow generally declines following dry periods and increases during wet periods, although the full impact of the precipitation on the streamflow is delayed by several years. This delay between precipitation and streamflow can be explained by interactions with groundwater storage, which we illustrate later in this paper.

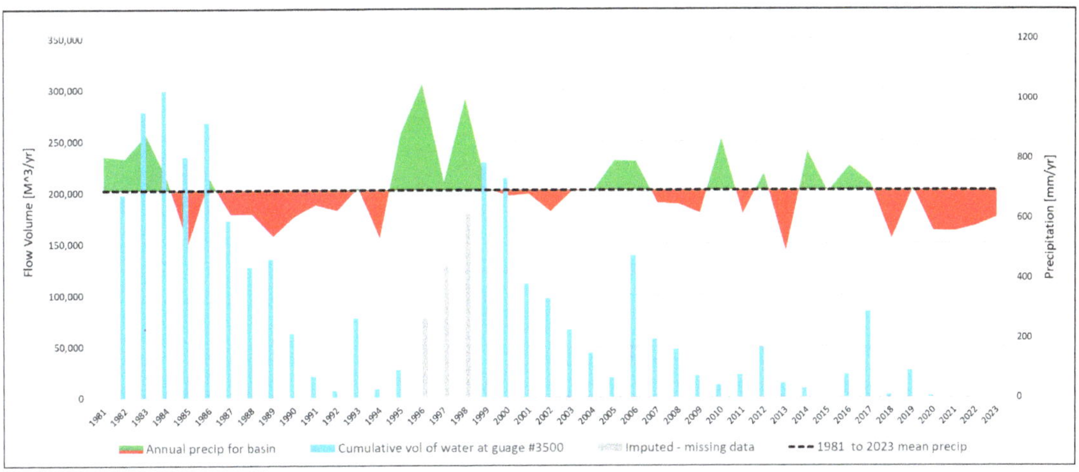

Figure 3. Cumulative flow at gage #3500 vs. cumulative precipitation by year, 1981–2023. Green values indicate above-average annual cumulative precipitation, and red values indicate below-average. Flow data for the years 1996–1998 are absent from the record due to deactivation of the gage; therefore, we show linearly imputed values as gray bars.

Uncoupled from a conventional rain–runoff relationship, the true headwaters of the Williamson River are at the Klamath Marsh. Isotropic analysis performed by Melady (2002) and Cummings (2007) confirmed distinct differences in the deuterium and oxygen isotope ratios in the surface and groundwater, along with distinguishable concentrations of magnesium, calcium, potassium, and sodium that demonstrate that the streamflow is from groundwater sources [5,7].

2.3. Drought Impacts in the Basin

Annual precipitation and streamflow data for the basin show two multi-year drought periods since 1981 (Figure 3). The first drought was from 1990 to 1994, culminating in extremely low streamflow. The second drought was longer, beginning in 2020 and continuing to the winter of 2023.

While the first drought was shorter, there were three drier than normal years prior to the 1990–1994 drought, which should be included as they affect groundwater storage. This first drought period culminated in extremely low flows that were temporarily relieved by a wetter than normal winter in 1993. On average, observed flow volumes during this first drought period ranged from between 29 and 75 percent lower than flows in the previous decade. Unfortunately, the gage was deactivated in October 1995 and subsequently reactivated in September 1998. After the gage was reactivated, the water volumes in 2000 and 2001 indicate that the aquifer is capable of being recharged and that the storage capacity of this system is not impaired by such impacts.

The second drought, from 2020 to 2023, was the result of similar low precipitation volumes. Figure 3 shows that yearly precipitation volume has been declining since about 1998, with 2006, 2007, 2011, 2013, 2015, and 2017 being well above average, but most years show a trend toward a declining average. The remaining years since 1998 are below average, with 2014 and post-2018 extremely low. This long-term decline in precipitation, starting in about 2008, has depleted the watershed entirely, with 8 April 2021 being the last date any surface water was measured leaving this watershed.

3. Datasets Used in the Study

3.1. Groundwater Level Monitoring Well Data

Both the USGS and the Oregon Water Resource Department (OWRD) maintain monitoring wells in the basin study area. We used 12 active monitoring wells (4 USGS, 8 OWRD) and 11 inactive wells (all OWRD), with various periods of activity since the 1960s. The depth of our sample wells averages 280 ft, with a range between 75 ft and 670 ft. Of the 15 wells with accessible logs, 13 are in the Klamath volcanic area, a locally confined highly porous aquifer [2]. The remaining two are in the Yonna formation (see Section 2.1). When active, more recent measurements have a resolution of 0.3 cm (0.01 ft), but older measurements are sometimes recorded with a coarser resolution of 3–30 cm (0.1–1.0 ft). Figure 4 shows the monitoring well locations used in this study. Well records prior to 1968 did not include enough data for the GWDM tools due to both the low number of data points and a low data resolution. After removing 2 wells with insufficient data, the dataset included 21 wells with 1312 water level records. We examined the water level data and eliminated four measurements that appeared to be outliers resulting either from recording errors or faulty depth measurements. The final dataset included 1308 measurements that we used in the analysis. The full dataset including well locations and the water level records is available in the HydroShare.org data repository [59] described in the Supplementary Materials section below.

3.2. Streamflow Data

Flow in the Williamson River is measured at the US Geological Survey (USGS) streamflow gage number 11493500 (gage #3500), located at approximately 42.740° N, −121.834° W; it is 0.5 km (0.31 mi) south of the topographic high known as Kirk Reef (Figure 2, area C). Installed in 1954, the gage records discharge on a daily timestep in cubic feet/s (CFS). We used these data to compute cumulative annual flow volumes as thousands of acre-feet (TAF) per year, which is the measurement commonly used by irrigators and water managers in the western United States. One acre-ft is approximately 1233.5 m^3, and one acre-ft per day is approximately 0.504 CFS (0.0143 m^3/s).

Figure 5 shows the cumulative annual flow per year in TAF (Figure 5a) and the number of days per year of non-zero flow at gage #3500 (Figure 5b). The gage was deactivated on

October 1995 and reactivated on September 1998, creating a 3-year gap in the data. For our analysis, we imputed streamflow values for this 3-year gap using linear interpolation to ensure that the missing values are not interpreted as zero flow in our analysis, since actual recorded zero flow values do occur in the record.

3.3. GLDAS Soil Moisture Data

To impute missing groundwater measurements at a selected well, the GWDM tools use the measured data at the well combined with the soil moisture dataset from the NASA Global Land Data Assimilation System (GLDAS), which is a global gridded dataset [60]. GLDAS consists of three separate land surface models: LSM, Noah, and VIC. We used the monthly averaged results from the Noah model, which has a resolution of 0.25 × 0.25 degrees. We downloaded these data from the Goddard Earth Sciences Data and Information Services Center (GES DISC) website [61].

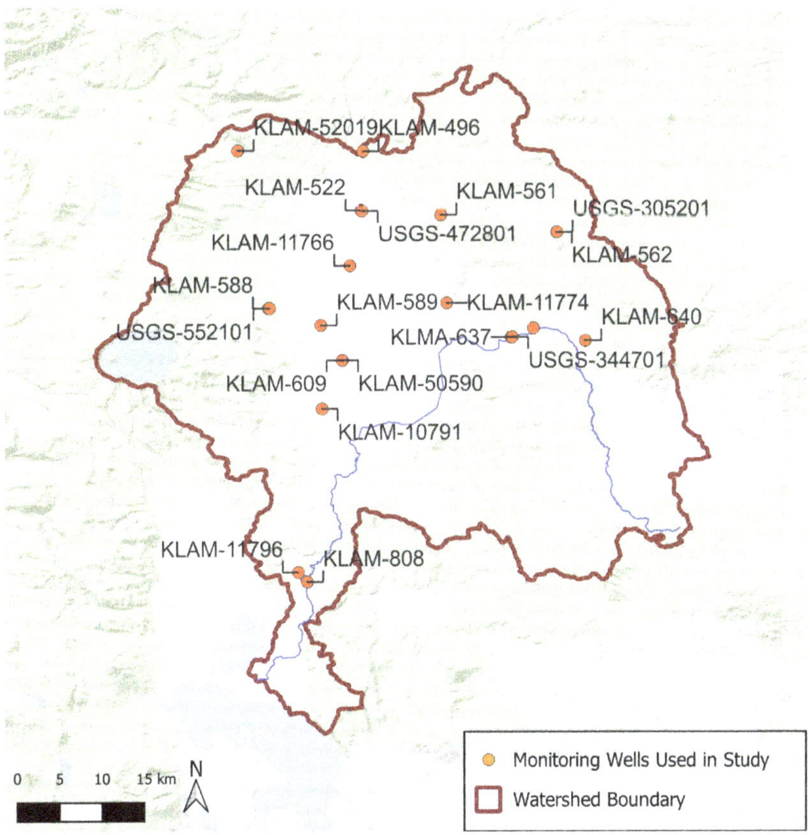

Figure 4. Locations of the various monitoring wells used in this study.

3.4. Precipitation Data

We used precipitation data from the CHIRPS gridded data product developed by the Climate Hazards Center at UC Santa Barbara in collaboration with the USGS Earth Resources Observation and Science (EROS) Center [62]. The CHIRPS dataset is derived from a variety of NASA and NOAA Earth observation sources and aggregated to a high-resolution global grid (0.05°) covering a period from 1981 to the present. We obtained CHIRPS data for the Williamson Basin, from the NASA SERVER ClimateSERV website [63].

To download the data, we first uploaded a GEOJSON file, which defined the boundaries of the basin. The website then provided a time series of daily rainfall data consisting of the daily average precipitation of the CHIRPS grid cells in the interior of the basin. We elected to use the CHIRPS dataset over data from individual rain/snow gages in the basin because our objective was to perform a groundwater storage analysis in a manner that aggregates both in situ data and Earth observations of the entire basin. The CHIRPS data represent a basin-scale, gridded dataset that is consistent with our basin-scale groundwater analysis.

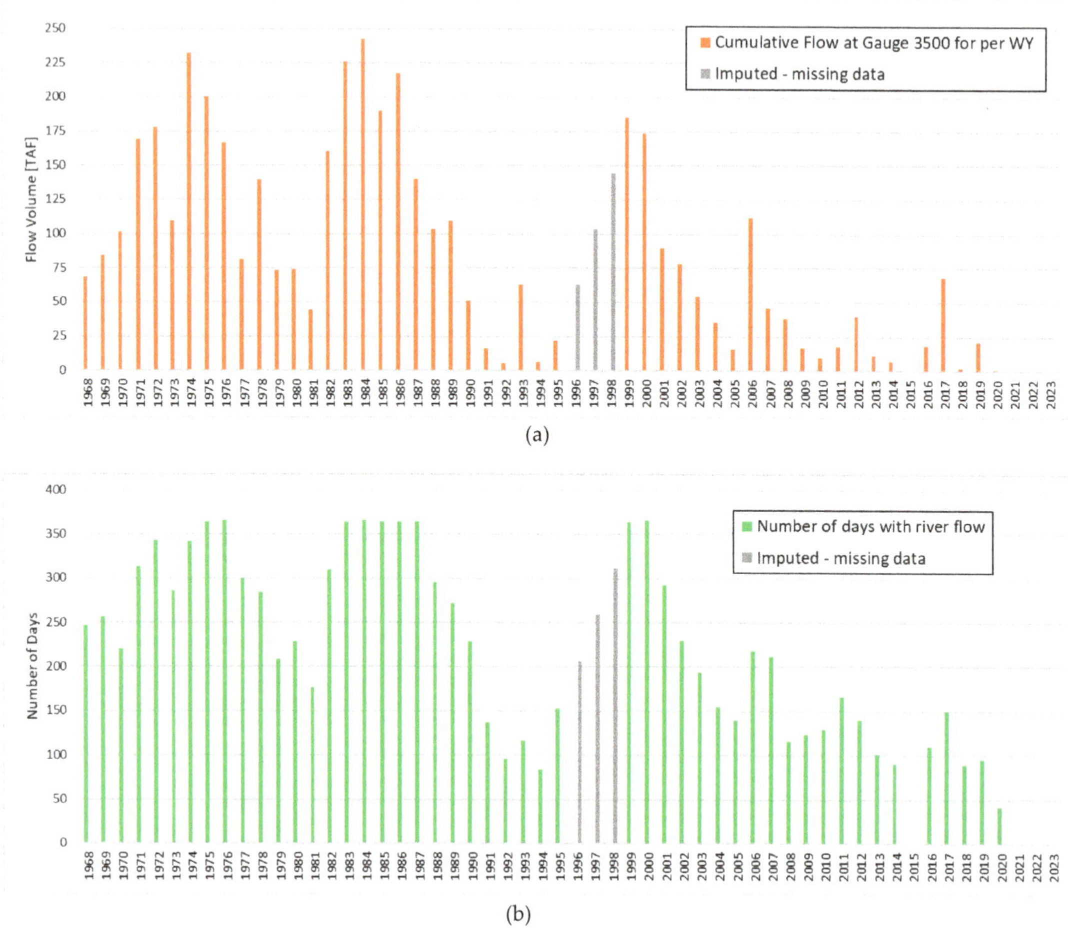

Figure 5. (**a**) Total annual flow volume and (**b**) number of days with measurable flow at gage #3500 per year from 1968 to 2023. Missing data were imputed using linear interpolation and are shown in gray.

4. Interpolated Groundwater Level Data Using GWDM

We generated an estimate of the historical groundwater storage changes in the Williamson Basin using the tools available within the GWDM platform [49]. GWDM uses a combination of in situ water level data and Earth observations in a multi-step process to estimate aquifer storage changes [48,50,51]. In the first step, the user imports well locations and historical water levels that define a water level time series for each well. However, it is common to have significant gaps in historical water level records at a well, ranging from several months to several years. Since computing the aquifer storage change requires a relatively complete water level record at each well, GWDM uses a machine learning algorithm to impute missing data for each well over the study period. The theoretical basis for this algorithm is that groundwater levels are highly correlated with drought indicators such as soil moisture; see Figure 6. During a drought, there is less groundwater recharge, and groundwater is pumped at a higher rate to meet agricultural, municipal, and industrial demands. During wet periods, there is a higher rate of recharge and less pumping. As a result, groundwater levels drop during drought periods and rise during wet periods. The machine learning algorithm learns these relationships, often non-linear, and uses them to impute missing values. The complete well records are then sampled at regular intervals and spatially interpolated to a set of time-varying rasters that are used to compute groundwater storage change vs. time. The entire process is described in more detail in [49–51,64]. GWDM uses an Extreme Learning Machine (ELM) model, which performs a linear analytical fit on all the available data; thus, there is no testing or training split. The model features include GLDAS results such as soil moisture, the flux and lags of soil moisture anomalies, and smoothed soil moisture averages over a 10-year period. As the GWMD model, features, and accuracy assessment methods are well defined in the literature and are published as an open-source tool, we do not repeat them here [49–51,64].

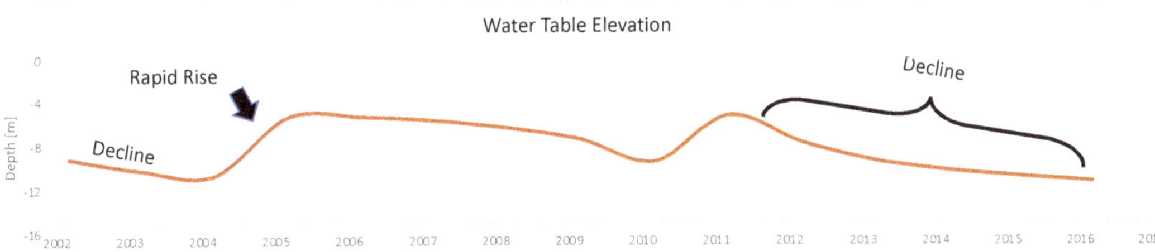

(a)

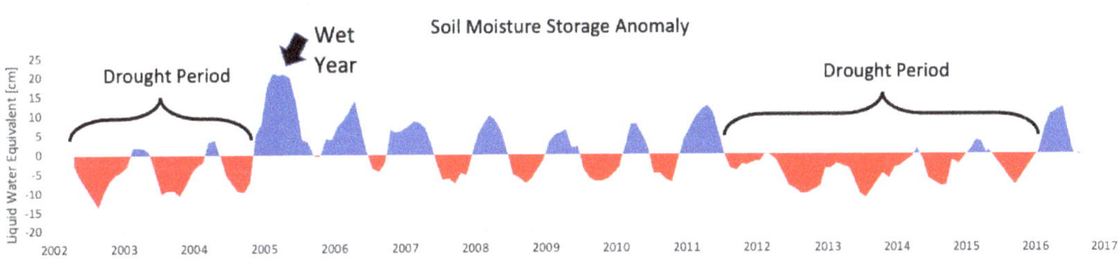

(b)

Figure 6. Correlation between groundwater levels and GLDAS soil moisture for a monitoring well near Beryl, UT, USA. (**a**) Water table elevation from 2002 to 2017. (**b**) GLDAS soil moisture anomaly for the same period (blue = positive anomaly, red = negative anomaly). Adapted from Evans et al. [50].

After the initial imputation, GWDM refines the imputed data using correlations among the existing well data and imputed results. This reduces uncertainty and provides a check on the imputed data. The GWDM tools include a set of Python scripts for cleaning and formatting data and for performing the imputation and volumetric analysis [49]. The wells, water levels time series, interpolated water level rasters, and groundwater storage time series can be uploaded and visualized in a web application [48,65].

To apply this process to the Williamson Basin, we used the well locations and associated water level measurements and created a water level time series for each well. Next, we used piecewise cubic hermite interpolating polynomial (PCHIP) interpolation to impute data for all gaps in the monthly water level records for each well. We retained these imputed values for gaps less than two years. We then deleted the PCHIP-interpolated values for gaps greater than or equal to two years while retaining 180 days of PCHIP-imputed data at the beginning and end of each of these gaps. At the end of this step, we have imputed the small gaps in the water level time series at each well, but the large gaps remain, with imputed data at the beginning and the end within 180 days of measured values.

Next, we used the GWDM tools to build a separate machine learning model for each well using the Extreme Learning Machine (ELM) algorithm, which directly solves the linear optimization problem without iterative training [48]. The model features for each well are the monthly GLDAS soil moisture anomalies corresponding to the centroid of the basin and the three-, five-, and ten-year moving average of the soil moisture anomaly for the period from 1968 to 2023. The moving averages capture longer-term trends in the data. Additional model features are the year and one-hot encoded month indices to capture seasonality and non-stationary trends in the data. We fit a separate ELM model for each well using the observed water levels as the dependent variable.

We used the ELM models to impute missing data to generate a complete water level record for each well from 1968 to 2023. Finally, we combined the measured, PCHIP, and EML datasets into a single water level dataset where the priority was the measured data, then the PCHIP data (short gaps), and then the ELM-imputed data for the longer gaps. In other words, for all months where there was either a measurement or PCHIP-interpolated value, we used that value. If no measurement or PCHIP value was available, we used the EML-imputed data value.

We sampled the time series at each well on yearly intervals from 1968 to 2023. We spatially interpolated these data to a 150 × 150 grid, which we then clipped to the basin boundaries, resulting in a yearly groundwater elevation map for the basin over the study period. We calculated the volume between each subsequent pair of groundwater elevation rasters and then multiplied the value, which represents the water level change, by a representative unconfined storage coefficient for the basin (0.1) to generate annual groundwater volume changes; then, we integrated these values to generate a groundwater storage curve for the basin.

5. Results

5.1. Correlation Between Precipitation and Flow Data

Before exploring the relationship between groundwater and streamflow in the basin, we first examined the relationship between precipitation and streamflow. We interpolated the annual cumulative data to estimate values at the start of each month. Cumulative annual flow data for three years, 1996–1998, were not available because the gage was deactivated during this period, so data in this period are all linearly interpolated from the values on each side of the gap. We normalized both datasets using standard score (z-score) normalization and found that a weak correlation exists between precipitation and flow on a year-to-year basis (Figure 7a). The Pearson correlation coefficient (r) of annual cumulative rainfall to annual cumulative flow by year is $r(n = 41) = 0.29$, $p = 0.06$, which shows that the two series are only slightly correlated. Nevertheless, streamflow generally declines following dry periods and increases during wet periods, but the full impact of the precipitation on the streamflow is delayed by several years.

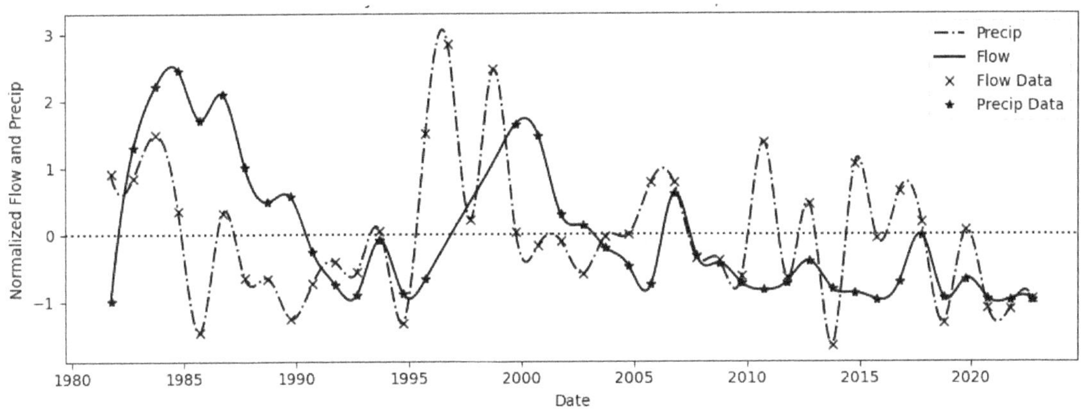

(a)

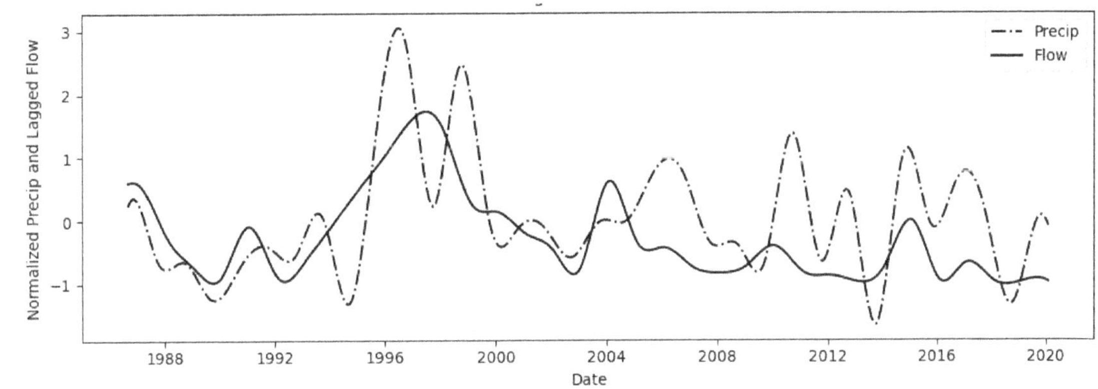

(b)

Figure 7. (**a**) Cumulative annual precipitation and flow data normalized and interpolated to monthly values using standard score or z-score normalization to plot both datasets on the same scale. Data were normalized to the study period, not the long-term average and variance. (**b**) Cumulative annual precipitation compared to cumulative annual flow with a lag of 32 months. The period from 1996 to 1998 has no annual flow data; data in this period were interpolated.

To better characterize the correlation between cumulative annual precipitation, we computed the Pearson correlation coefficient using monthly lags from −60 to +60 lags (±5 years) between the two series. We found that the highest correlation occurs at a lag of −32 months, with $r = 0.525$, $p < 0.001$ (Figure 8). The computed p-value is essentially zero but is based on the interpolated monthly values, so it is not representative of the data. Our analysis showed that there is not a sharp peak in the correlation values, but a relatively smooth transition. This indicates that the lag that produces the highest correlation probably changed over the 43-year study period. We would expect a longer lag at low groundwater levels and a shorter lag at high groundwater levels. Figure 7b shows cumulative annual precipitation data plotted with cumulative annual flow data lagged by 32 months. In general, these two datasets are visually correlated. The no-data period, from 1996 to 1998, is not as visually correlated, as the interpolated data do not follow precipitation.

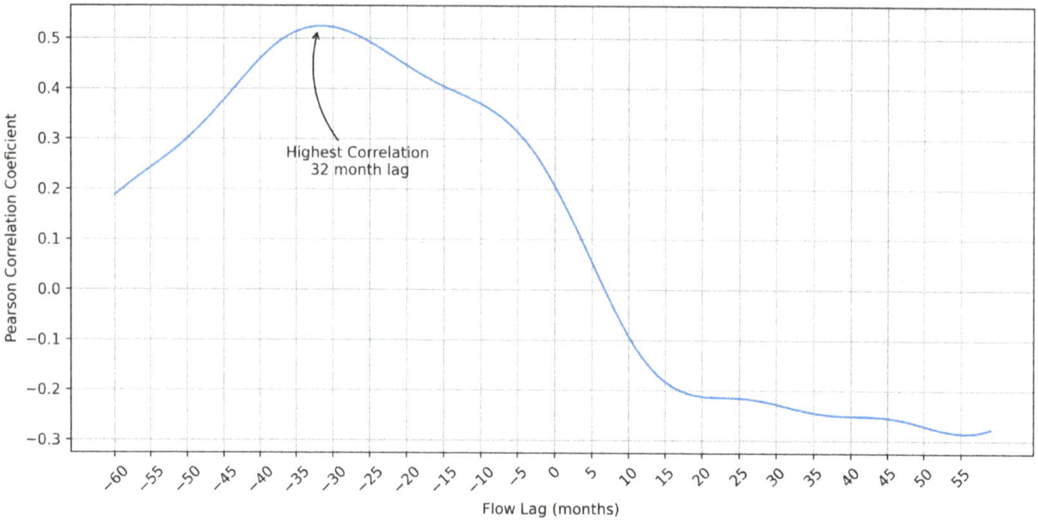

Figure 8. Lagged correlation of cumulative annual precipitation and cumulative annual stream with lags from −60 to +60 months. The highest correlation, $r = 0.525$, occurred at lag = −32 months.

5.2. Groundwater Storage Change

To better understand the correlation between groundwater and streamflow, we derived the groundwater storage change using the GWDM tools. The monthly GLDAS soil moisture (soilw) dataset for the GLDAS grid cell containing the centroid of the Williamson Basin and the 3, 5, and 10 yr moving averages of the soil moisture are shown in Figure 9.

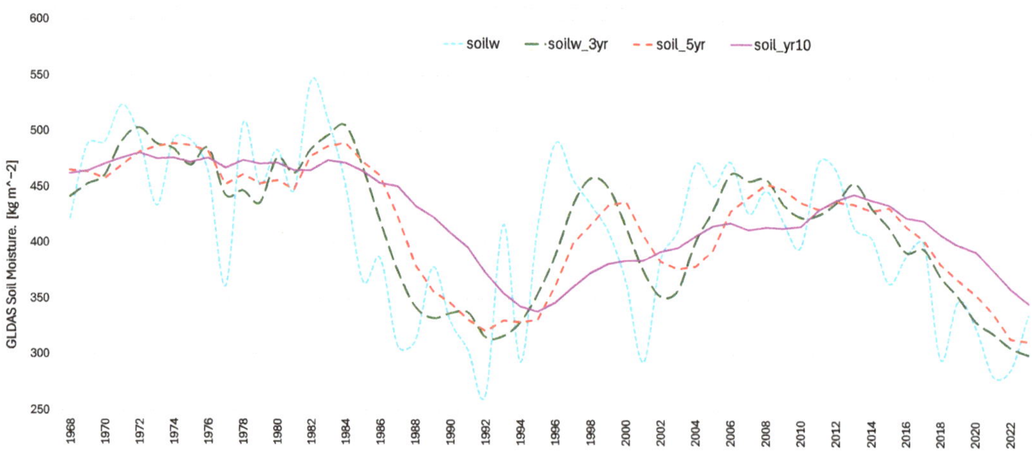

Figure 9. GLDAS soil moisture datasets for the Williamson Basin.

Using PCHIP interpolation and the ELM models for each well, we generated a complete set of water levels for each well. Figure 10 shows an example set of measured and imputed water level time series data for well KLAM0000496 in the Williamson Basin. The "combined" trace in the figure represents the data we used for this study.

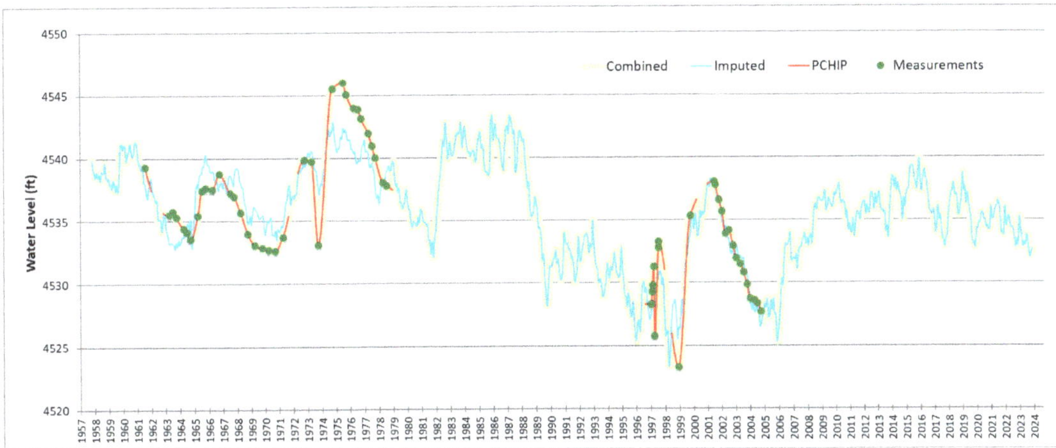

Figure 10. An example of actual water level measurements, short-term PCHIP-interpolated values, and ELM-imputed values to create a combined dataset for the KLAM0000496 well in the Williamson Basin. Each well in the dataset was put through the same analysis to create PCHIP-interpolated values and ELM imputed values to generate a full time series. Well KLAM0000496 is the Klamath volcanic aquifer and showcases high fluctuations in water levels. Fluctuations up to 3.6 m (12 ft) are common in wells near the Cascade Range [9].

We used the generated water level time series, for each well, to generate a groundwater storage curve for the basin as shown in Figure 11. The volumetric storage change values are relative to the starting value of zero in 1968. Figure 11 shows that groundwater storage was relatively constant from 1968 through 1987 and then dropped significantly for 7 years before partially rebounding from 1994 to 2000. This rebound was followed by a decline from 2000 to 2005 and then a steeper decline starting in 2015 that has continued through the end of 2023. These data show that the basin has lost approximately 2000 million m^3 (1600 TAF) over the 55-year study period. Soil moisture shows a similar decline (Figure 9), indicating both processes are driven by climatic conditions.

Figure 11. Net groundwater storage change for the Williamson Basin from 1968 to 2023 with 1968 as the datum; green indicates values above the datum and red below. We hypothesize that groundwater pumping is the reason for the consistently lower volume of water in the aquifer when compared to the 1970s/80s. Pumping prevents the aquifer from fully recharging. Groundwater pumping began in the 2000s in response to the previous dry years along with mandates to reduce surface water usage [6,66]. However, the volume of water removed cannot be quantified at this time.

5.3. Groundwater–Streamflow Correlation

Figure 12a compares the groundwater storage curve to the cumulative streamflow data at gage 3500. Visually, there is a strong correlation between the decline in groundwater storage and declining streamflow from 1985 to 1995, followed by a rebound in both 1985 and 2000. After the rebound, both streamflow and storage declined through 2005, which was followed by a steeper decline in groundwater storage starting in 2015. This decline in groundwater storage is followed by a complete cessation of streamflow in 2021. The Pearson correlation coefficient between the groundwater storage and streamflow is $r(\text{n} = 54) = 0.64$, $p < 0.01$, relatively well correlated.

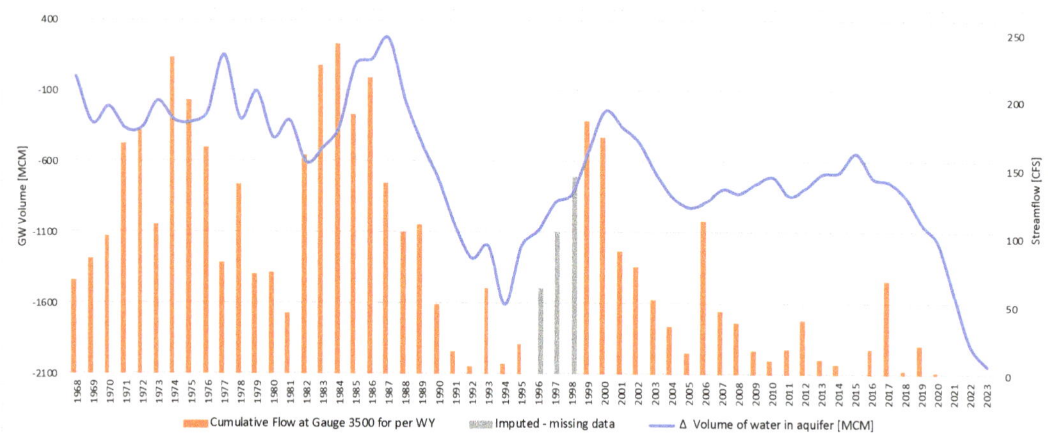

(**a**)

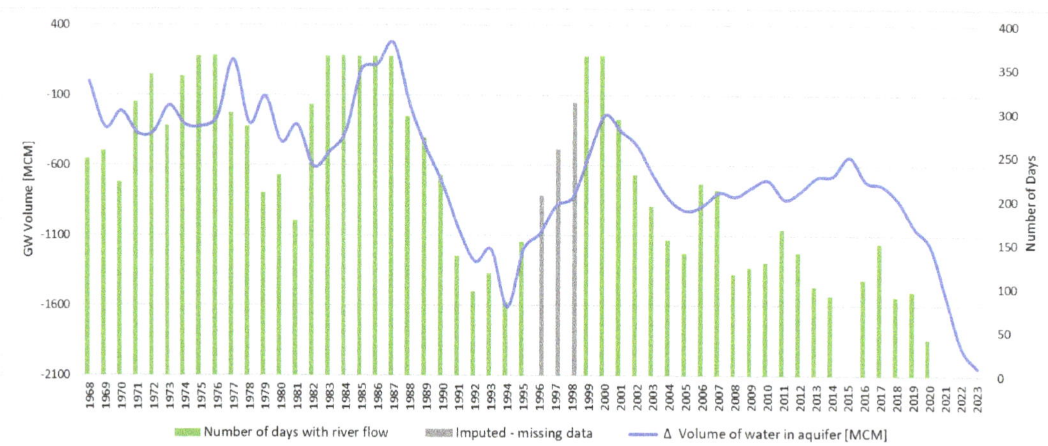

(**b**)

Figure 12. Groundwater storage change in the Williamson Basin vs. (**a**) measured annual volume of water at USGS gage #3500 and (**b**) number of days with measurable flow at gage #3500 per year from 1968 to 2023. Gray bars indicated interpolated data.

Figure 12b shows the groundwater storage and the days per year with non-zero streamflow. The same overall behavior is observed, but groundwater storage is even more strongly correlated with flow days, resulting in a Pearson correlation coefficient of $r(\text{n} = 54) = 0.79, p < 0.01$.

5.4. Groundwater Storage Change and Rainfall Correlation

Finally, we examined the relationship between groundwater storage change and precipitation. We started the comparison in 1981 as opposed to 1968, as the CHIRPS data were not available prior to 1981. Figure 13a shows groundwater storage change versus cumulative annual precipitation along with the 3- and 5-year moving average of the cumulative precipitation. The figure shows that long-term changes in groundwater storage mirror fluctuations in precipitation, particularly the 3 and 5 yr moving averages, but the peaks and valleys are offset. We evaluated different lags and found that, if we lag groundwater storage by 3 years, we achieved a high degree of correlation with both the 3 and 5 yr precipitation curves as shown in Figure 13b. This matches the lag we found between precipitation and streamflow in Section 5.1.

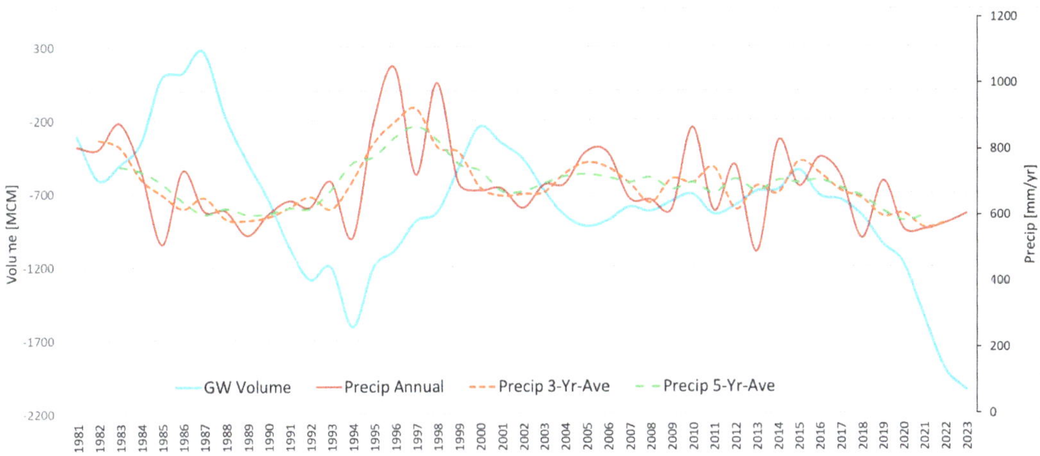

(a)

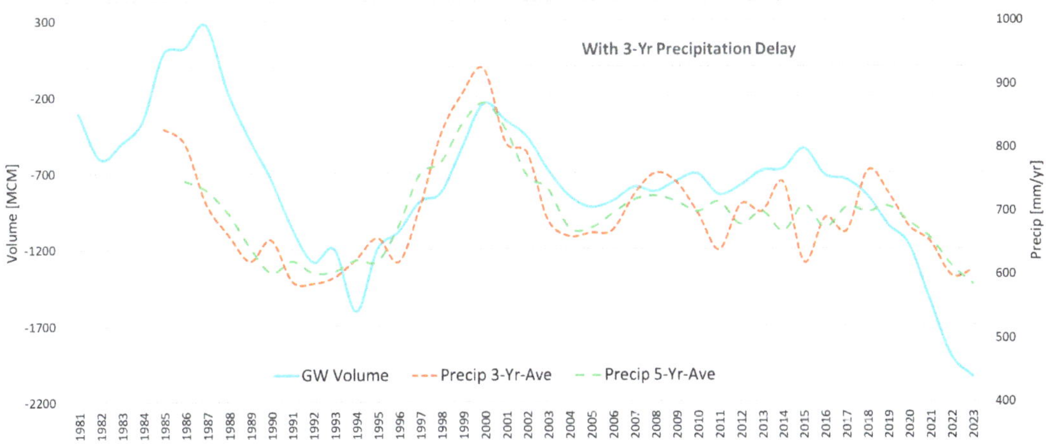

(b)

Figure 13. Groundwater storage change in the Williamson Basin from 1981 to 2023 vs. (**a**) cumulative annual precipitation and 3 and 5 yr moving average annual precipitation and (**b**) 3 and 5 yr moving average annual precipitation with a 3 yr lagged groundwater storage change.

The Pearson correlation coefficient for lagged groundwater storage and 3 yr and 5 yr moving average precipitation are $r(n = 37) = 0.60$, $p < 0.01$, and $r(n = 36) = 0.64$, $p < 0.01$, respectively. This indicates that it takes approximately 3 years for precipitation to fully percolate through the vadose zones and flow into the basin to effectively recharge the groundwater in the basin aquifers. Even though the precipitation in the basin started to rebound in 2021, as shown by the solid red line in Figure 13a, the groundwater still has not started to rebound, as it is most closely correlated with the 3 and 5 yr moving average of precipitation with a 3 yr delay. Once groundwater in the basin has rebounded to a high enough level to feed the seeps and springs at the head of the Williamson River, we expect the flow to resume.

6. Discussion

Due to the geology of the Williamson Basin, river flow is uncoupled from a conventional rain–runoff relationship with little short-term relationship between precipitation and flow. Studies have shown that the headwaters of the Williamson River are at the Klamath Marsh and are related to groundwater rather than the tributaries that feed the marsh. Isotopic data have shown distinct differences in deuterium and oxygen isotope ratios in the surface and groundwater, along with distinguishable concentrations of magnesium, calcium, potassium, and sodium, supporting this conclusion [1,6]. These studies, combined with the lag between precipitation and streamflow demonstrated in Section 5.1, indicate that flow in the Williamson River is governed by groundwater. Lagged correlation with precipitation, streamflow, and groundwater storage indicates that precipitation in this basin recharges groundwater over about a 3-year time period. However, while precipitation and recharge are coupled and demonstrate a correlation in time, the correlation between groundwater and streamflow may be more complex. A simple analogy would be a bathtub; enough water must fill the tub before it can overflow, and in a similar manner, we propose that groundwater underneath the National Wildlife Refuge needs to reach a certain volume before it can discharge through the various springs and seeps. Groundwater levels below this level result in low-flow or no-flow conditions from this basin.

The general relationship between the groundwater storage volume and Williamson River flow has traditionally been known to hydrogeologists and those with local knowledge, but a detailed characterization of basin processes with a method to estimate long-term changes in groundwater storage was previously unavailable. This is because examining water levels at individual groundwater monitoring well locations does not capture the full behavior of the groundwater–surface water interaction, as individual wells may not provide insight into aquifer storage due to incomplete records and local behavior at a well.

The tools we used in this study provide methods to address these issues. The imputation of groundwater level data for each well using correlations between well data and Earth observations addresses sparse data at the wells. The spatial interpolation and mapping algorithm uses these individual well data and computes a time series of aquifer level maps that can be integrated to provide aquifer volume changes. Computing these storage changes and integrating them over time provides the basis for the analysis we presented in this paper. These tools provide insights that allow us to better characterize the relationship between precipitation, groundwater, and streamflow in the Klamath Basin and the effects of climate change on this area. This is a novel approach, as the majority of groundwater analysis tools do not compute groundwater changes on a yearly timestep [52]. Our tools allow us to use the available, but gap-filled, groundwater level data in the Klamath Basin along with precipitation records to characterize the precipitation–groundwater–streamflow interactions in a manner that was previously not possible. This hydrologic system description and understanding allows managers and others to evaluate and gage the impact of droughts and anthropogenic actions on the Klamath Basin.

Our analysis confirms that the overall volume of water within the aquifer is strongly correlated with streamflow leaving the watershed. As illustrated by the drought in the 1990s and the current drought, once the groundwater storage drops approximately 1800 MCM (1500 TAF) below the 1968 baseline, streamflow in the basin ceases completely. Due to the multi-year delay between precipitation and groundwater storage, even though precipitation has rebounded in the last three years, the groundwater has not yet responded, as the groundwater storage has not reached the critical level. Our analysis suggests that it should rebound soon. If precipitation continues at recent rates, we should expect streamflow to return in the next 3–5 years.

More concerning is the trend seen in the aquifer volume since 2001. Combined with the continuous decline in streamflow shown in Figure 12a, this indicates an overall drop in the total volume of groundwater within the basin. This trend is possibly anthropogenic. It coincides with actions taken by various federal agencies as part of a biological consultation. As a result of the National Marine Fisheries Service (NMFS)'s 2002 biological opinion on the endangered Chinook Salomon (*Oncorhynchus tshawytscha*), the Bureau of Reclamation reduced demand and stored extra water for the endangered species. The goal was to create a "water bank and water supply to provide flows to the Klamath River below Iron Gate Dam to improve coho salmon habitat" [67]. Designed to save water for habitat flows, up to 100 TAF of water was to be conserved in Upper Klamath Lake. Conservation was achieved through land idling, increased storage of surface water, and the encouragement of groundwater usage to supplement/replace diverted live flow or deliveries of project water [67,68]. These policies may have contributed to the depletion of groundwater storage.

Supporting this theory is the sharp drop in groundwater in 2020 and the resulting no-flow conditions at gage #3500. Using the CHIRPS data, we can roughly estimate that 3468 MCM of water (2811 TAF) entered this watershed between 2020 and 2023, with only 1.6 MCM (1.3 TAF) of outflow measured during the same 4-year period. This coincides with a severe drought and an increase in surface water restrictions, leading to increased groundwater consumption. Further research is needed to separate out anthropogenic pumping impacts on this watershed.

On a positive note, the groundwater storage increase from 1994 to 2000 illustrates that, with sufficient precipitation, the aquifer storage is capable of rebounding substantially. This understanding can and will assist in management decisions. This knowledge allows managers to properly manage a high water supply forecast from an above-normal snowpack, as water managers can adjust their forecast with awareness about aquifer recharge and storage. Biologists and natural resource managers, aware of the timeframe and water availability, can adjust management and conservation plans for aquatic species [6,66].

As with all imputation methodologies, the resulting data inherently possess a degree of error and uncertainty. The GWDM framework generates an initial model and imputes results for each individual well, iteratively refining these outputs by incorporating data from all wells within the aquifer system. This iterative process minimizes uncertainty and significantly reduces large deviations. Subsequently, spatial interpolation is applied to the groundwater elevation data at each temporal step, meaning that each well contributes only to the localized portion of the interpolated surface. Comparative studies of published GWDM results with established United States Geological Survey (USGS) models have demonstrated strong concordance between methodologies [49–51,64]. While the uncertainty associated with the imputed data in this study remains unquantified, the primary objective is to assess long-term trends and their correlation with processes such as precipitation and streamflow. Despite the inherent uncertainties, the long-term trends have reduced variability. Though unquantified uncertainty exists, our analysis of groundwater storage and how it is related to precipitation and stream flow is valid.

7. Conclusions

The primary objective of this research was to use a groundwater storage analysis tool to characterize the time history of groundwater storage in the Williamson Basin by supplementing sparse in situ data with Earth observations. The resulting groundwater storage data allowed us to gain new insights into the complex relationship between precipitation, groundwater, and streamflow in the basin. Using the data from this analysis, we have demonstrated that both streamflow quantities and the number of days per year with streamflow are highly correlated with the groundwater storage levels, and once the groundwater storage drops below 1800 MCM (1500 TAF) relative to the 1968 baseline, the streamflow in the basin completely ceases. We computed that there is a 3 yr lag between the 3 and 5 yr moving average precipitation in the basin and groundwater storage once this threshold is reached. This means that, even though precipitation has started to rebound in the basin, it will take several more years, more than the 3-year lag, of elevated precipitation before the streams begin to flow again. This is because the time required includes both the 3-year lag and the time required to recharge the aquifer to the critical level. The high level of correlation we found between precipitation, groundwater storage, and streamflow validates the use of the new groundwater storage tools that we applied. This approach represents a new technique that water managers can use in groundwater-dependent basins to better understand and more effectively manage precious water resources.

The methods and tools demonstrated in this study, particularly the GWDM tools, the associated framework, and the use of globally (and freely) available GLDAS data, have significant potential for application in other basins and locations. This approach is especially valuable for regions with sparse measured groundwater data, as it enables the integration of Earth observations to address gaps in data and refine groundwater storage estimations. By using publicly accessible resources and computational techniques, the framework can be adapted to analyze precipitation–groundwater–streamflow interactions in diverse hydrologic systems worldwide. Its ability to characterize and incorporate long-term trends and manage uncertainties makes it an effective tool for water managers and researchers dealing with groundwater-dependent basins, offering insights to improve resource management even in data-limited regions.

Supplementary Materials: The groundwater data used in this study, including well locations and water level time series, can be downloaded at http://www.hydroshare.org/resource/1d9791b2ef74454d8b335aab72d9bc97 (accessed on 14 November 2024).

Author Contributions: Conceptualization, D.S., N.L.J. and G.P.W.; methodology, D.S., N.L.J. and G.P.W.; software, N.L.J. and G.P.W.; data curation, D.S. and N.L.J.; writing—original draft preparation, D.S. and N.L.J.; writing—review and editing, G.P.W.; visualization, D.S., N.L.J. and G.P.W. All authors have read and agreed to the published version of the manuscript.

Funding: This research received no external funding.

Data Availability Statement: All datasets used in the creation of this research paper are freely available on the internet through Federal and State environmental and research agencies. Federally run monitoring well logs can be found through the US Geological Survey groundwater mapper website (https://waterdata.usgs.gov/nwis/gw (accessed on 14 November 2024)) and by entering the associated well number. Other monitoring wells' data can be accessed at Oregon's Water Resource Department (https://www.oregon.gov/owrd/pages/index.aspx (accessed on 14 November 2024)) by locating the well by county (Klamath) and the well ID number. All flow data are available through the US Geological Survey (https://waterdata.usgs.gov/nwis/sw (accessed on 14 November 2024)) by entering the identification number of the gage used.

Conflicts of Interest: The authors declare no conflicts of interest.

References

1. Ochoa, C.G.; Jarvis, W.T.; Hall, J. A Hydrogeologic Framework for Understanding Surface Water and Groundwater Interactions in a Watershed System in the Willamette Basin in Western Oregon, USA. *Geosciences* **2022**, *12*, 109. [CrossRef]
2. Conaway, J.S. Hydrogeology and Paleohydrology in the Williamson River Basin, Klamath County Oregon. Master's Thesis, Portland State University, Portland, OR, USA, 2000.
3. Mayer, T.; Wurster, F.; Craver, D. *Klamath Marsh Hydrology and Water Rights*; USFWS: Washington, DC, USA, 2007; p. 22.
4. Leonard, A.R.; Harris, A.B. *Ground Water in Selected Areas in the Klamath Basin, Oregon*; Oregon Water Science Center: Portland, OR, USA, 1974; p. 104.
5. Melady, J. Hydrogeologic Investigation of the Klamath Marsh, Klamath County, Oregon. Master's Thesis, Portland State University, Portland, OR, USA, 2002.
6. Gannett, M.W.; Lite, K.E., Jr.; Marche, J.L.L.; Fisher, B.J.; Polette, D.J. *Ground-Water Hydrology of the Upper Klamath Basin, Oregon and California*; Geological Survey (U.S.): Reston, VA, USA, 2007.
7. Cummings, M.L. Hydrogeology of the Williamson River Basin, Upper Klamath Basin, Klamath County, Oregon. Master's Thesis, Portland State University, Portland, OR, USA, 2007; p. 70.
8. Stannard, D.I.; Gannett, M.W.; Polette, D.J.; Cameron, J.M.; Waibel, M.S.; Spears, J.M. *Evapotranspiration from Wetland and Open-Water Sites at Upper Klamath Lake, Oregon, 2008–2010*; USGS: Reston, VA, USA, 2013; p. 66.
9. Gannett, M.W.; Wagner, B.J.; Lite, K.E., Jr. *Groundwater Simulation and Management Models for the Upper Klamath Basin, Oregon and California*; U.S. Geological Survey: Reston, VA, USA, 2012.
10. Risley, J.C. *Using the Precipitation-Runoff Modeling System to Predict Seasonal Water Availability in the Upper Klamath River Basin, Oregon and California*; U.S. Geological Survey: Reston, VA, USA, 2019.
11. Markstrom, S.L.; Regan, R.S.; Hay, L.E.; Viger, R.J.; Webb, R.M.; Payn, R.A.; LaFontaine, J.H. *PRMS-IV, the Precipitation-Runoff Modeling System, Version 4*; U.S. Geological Survey: Reston, VA, USA, 2015.
12. Chang, H.; Jones, J. Climate Change and Freshwater Resources in Oregon. In *Oregon Climate Assessment Report*; Oregon Climate Change Research Institute: Portland, OR, USA, 2010.
13. Carroll, R.W.H.; Niswonger, R.G.; Ulrich, C.; Varadharajan, C.; Siirila-Woodburn, E.R.; Williams, K.H. Declining Groundwater Storage Expected to Amplify Mountain Streamflow Reductions in a Warmer World. *Nat. Water* **2024**, *2*, 419–433. [CrossRef]
14. Mayer, T.D.; Naman, S.W. Streamflow Response to Climate as Influenced by Geology and Elevation1. *JAWRA J. Am. Water Resour. Assoc.* **2011**, *47*, 724–738. [CrossRef]
15. Hess, G.W.; Stonewall, A.J. *Comparison of Historical Streamflows to 2013 Streamflows in the Williamson, Sprague, and Wood Rivers, Upper Klamath Lake Basin, Oregon*; U.S. Geological Survey: Reston, VA, USA, 2014.
16. Tague, C.; Grant, G.; Farrell, M.; Choate, J.; Jefferson, A. Deep Groundwater Mediates Streamflow Response to Climate Warming in the Oregon Cascades. *Clim. Change* **2008**, *86*, 189–210. [CrossRef]
17. Brutsaert, W. Long-Term Groundwater Storage Trends Estimated from Streamflow Records: Climatic Perspective. *Water Resour. Res.* **2008**, *44*. [CrossRef]
18. Dudley, R.W.; Hodgkins, G.A. Historical Groundwater Trends in Northern New England and Relations with Streamflow and Climatic Variables. *JAWRA J. Am. Water Resour. Assoc.* **2013**, *49*, 1198–1212. [CrossRef]
19. Huntington, J.L.; Niswonger, R.G. Role of Surface-Water and Groundwater Interactions on Projected Summertime Streamflow in Snow Dominated Regions: An Integrated Modeling Approach. *Water Resour. Res.* **2012**, *48*, W11524. [CrossRef]
20. Ayers, J.R.; Villarini, G.; Schilling, K.; Jones, C.; Brookfield, A.; Zipper, S.C.; Farmer, W.H. The Role of Climate in Monthly Baseflow Changes across the Continental United States. *J. Hydrol. Eng.* **2022**, *27*, 04022006. [CrossRef]
21. Arnold, J.G.; Allen, P.M. Automated Methods for Estimating Baseflow and Ground Water Recharge from Streamflow Records. *JAWRA J. Am. Water Resour. Assoc.* **1999**, *35*, 411–424. [CrossRef]
22. Winter, T.C. *Ground Water and Surface Water: A Single Resource*; Diane Publishing: Collingdale, PA, USA, 2000.
23. The NCAR WRF-Hydro® Modeling System Technical Description — WRF-Hydro Modeling System 5.4.0 Documentation. Available online: https://wrf-hydro.readthedocs.io/en/latest/ (accessed on 4 June 2025).
24. NOAA. The National Water Model. Available online: https://water.noaa.gov/about/nwm (accessed on 8 July 2024).
25. Zealand, C.M.; Burn, D.H.; Simonovic, S.P. Short Term Streamflow Forecasting Using Artificial Neural Networks. *J. Hydrol.* **1999**, *214*, 32–48. [CrossRef]
26. Day, G.N. Extended Streamflow Forecasting Using NWSRFS. *J. Water Resour. Plan. Manag.* **1985**, *111*, 157–170. [CrossRef]
27. Boyle, D.P.; Gupta, H.V.; Sorooshian, S.; Koren, V.; Zhang, Z.; Smith, M. Toward Improved Streamflow Forecasts: Value of Semidistributed Modeling. *Water Resour. Res.* **2001**, *37*, 2749–2759. [CrossRef]
28. Snow, A.D.; Christensen, S.D.; Swain, N.R.; Nelson, E.J.; Ames, D.P.; Jones, N.L.; Ding, D.; Noman, N.S.; David, C.H.; Pappenberger, F.; et al. A High-Resolution National-Scale Hydrologic Forecast System from a Global Ensemble Land Surface Model. *JAWRA J. Am. Water Resour. Assoc.* **2016**, *52*, 950–964. [CrossRef] [PubMed]

29. Hapuarachchi, H.A.P.; Wang, Q.J.; Pagano, T.C. A Review of Advances in Flash Flood Forecasting. *Hydrol. Process.* **2011**, *25*, 2771–2784. [CrossRef]
30. Markstrom, S.L.; Niswonger, R.G.; Regan, R.S.; Prudic, D.E.; Barlow, P.M. GSFLOW—Coupled Ground-Water and Surface-Water Flow Model Based on the Integration of the Precipitation-Runoff Modeling System (PRMS) and the Modular Ground-Water Flow Model (MODFLOW-2005). Available online: https://pubs.usgs.gov/tm/tm6d1/ (accessed on 8 July 2024).
31. Hunt, R.J.; Walker, J.F.; Selbig, W.R.; Westenbroek, S.M.; Regan, R.S. *Simulation of Climate-Change Effects on Streamflow, Lake Water Budgets, and Stream Temperature Using GSFLOW and SNTEMP, Trout Lake Watershed, Wisconsin*; U.S. Geological Survey: Reston, VA, USA, 2013.
32. Ntona, M.M.; Busico, G.; Mastrocicco, M.; Kazakis, N. Modeling Groundwater and Surface Water Interaction: An Overview of Current Status and Future Challenges. *Sci. Total Environ.* **2022**, *846*, 157355. [CrossRef]
33. Hughes, J.D.; Petrone, K.C.; Silberstein, R.P. Drought, Groundwater Storage and Stream Flow Decline in Southwestern Australia. *Geophys. Res. Lett.* **2012**, *39*. [CrossRef]
34. Zipper, S.; Brookfield, A.; Ajami, H.; Ayers, J.R.; Beightel, C.; Fienen, M.N.; Gleeson, T.; Hammond, J.; Hill, M.; Kendall, A.D.; et al. Streamflow Depletion Caused by Groundwater Pumping: Fundamental Research Priorities for Management-Relevant Science. *Water Resour. Res.* **2024**, *60*, e2023WR035727. [CrossRef]
35. Rodell, M.; Famiglietti, J.S. The Potential for Satellite-Based Monitoring of Groundwater Storage Changes Using GRACE: The High Plains Aquifer, Central US. *J. Hydrol.* **2002**, *263*, 245–256. [CrossRef]
36. NASA. JPL Monthly Mass Grids—Global Mascons (JPL RL06.1_v03) | Get Data. Available online: https://grace.jpl.nasa.gov/data/get-data/jpl_global_mascons (accessed on 9 January 2024).
37. NASA. *Studying the Earth's Gravity from Space: The Gravity Recovery and Climate Experiment (GRACE)*; NASA Facts—FS-2002-1-029-GSFC; NASA: Washington, DC, USA, 2002.
38. McStraw, T.C.; Pulla, S.T.; Jones, N.L.; Williams, G.P.; David, C.H.; Nelson, J.E.; Ames, D.P. An Open-Source Web Application for Regional Analysis of GRACE Groundwater Data and Engaging Stakeholders in Groundwater Management. *JAWRA J. Am. Water Resour. Assoc.* **2022**, *58*, 1002–1016. [CrossRef]
39. Scanlon, B.R.; Faunt, C.C.; Longuevergne, L.; Reedy, C.; Alley, W.M.; McGuire, V.L.; McMahon, B. Groundwater Depletion and Sustainability of Irrigation in the US High Plains and Central Valley. *Proc. Natl. Acad. Sci. USA* **2012**, *109*, 9320–9325. [CrossRef]
40. Chen, J.; Famigliett, J.S.; Scanlon, B.R.; Rodell, M. Groundwater Storage Changes: Present Status from GRACE Observations. *Surv. Geophys.* **2015**, *37*, 397–417. [CrossRef]
41. Rodell, M.; Velicogna, I.; Famiglietti, J.S. Satellite-Based Estimates of Groundwater Depletion in India. *Nature* **2009**, *460*, 999–1002. [CrossRef] [PubMed]
42. Reager, J.T.; Famiglietti, J.S. Characteristic Mega-Basin Water Storage Behavior Using GRACE. *Water Resour. Res.* **2013**, *49*, 3314–3329. [CrossRef] [PubMed]
43. Huang, Z.; Pan, Y.; Gong, H.; Yeh, J.-F.; Li, X.; Zhou, D.; Zhao, W. Subregional-Scale Groundwater Depletion Detected by GRACE for Both Shallow and Deep Aquifers in North China Plain. *Geophys. Res. Lett.* **2015**, *42*, 1791–1799. [CrossRef]
44. Alshehri, F.; Mohamed, A. Analysis of Groundwater Storage Fluctuations Using GRACE and Remote Sensing Data in Wadi As-Sirhan, Northern Saudi Arabia. *Water* **2023**, *15*, 282. [CrossRef]
45. Purdy, A.J.; David, C.H.; Sikder, M.S.; Reager, J.T.; Chandanpurkar, H.A.; Jones, N.L.; Matin, M.A. An Open-Source Tool to Facilitate the Processing of GRACE Observations and GLDAS Outputs: An Evaluation in Bangladesh. *Front. Environ. Sci.* **2019**, *7*, 155. [CrossRef]
46. Kuss, A.; Brandt, W.T.; Randall, J.; Floyd, B.; Bourai, A.; Newcomer, M.; Schmidt, C. Comparison of Changes in Groundwater Storage Using GRACE Data and a Hydrological Model in California's Central Valley. In Proceedings of the ASPRS 2012 Annual Conference, Sacramento, CA, USA, 19–23 March 2012.
47. Zhang, X. Impacts of Water Resources Management on Land Water Storage in the Lower Lancang River Basin: Insights from Multi-Mission Earth Observations. *Remote Sens.* **2023**, *15*, 1747. [CrossRef]
48. Evans, S.W.; Jones, N.L.; Williams, G.P.; Ames, D.P.; Nelson, E.J. Groundwater Level Mapping Tool: An Open Source Web Application for Assessing Groundwater Sustainability. *Environ. Model. Softw.* **2020**, *131*, 104782. [CrossRef]
49. Jones, N.L. GWDM—Ground Water Data Mapper 2.0 Documentation. Available online: https://gwdm.readthedocs.io/en/latest/ (accessed on 22 March 2024).
50. Evans, S.; Williams, G.P.; Jones, N.L.; Ames, D.P.; Nelson, E.J. Exploiting Earth Observation Data to Impute Groundwater Level Measurements with an Extreme Learning Machine. *Remote Sens.* **2020**, *12*, 2044. [CrossRef]
51. Ramirez, S.G.; Williams, G.P.; Jones, N.L. Groundwater Level Data Imputation Using Machine Learning and Remote Earth Observations Using Inductive Bias. *Remote Sens.* **2022**, *14*, 5509. [CrossRef]
52. Tao, H.; Hameed, M.M.; Marhoon, H.A.; Zounemat-Kermani, M.; Heddam, S.; Kim, S.; Sulaiman, S.O.; Tan, M.L.; Sa'adi, Z.; Mehr, A.D.; et al. Groundwater Level Prediction Using Machine Learning Models: A Comprehensive Review. *Neurocomputing* **2022**, *489*, 271–308. [CrossRef]

53. Stevens, M.D.; Ramirez, S.G.; Martin, E.-M.H.; Jones, N.L.; Williams, G.P.; Adams, K.H.; Ames, D.P.; Pulla, S.T. Groundwater Storage Loss in the Central Valley Analysis Using a Novel Method Based on In Situ Data Compared to GRACE-Derived Data. *Environ. Model. Softw.* **2025**, *186*, 106368. [CrossRef]
54. Swenson, S.; Wahr, J. Methods for Inferring Regional Surface-Mass Anomalies from Gravity Recovery and Climate Experiment (GRACE) Measurements of Time-Variable Gravity. *J. Geophys. Res. Solid Earth* **2002**, *107*, ETG 3-1–ETG 3-13. [CrossRef]
55. Vasco, D.W.; Kim, K.H.; Farr, T.G.; Reager, J.T.; Bekaert, D.; Sangha, S.S.; Rutqvist, J.; Beaudoing, H.K. Using Sentinel-1 and GRACE Satellite Data to Monitor the Hydrological Variations within the Tulare Basin, California. *Sci. Rep.* **2022**, *12*, 3867. [CrossRef]
56. Famiglietti, J.; Lo, M.; Ho, S.; Bethune, J.; Anderson, K.; Syed, T.; Swenson, S.; De Linage, C.; Rodell, M. Satellites Measure Recent Rates of Groundwater Depletion in California's Central Valley. *Geophys. Res. Lett.* **2011**, *38*. [CrossRef]
57. Newcomb, R.C.; Hart, D.H. *Preliminary Report on the Ground-Water Resources of the Klamath River Basin, Oregon*; U.S. Geological Survey: Reston, VA, USA, 1958.
58. Beck, H.E.; Zimmermann, N.E.; McVicar, T.R.; Vergopolan, N.; Berg, A.; Wood, E.F. Present and Future Köppen-Geiger Climate Classification Maps at 1-Km Resolution (With Publisher Correction). *Sci. Data* **2020**, *7*, 274. [CrossRef]
59. Jones, N.; Williams, G.; Daniel, S. 2024 Oregon Williamson Basin Groundwater Study Data. Available online: http://www.hydroshare.org/resource/1d9791b2ef74454d8b335aab72d9bc97 (accessed on 14 November 2024).
60. NASA. Global Land Data Assimilation System (GLSDAS). Available online: https://ldas.gsfc.nasa.gov/gldas (accessed on 10 July 2024).
61. NASA. Goddard Earth Sciences Data and Information Services Center (GES DISC). Available online: https://disc.gsfc.nasa.gov/ (accessed on 10 July 2024).
62. UCSB. CHIRPS: Rainfall Estimates from Rain Gauge and Satellite Observations | Climate Hazards Center—UC Santa Barbara. Available online: https://www.chc.ucsb.edu/data/chirps (accessed on 10 July 2024).
63. NASA. SERVIR ClimateSERV 2.0—Data and Tools for Sustainable Development. Available online: https://climateserv.servirglobal.net/ (accessed on 10 July 2024).
64. Ramirez, S.G.; Williams, G.P.; Jones, N.L.; Ames, D.P.; Radebaugh, J. Improving Groundwater Imputation through Iterative Refinement Using Spatial and Temporal Correlations from In Situ Data with Machine Learning. *Water* **2023**, *15*, 1236. [CrossRef]
65. Nelson, E.J.; Pulla, S.T.; Matin, M.A.; Shakya, K.; Jones, N.; Ames, D.P.; Ellenburg, W.L.; Markert, K.N.; David, C.H.; Zaitchik, B.F. Enabling Stakeholder Decision-Making with Earth Observation and Modeling Data Using Tethys Platform. *Front. Environ. Sci.* **2019**, *7*, 148. [CrossRef]
66. NOAA. Fisheries 2002 Klamath Project Biological Opinion | NOAA Fisheries. Available online: https://www.fisheries.noaa.gov/resource/document/2002-klamath-project-biological-opinion (accessed on 27 August 2024).
67. U.S. National Marine Fisheries Service. *2002 Klamath Project Biological Opinion*; U.S. National Marine Fisheries Service: Silver Spring, MD, USA, 2002.
68. Gannet, M.W.; Lite, K.E., Jr.; la Marche, J.L.; Polette, D.J. *Ground-Water Hydrology of the Upper Klamath Basin, Oregon and California: U.S. Geological Survey Scientific Investigations Reportentific Investigations Report*; Scientific Investigations Report; Geological Survey (US): Reston, VA, USA, 2007; p. 84.

hydrology

MDPI

Article

Assessment of the Impact of Climate Change on Dam Hydrological Safety by Using a Stochastic Rainfall Generator

Enrique Soriano [1], Luis Mediero [1,*], Andrea Petroselli [2], Davide Luciano De Luca [3], Ciro Apollonio [2] and Salvatore Grimaldi [4]

1 Department of Civil Engineering: Hydraulics, Energy and Environment, Universidad Politécnica de Madrid, 28040 Madrid, Spain; e.soriano@upm.es
2 Department of Agriculture and Forest Sciences (DAFNE), Tuscia University, 01100 Viterbo, Italy; petro@unitus.it (A.P.); ciro.apollonio@unitus.it (C.A.)
3 Department of Civil, Building and Environmental Engineering, Sapienza University of Rome, 00185 Roma, Italy; davideluciano.deluca@uniroma1.it
4 Department for Innovation in Biological, Agro-Food and Forest Systems (DIBAF), Tuscia University, 01100 Viterbo, Italy; salvatore.grimaldi@unitus.it
* Correspondence: luis.mediero@upm.es; Tel.: +34-910-67-43-76

Abstract: Dam breaks can lead to important economic and human losses. Design floods, which are useful to assess possible dam breaks, are usually estimated through statistical analysis of rainfall and streamflow observed data. However, such available samples are commonly limited and, consequently, high uncertainties are associated with the design flood estimates. In addition, climate change is expected to increase the frequency and magnitude of extreme rainfall and flood events in the future. Therefore, a methodology based on a stochastic rainfall generator is proposed to assess hydrological dam safety by considering climate change. We selected the Eugui Dam on the Arga river in the north of Spain as a case study that has a spillway operated by gates with a maximum capacity of 270 m^3/s. The stochastic rainfall generator STORAGE is used to simulate long time series of 15-min precipitation in both current and future climate conditions. Precipitation projections of 12 climate modeling chains, related to the usual three 30-year periods (2011–2024; 2041–2070 and 2071–2100) and two emission scenarios of AR5 (RCP 4.5 and 8.5), are used to consider climate change in the STORAGE model. The simulated precipitation time series are transformed into runoff time series by using the continuous COSMO4SUB hydrological model, supplying continuous 15-min runoff time series as output. Annual maximum flood hydrographs are selected and considered as inflows to the Eugui Reservoir. The Volume Evaluation Method is applied to simulate the operation of the Eugui Dam spillway gates, obtaining maximum water levels in the reservoir and outflow hydrographs. The results show that the peak outflows at the Eugui Dam will be lower in the future. Therefore, maximum reservoir water levels will not increase in the future. The methodology proposed could allow practitioners and dam managers to check the hydrological dam safety requirements, accounting for climate change.

Keywords: climate change; floods; hydrological dam safety; rainfall-runoff model; stochastic rainfall generator

1. Introduction

Dams offer great benefits to society, such as water storage for supplying demands, generation of hydropower electricity, flood control, and reservoirs for recreational uses [1].

However, dams require a high safety level, as the human and economic losses caused by their failure could be catastrophic [2].

European research programs, such as Horizon 2020, have prioritized flood risk protection and territory management in line with the Flood Directive 2007/60/EC. This directive mandates coordinated efforts among governments and institutions, such as basin authorities and civil protection agencies. Research initiatives focused on understanding flood dynamics, risk assessment, early warning systems, natural and structural defenses, and adaptation strategies, among others. The findings inform policies and actions to mitigate flood risks and protect communities and ecosystems, ensuring compliance with European legislation and fostering resilience against future challenges [3]. According to the International Commission on Large Dams [4], overtopping causes about 35% of all embankment dam failures. Therefore, it is crucial to correctly estimate flood design hydrographs in hydrological dam safety analyses. They are usually obtained by hydrometeorological approaches, using observed precipitation data and rainfall-runoff models, as observed streamflow time series are either unavailable or short at most sites [5]. In addition, despite the greater availability of precipitation data than streamflow data, they are usually recorded at a daily scale, while sub-daily precipitation observations are scarce [6].

In this context, the use of stochastic rainfall generators (SRGs) appears helpful to generate long sub-daily rainfall time series [7,8]. SRGs generally present a simple mathematical formulation and low computational costs. Therefore, large ensembles of long time series can be quickly obtained [9]. Moreover, an SRG can be easily used for obtaining precipitation time series in changing conditions [10] that can be representative of future rainfall in a changing climate on hydrological scales, which are finer than the spatial and time scale supplied by Regional Climate Models (RCMs). More specifically, RCM outputs are mainly available on a daily scale and are averaged over large spatial resolutions, so they require additional statistical downscaling or bias correction methods for including them in hydrological analyses [11,12]. Only very recent RCM outputs regarded high resolutions (hourly) and small spatial scales [13,14]. As dam spillways are usually designed for high return periods (for example, 500–10,000 years in Spain) and observed precipitation data time series are usually in the order of 50 years, SGRs can reduce the uncertainties associated with such large return period estimates.

A variety of approaches for rainfall simulation were introduced in the last twenty years [15–17]. A summary is available in [18], where a performance comparison is provided.

In addition, recent studies indicate that climate change has modified the magnitude and timing of floods in Europe [19]. The number of floods exceeding certain severity and magnitude thresholds increased in the period 1985–2009 [20]. Furthermore, climate projections point to an increase in flood magnitude and frequency in some areas [21]. Therefore, several studies have been carried out recently to assess the expected influence of climate change on floods, although they are usually focused on either monthly or annual time scales [22]. However, few studies have analyzed the impact of climate change on extreme events, such as floods. Giuntoli et al. (2015) identified future changes in the frequency of high flows on a global scale [23]. Dankers et al. (2014) found an increasing trend in the 30-year return level of the 5-day mean peak flow at half of the grid points considered at the global scale [24]. Roudier et al. (2016) found that the 10- and 100-year floods will increase over most of Europe [25]. In Europe, the frequency of flood magnitudes above the 100-year flood quantiles is expected to double within three decades [26]. Lehner et al. (2006) also studied the change in 100-year flood recurrence, finding that northern to northeastern Europe are the regions most prone to a rise in flood frequencies [27].

In Spain, there are more than 1200 large dams. Therefore, Spain is the European Union country with the highest number of dams, as well as the fifth in the world. Consequently, assessing the influence of climate change on hydrological dam safety is a crucial concern in this country. In Spain, there are two sources of climate change projections: (i) the data provided by the Spanish Met Office (AEMET, Agencia Estatal de Meteorología in Spanish) and (ii) the projections provided by the Coordinated Regional Climate Downscaling Experiment (CORDEX) [28]. Garijo et al. (2018) observed that the CORDEX precipitation projections perform better for studies focused on extreme events than the AEMET projections in Spain, which tend to smooth extremes in time series of precipitation projections [29]. Therefore, the climate change projections provided by CORDEX will be used in this study, as dam design floods are usually associated with extreme events.

SRGs have been used in recent years to downscale global climate model results [30,31] and to increase the temporal resolution of climate change data [32]. In addition, SRGs have also been used to generate synthetic rainfall time series to assess hydrological dam safety [33–35]. However, SRGs cannot supply information on the physical phenomenon in flood events. Therefore, such results are improved by coupling SRGs with rainfall-runoff models, as catchment response in flood events can be calibrated with observations.

This study presents a methodology to assess how climate change could affect hydrological dam safety in the future by using a stochastic rainfall generator, a continuous rainfall-runoff model, and a method to identify the best operation of spillway gates in a dam. Influence of climate change on hydrological dam safety is limited to expected changes in precipitation. Possible changes in land uses, such as changes in urbanization extents, deforestation or reforestation, are not considered in the study.

The methodology is applied to the Eugui Dam in Spain. The structure of the paper is as follows. First, a description of models, a case study and available data used in the study is presented. Then, the results of the maximum reservoir water levels obtained with the previous procedure are commented. Finally, conclusions are presented.

2. Models

Three models are used in this study. First, the STORAGE stochastic rainfall generation model [36] that is based on the Neymann–Scott Rectangular Pulse Model. This model allows for simulation of long synthetic rainfall time series at several sub-daily scales. Second, the COSMO4SUB model is used for simulating rainfall-runoff processes. It is a continuous model, tailored for small and ungauged basins, that uses a high-resolution digital terrain model (DTM), land use data and precipitation data generated by the STORAGE model as input data. It can generate continuous runoff time series as output data from which annual maximum hydrographs can be extracted. Finally, the Volumetric Evaluation Method (MEV) [37] is used to simulate the operation of the spillway gates in flood events. The MEV method specifies when the gates should be opened and closed to keep the dam safe and reach the target final reservoir water level.

2.1. STORAGE

The STORAGE (STOchastic RAinfall Generation) model is an SRG model that can generate long and high-resolution rainfall time series based on observations at rain-gauging stations, both in stationary and changing climate contexts. STORAGE is a modified version of the basic Neyman–Scott Rectangular Pulses (NSRP) SRG model that is characterized by two main innovations compared with the NSRP SRG model. First, the parameter estimation is carried out by using a set of data time series (annual maximum rainfall, annual and monthly cumulative rainfall, and annual number of wet days), which are usually longer than observed high-resolution time series that are commonly considered for the calibration of SRGs. A day is considered wet when rainfall is higher than 1 mm. Second, the seasonality is modeled using a series of goniometric functions. This approach makes STORAGE strongly parsimonious with respect to basic versions of NSRP models, in which a user should estimate a specific parametric set for each month or season. For specific details of the STORAGE model, the reader can refer to [36]. The current version of the STORAGE model has a limitation in simulating synthetic rainfall time series for periods of 500 years.

The STORAGE model has been applied under changing climate conditions by using a set of climate change projections. The STORAGE model parameters are derived from the given 30-year periods in such climate projections, in which stationarity is assumed. Time series for periods of 500 years are simulated, composing any 30-year ensemble.

2.2. COSMO4SUB

COSMO4SUB (Continuous Simulation Model For Small and Ungauged Basin) [38,39] is a continuous rainfall-runoff model that is composed of three modules: (i) the rainfall simulation; (ii) the excess rainfall estimation; (iii) the transformation of excess rainfall into runoff.

COSMO4SUB is a simplified rainfall-runoff model that aims at reproducing rainfall-runoff transformation processes in small catchments that are characterized by either small concentration or response times. Slow processes like baseflow, evaporation and snow are not considered in the model. Therefore, the continuous transformation is limited to the surface runoff, which is estimated by applying the curve number method. The antecedent moisture condition is estimated based on the cumulative rainfall in the five days preceding a given rainfall event. COSMO4SUB is based on the instantaneous unit hydrograph (IUH) theory. Consequently, rainfall should be homogeneous within the catchment. Therefore, COSMO4SUB can be applied to small catchments, usually with a contributing area smaller than 200 km^2.

The rainfall simulation module uses sub-daily precipitation time series that can be obtained from either observations, if data recorded at gauging stations are available, or from synthetic time series generated by models like STORAGE. In this study, the rainfall time series previously generated by the STORAGE model will be used as input data.

For the excess rainfall estimation module, the continuous version of Curve Number (CN) for Green-Ampt (CN4GA) was adopted. CN4GA is a mixed and automatic approach combining the empirical equation of the NRCS CN method [40] and the Green-Ampt physically based infiltration equation [41]. CN4GA parameters are essentially the Curve Number (CN) in the AMC-II condition at the catchment, which considers an average soil moisture content at the beginning of the storm, and the rainfall event separation time (Ts), which basically represents the dry (or almost dry) period that should be waited between two storms so that the initial abstraction becomes effective again. In this study, a Ts value of 24 h is considered. The continuous CN4GA procedure is able to automatically assess the AMC condition for a given event based on the amount of rainfall occurred in the five antecedent days, modifying the CN-II value for an average soil moisture condition into either CN-I (dry soil) or CN-III (wet soil) according to the original formulation of the CN method [41].

In the third and last step, the excess rainfall concentration and diffusion processes are simulated by applying the WFIUH (Width Function-Based Instantaneous Unit Hydrograph), which is automatically calculated from DTM flow paths and from the time of concentration (t_c) estimate, which characterizes the response time in the catchment, allowing determining its flow travel time distribution. In detail, in each hillslope cell, the surface flow velocity is calculated based on slope and land cover employing empirical formulas. Conversely, in the river network cells the surface flow velocity is calibrated ensuring that the projection of the WFIUH center of mass on the temporal axis is equal to the catchment lag time, which is expressed as 60% of t_c.

The application of COSMO4SUB allows for simulating long streamflow time series. In this study, we selected the annual maximum peak discharges, reconstructing the relationship between peak discharge and return period.

2.3. Volumetric Evaluation Method

The Volumetric Evaluation Method (VEM) aims to identify the decision on the operation of spillway gates from the data available in real time. The inflow hydrograph, reservoir curve, and existing dam freeboard available at the given time step are considered as input. The method is based on three assumptions: (i) the outflow has to be equal or lower than the inflow, (ii) increasing inflows will increase outflows, and (iii) the higher the reservoir water level, the higher the increase percentage of outflow with respect to inflow.

The application of the model is iterative in each time step and can be used on a general scale without the need for any other type of control. It is valid for any reservoir. The method provides the outflow from the reservoir water level, ensuring the best utilization of the available freeboard from the existing data at every time step. The method starts by calculating the number of time steps that would be needed to fill the reservoir given the current inflow and reservoir storage (Equation (1)).

$$n = \frac{S_a}{(I - O)\cdot \Delta t} \tag{1}$$

where n is the number of time steps to fill the reservoir, I is the inflow, O is the outflow, S_a is the available storage volume to the dam freeboard, and Δt is the time step.

The outflow in each time step is then obtained with Equation (2)

$$O_t = O_{t-1} + \Delta O = O_{t-1} + \frac{(I - O_{t-1})^2 \Delta t}{S_a} \tag{2}$$

where O_t is the outflow at time step t, O_{t-1} is the outflow at time step $t-1$, ΔO is the increase in outflow between time steps t and $t-1$, S_a is the available storage volume to the dam freeboard, and Δt is the time step.

The method calculates the outflow to be discharged at the next time step depending on the inflow hydrograph and reservoir water level. The method considers if the inflow hydrograph is in its rising limb and the inflow discharge is increasing or in its descending limb and the inflow discharge is decreasing. In addition, the method considers two options for the reservoir water level: (i) the reservoir is filling and the reservoir water level is increasing, and (ii) the reservoir is emptying and the reservoir water level is decreasing. Depending on the situation in which the reservoir is at a given time step, in terms of the inflow hydrograph and water reservoir level changes at the last time step, a condition is applied to Equation (2) to obtain the outflow in such a time step [37] (Girón et al., 1988).

3. Case Study and Data

The Eugui Dam has been selected as case study, given it has a small–medium catchment size, its data availability and its importance regarding water supply and flood defense in the downstream city of Pamplona. It is located in the region of Navarre in northern Spain (Figure 1). The dam is on the Arga River, which belongs to the Ebro River Basin Authority. The draining catchment area of the Eugui Reservoir is 69 km^2. The reservoir at the top of its conservation pool has a volume of 22 hm^3 and an area of 123 ha. It is a concrete dam with a height of 51 m and a crest length of 252 m that was built in 1972. Its foundation elevation is 579 m a.s.l. The dam spillway is operated by gates with a maximum capacity of 270 m^3/s.

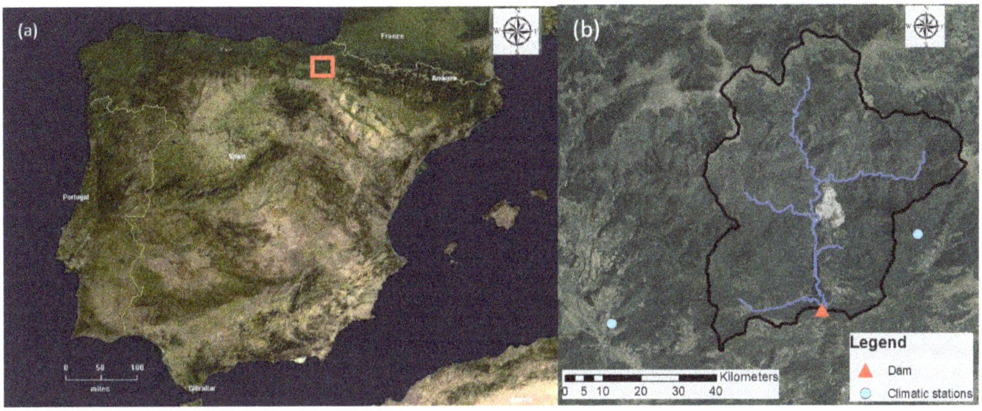

Figure 1. (**a**) Location of the case study in northern Spain (red square). (**b**) Eugui Dam catchment (red triangle) and stations of climate projections (blue circles).

The observed streamflow data required for COSMO4SUB model calibration were obtained from a 15-min gauging station located at the dam that provides reservoir water levels and outflow releases from the dam. Therefore, inflow discharges can be obtained by applying mass conservation. In the dam, there is also a rain-gauging station that provides 15-min data that has also been used to perform the calibration of the STORAGE and the COSMO4SUB models. These data were supplied by the Ebro real-time SAIH system. The 2008–2020 period is used for the calibration.

The Spanish National Geographic Institute (IGN, Instituto Geográfico Nacional in Spanish) has provided a high-resolution DTM with a cell size of two meters that has been used as input data for the COSMO4SUB rainfall-runoff model. In addition, land-use data was also used to obtain the initial estimate of CN for the COSMO4SUB. In this study, land-use data have been supplied by the Corine Land Cover (CLC) 2018 [42] (EEA, 2018). The initial values of CN are estimated based on CLC maps by using the model that implements linked look-up tables, according to the original CN formulation, land cover and CLC.

Climate change projections have been obtained from 12 GCM-RCM modeling chains of the EURO-CORDEX program (Table 1). The projections are supplied in cells of 0.11°, four of which were used in this study (Figure 1b). The reference period is 1971–2004 and the future period is 2011–2100. Three subperiods are considered in the future period: 2011–2040, 2041–2070 and 2071–2100. Climate projections used in the study consider two representative concentration pathways (RCP): RCP 4.5 and RCP 8.5.

Table 1. Climate change modeling chains used in the study, including global climate models (GCMs) and regional climate models (RCMs).

Acronym	GCM	RCM
ICH-CCL	ICHEC-EC-EARTH	CCLM4-8-17
MPI-CCL	MPI-ESM-LR	CCLM4-8-17
MOH-RAC	MOHC-HadGEM2-ES	RACMO22E
CNR-CCL	CNRM-CMS	CCLM4-8-17
ICH-RAC	ICHEC-EC-EARTH	RACMO22E
MOH-CCL	MOHC-HadGEM2-ES	CCLM4-8-17
IPS-WRF	IPSL-CMSA-MR	WRF331F

Table 1. *Cont.*

Acronym	GCM	RCM
IPS-RCA	IPSL-CM5A-MR	RCA4
MOH-RCA	MOHC-HadGEM2-ES	RCA4
ICH-RCA	ICHEC-EC-EARTH	RCA4
CNR-RCA	CNRM-CM5	RCA
MPI-RCA	MPI-ESM-LR	RCA4

4. Methodology

The methodology proposed to estimate the expected changes in hydrological dam safety due to climate change consists of four steps (Figure 2): (i) stochastic rainfall generation of long time series with the STORAGE model by using climate change projections, (ii) calibration of the COSMO4SUB hydrological model, (iii) generation of flood hydrographs using the calibrated COSMO4SUB model with the rainfall time series generated by STORAGE as input data, and (iv) application of the VEM method to obtain the expected maximum water reservoir levels in flood events considering the effect of climate change.

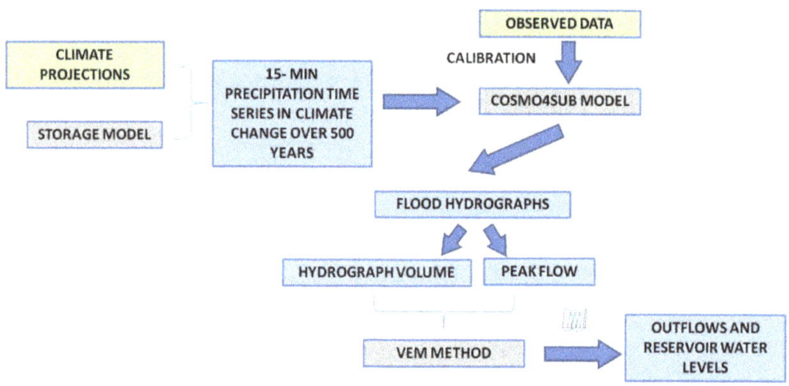

Figure 2. Diagram of the methodology used in the study.

4.1. Rainfall Generation with the STORAGE Model

Synthetic rainfall time series for a period of 500 years have been generated with a 15-min time step for each of the three subperiods considered in the future period (2011–2040, 2041–2070 and 2071–2100), based on climate change projections by using the STORAGE model. Rainfall is considered to be uniformly distributed in space throughout the catchment, as this is one of the hypotheses of the IUH method that is used in the COSMO4SUB model. In addition, the IUH method can be used in catchments with a contributing area smaller than 150–200 km^2, such as the case of the Eugui Dam.

The STORAGE model is useful when high-resolution observed data is not available for model calibration, such as sub-daily time series of precipitation and streamflow. In these, cases, statistics like mean annual and seasonal precipitation and parameters of intensity-duration-frequency (IDF) curves related to given return periods can be used as input data of the STORAGE model. In this study, climate change projections have a daily scale and the sub-daily parameters of the STORAGE model cannot be calibrated. Therefore, sub-daily parameters have not been considered. Only the parameters with a temporal scale equal or greater than daily have been introduced: mean annual precipitation (MAP) (Figure 3), the mean seasonal precipitation associated with December–January–February (DJF), March–April–May (MAM), June–July–August (JJA) and September–October–November (SON) seasons (Figure 4), and the mean annual number of wet days (MANWD) per year (Figure 3). Stochastic rainfall time series have been generated for the 12 climate modeling chains (Table 1) in both the reference and future periods. In the three subperiods in the future period, the two emission scenarios (RCP 4.5 and RCP 8.5) are considered.

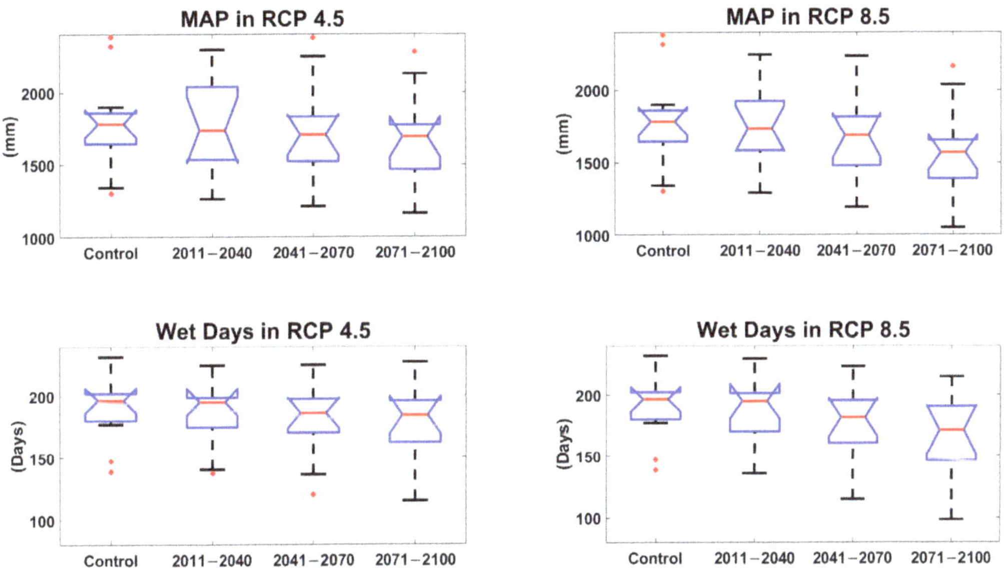

Figure 3. Mean annual precipitation (MAP) (**first row**) and wet days (**second row**) in control and the three future periods considered for RCP 4.5 (**first column**) and RCP 8.5 (**second column**). Red dots represent outliers.

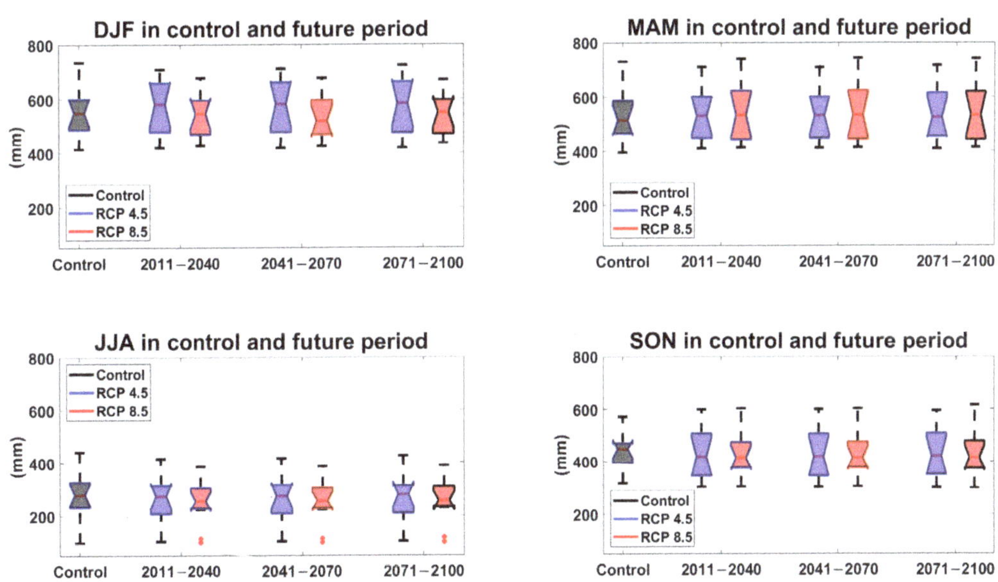

Figure 4. Mean seasonal precipitation in the control and three future periods for RCP 4.5 and RCP 8.5. Red dots represent outliers.

Regarding the sensitivity of key model parameters in the STORAGE model, MANWD influences all the parameters related to waiting times, as these increase when the MANWD value increases. In addition, the observed mean annual and seasonal precipitation (MAP, DIF, MAM, JJA and SON) and the IDF parameters influence intensity and duration of the rectangular bursts, as they increase when all such summary statistics increase.

In Figure 3, it can be seen that less MAP is expected in the future, as well as a smaller number of wet days. Therefore, drier years are expected in the future. This is especially evident in the last future period and particularly for the RCP 8.5.

On the other hand, there is no clear pattern of change for the seasonal precipitation in the four seasons considered, as it remains similar in the three future periods. For the RCP 4.5 emission scenario, higher rainfall is expected in all three future periods in DJF (Figure 4). Similar values in the future are expected in MAM and JJA, indicating no changes in seasonal precipitation. Finally, lower values are expected in SON. In the RCP 8.5 emission scenario, the lowest seasonal precipitation in DJF is expected in the period 2041–2070. However, seasonal precipitation in the future periods remains similar in the other three seasons. In MAM, the seasonal precipitation is expected to be slightly higher in the future. In the summer period (JJA), lower values are expected for the 12 climate modeling chains. Finally, in SON slightly lower seasonal precipitation is expected in the future.

The 15-min rainfall time series for periods of 500 years obtained from the STORAGE model have been used as input data in the COSMO4SUB model in order to obtain flood hydrographs. Six time series of 500 years have been generated for each of the twelve climate modeling chains (Table 1), corresponding to the three subperiods considered in the future period and the two emission scenarios considered (RCP 4.5 and RCP 8.5) in each subperiod. Therefore, a total of 72 time series were simulated.

Stationarity has been assumed in the 500-year synthetic series that have been generated from the rainfall indicators (Figures 3–5) in each 30-year period for each combination of climatic modeling chain, period in the future and RCP. Therefore, stationarity has been assumed in the time series of 30 years generated by the climate model chains in each subperiod considered in the future period. Usually, such short periods generated by climate model chains are not enough when assessing expected changes in high-return period events, as both flood and precipitation quantile estimates for such high return periods show a high uncertainty given the short time series used. Consequently, SRGs are used in this study to extend the short time series generated by climate modeling chains in order to reduce uncertainty when assessing expected changes for high return periods.

However, a more rigorous approach would consider specific time-varying functions of STORAGE parameters based on changes identified in the precipitation climate projections in each period to generate a large ensemble of a 500-year series. However, the approach used in the study does not consider time-varying functions in each period, as the uncertainties due to the definition of the mathematical expression of such time-varying functions for parameters and their extrapolation after 2100 in the 500-year periods would make the rigorous approach hard to apply. In addition, the high uncertainties associated with such an approach would make it difficult to identify the expected changes in hydrological dam safety expected in the future.

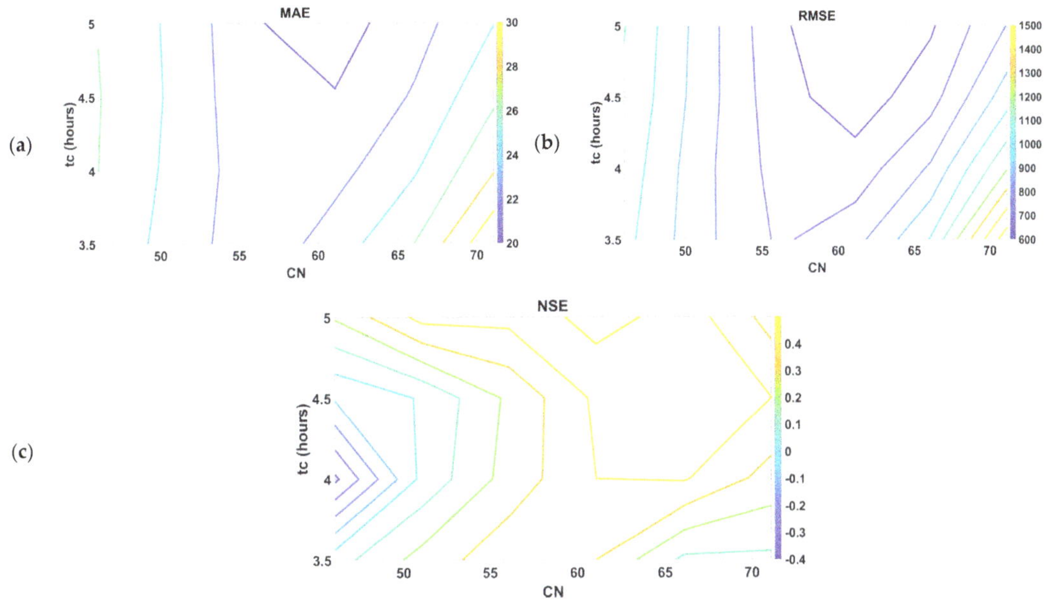

Figure 5. Results of the objective functions used for the calibration of the COSMO4SUB model in terms of CN and t_c values with data observed in the period 2008–2020: (**a**) mean absolute error (**b**) root mean square error (**c**) Nash–Sutcliffe efficiency coefficient.

4.2. Calibration and Runoff Generation with COSMO4SUB Model

The COSMO4SUB model is used to simulate runoff time series at the catchment outlet by using the 15-min rainfall time series generated by the STORAGE model as input data. Therefore, 15-min streamflow time series of 500 years for each climate modeling chain, 30-year period and emission scenario are obtained. Such streamflow time series are considered as inflow to the Eugui Reservoir.

The CN and t_c values can be considered optimal when derived from observed data. However, their estimation in ungauged catchments rely on geomorphological and land-use information, introducing significant uncertainty. Previous studies have explored a wide range of CN values between 30 and 75 and t_c values to assess model sensitivity and evaluate their impact on hydrograph variability and parameter influence. Such combinations would encompass all possible hydrographs the model can generate for a given catchment. While the CN range reflects land covers from brush to gravel surfaces, t_c shifts the timing of the WFIUH-1par peak, yielding diverse hydrological responses. The resulting hydrographs display substantial variability, confirming the model flexibility with few input parameters. As expected, increasing t_c reduces peak discharge, though this effect is diminished for CN values smaller than 50, indicating a stronger control by infiltration processes. Flood hydrograph volume shows minimal sensitivity to t_c. In addition, for CN values greater than 65, hydrograph volume variability also decreases, highlighting that CN plays a dominant role. Hydrograph flood duration increases with t_c, especially for CN values greater than 50, likely due to reduced infiltration and prolonged runoff generation.

First, the model parameter values are calibrated with observations. The parameters that have been considered in the COSMO4SUB model calibration are the curve number (CN) and t_c. t_c values between four and six hours and CN values between 45 and 75 were explored. The period 2008–2020 was used for the model calibration. The model calibration concentrated on high return period results, as the study is focused on hydrological dam safety and such return periods are needed. A multi-objective function was not required as the three model parameters supplied the same results when the variables were varied. CLC maps were used to obtain the initial values of the CN parameter.

Several simulations have been performed with a set of combinations of values for these two model parameters. The mean absolute error (MAE), root mean square error (RMSE) and the Nash–Sutcliffe efficiency coefficient (NSE) have been calculated for each combination of parameter values (Equations (3)–(5)).

$$MAE = \frac{\sum_i^N |Q_i - \hat{Q}_i|}{N} \tag{3}$$

$$RMSE = \sqrt{\frac{\sum_i^N |Q_i - \hat{Q}_i|^2}{N}} \tag{4}$$

$$NSE = \frac{\sum_i^N |\hat{Q}_i - Q_i|^2}{\sum_i^N \left|\hat{Q}_i - \overline{\hat{Q}}\right|^2} \tag{5}$$

where Q_i is the streamflow prediction at time step i, $\hat{Q}_i$ is the observed streamflow value at time step i, $\overline{\hat{Q}}$ is the mean of observed streamflow values in the period considered, and N is the sample size.

The hydrographs that generate the annual maximum peak flows for each model, scenario and period are extracted from such streamflow series. These hydrographs are the input data in the MEV model to obtain the outflow hydrographs and maximum reservoir water levels, considering an optimal gate operation strategy. In this study, both the initial and final reservoir water levels are the top level of the conservation pool.

Finally, return periods are assigned to the peak outflows and maximum reservoir water levels by using the Gringorten formula (Equation (6)), analyzing the changes between the future and baseline periods.

$$P_i = \frac{i - 0.44}{n + 0.12} \tag{6}$$

where n is the number of observations, i is the i-th observation in increasing order, and P_i is the estimate of the cumulative frequency of the i-th term [43].

Both STORAGE and COSMO4SUB have already been validated using observed datasets across diverse hydrological and meteorological conditions, demonstrating consistent and reliable performance. Both models incorporate physically meaningful parameters and boundary conditions, which inherently constrain their behavior and reduce the risk of overfitting unlike purely data-driven models that benefit more significantly from statistical resampling methods.

5. Results

This section presents the results obtained by applying the methodology proposed. First, the results of the COSMO4SUB model calibration are presented, and the synthetic runoff time series generated at the Eugui Resevoir are analyzed. Second, expected changes in peak outflow and maximum reservoir water levels in Eugui Dam are assessed.

5.1. Generation of Runoff Time Series

First, the calibrated values of the COSMO4SUB model parameters are obtained, using the functions described in the previous section. The three objective functions have been calculated to quantify the errors between simulations and observations. The curve number and time of concentration value combinations with the lowest errors were selected. A CN value of 65 and a t_c value of five hours have been considered (Figure 5).

With the calibrated parameter values of the COSMO4SUB model and the synthetic rainfall time series of 500 years generated by the STORAGE stochastic model, streamflow time series of 500 years at the Eugui Reservoir are generated for each climate modeling chain, scenario and period with the COSMO4SUB model (Figures 6 and 7). Figures 6 and 7 present the results by using empirical cumulative distribution functions (ECDF) to visualize the cumulative distribution of the flow series for the baseline period and the two emission scenarios in the future, considering the three selected periods. The graph shows the ECDFs for the 12 climate modeling chains considered in the study using thinner grey lines. The median values of the 12 climate modeling chains are presented with a wider black line. The baseline period presents the highest and most spread cumulative flow distribution, as it shows a more horizontal shape. However, in both emission scenarios in the future, the curves are more vertical, i.e., the flow discharges are spread over a smaller range, between 50 and 200 m^3/s.

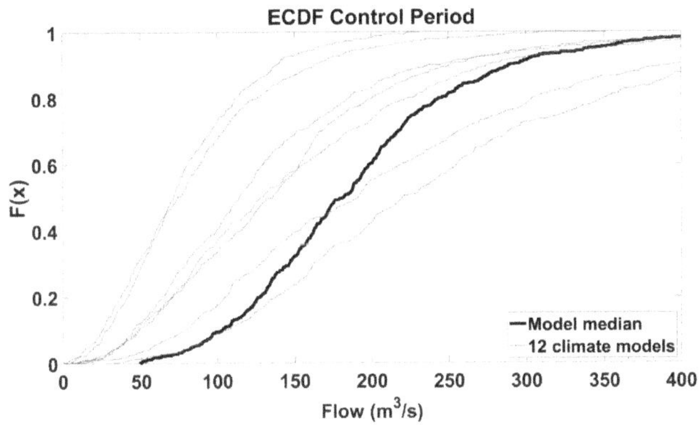

Figure 6. Empirical cumulative distribution function of peak inflows at the Eugui Resevoir in the reference period for the 12 climate modeling chains (grey lines) and the median (black line).

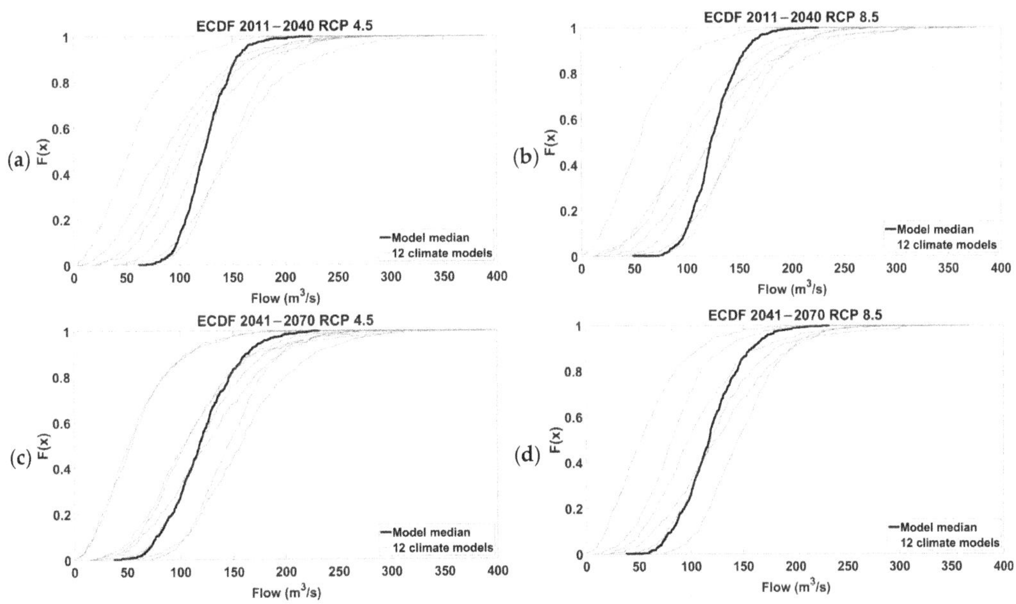

Figure 7. *Cont.*

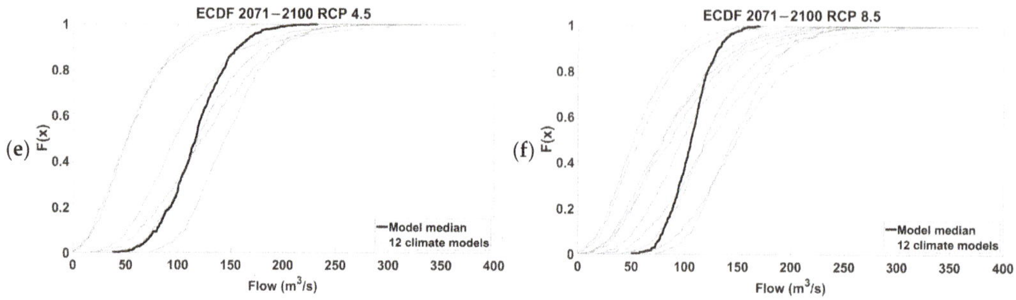

Figure 7. Empirical cumulative distribution function of peak inflows at the Eugui Resevoir in the future period for the 12 climate modeling chains (grey lines) and the median (black line). (**a**) The 2011–2040 period and RCP 4.5. (**b**) The 2011–2040 period and RCP 8.5 (**c**) The 2041–2070 period and RCP 4.5. (**d**) The 2041–2070 period and RCP 8.5. (**e**) The 2071–2100 period and RCP 4.5. (**f**) The 2071–2100 period and RCP 8.5.

It has been observed that the median values of annual maximum peak flows in the future periods will be lower than in the baseline period (Figure 8). No flow trends are observed with respect to time periods or emission scenarios. It seems that the average streamflow and precipitation depend exclusively on the selected model.

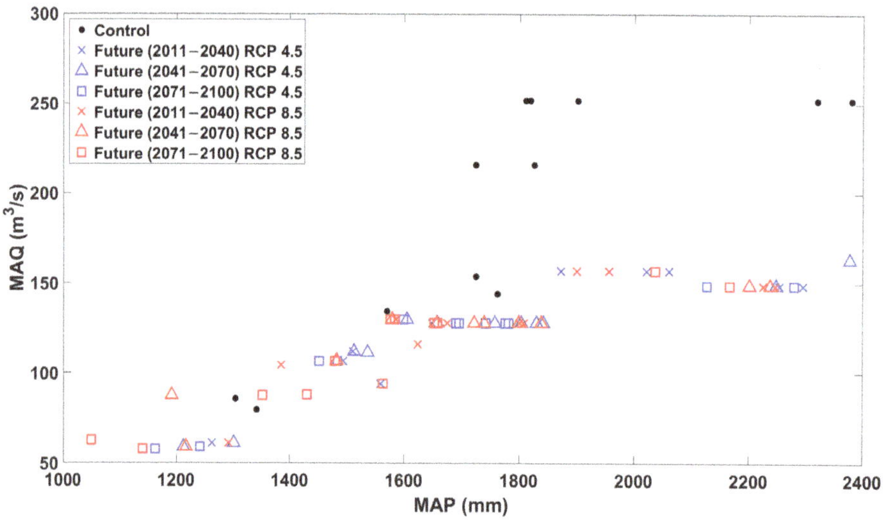

Figure 8. Mean values of annual maximum peak flow time series (MAQ) against annual mean precipitation (MAP) in control (black), future RCP 4.5 (blue) and future RCP 8.5 (red) for the 12 climate modeling chains.

5.2. Expected Changes in Peak Outflow and Maximum Reservoir Water Level Curves

The annual maximum flood hydrographs for the 500-year periods in each case are considered as inflow to the Eugui Dam. The VEM is applied to obtain the outflow hydrographs and reservoir storage volumes and water levels.

The reservoir has been assumed to be completely full at the beginning of each flood event to consider the most pessimistic situation. The median peak outflow frequency curve is expected to be lower in both future emission scenarios and in the three future periods considered. In the first future period (2011–2040), annual maximum peak outflows are expected to be lower than in the baseline period considered as the present situation (Figure 9). Annual maximum peak outflows will be similar in the two emission scenarios (Figure S1). The largest differences were found for the higher return periods. High return periods up to 1000 years have been considered in the study, as they are usually used in spillway design. The lowest annual maximum peak outflows are found for the emission scenario RCP 4.5. The 33rd and 67th percentiles present greater differences in the RCP 4.5 emission scenario, while in the RCP 8.5, they are closer to the median.

(a)

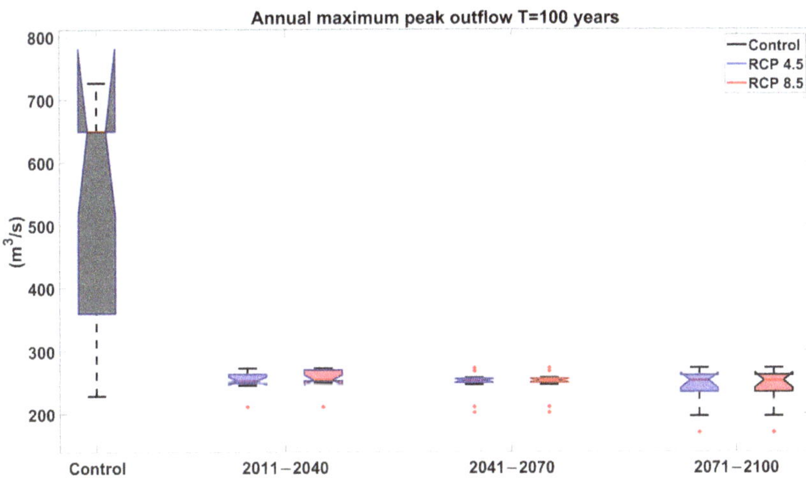

(b)

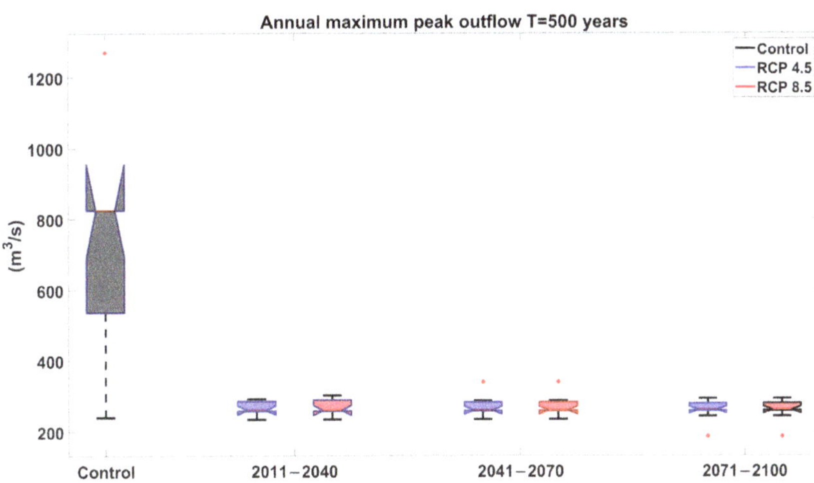

Figure 9. Annual maximum peak outflow frequency box plotted by using the Gringorten formula. (**a**) Results for the 100-year return period, (**b**) results for the 500-year return period. Red dots represent outliers.

For annual maximum reservoir water levels, a constant value is obtained in the case of the median of the 12 climate modeling chains (Figure 10), as the spillway is operated by gates and the VEM method tries to open gates gradually to keep the initial water level, which is fixed as the top of the conservation pool for as long as possible, while the inflow discharge can be discharged, opening the gates partially and maintaining the initial water level (Figure S2). Therefore, the reservoir water level is not increased, while the inflows are lower than the maximum outflow discharge that can be spilled under the partially opened gates with a reservoir water level equal to the top of the conservation pool. However, for given climate modeling chains, an increase in the maximum reservoir water levels can be observed for return periods greater than 25 years.

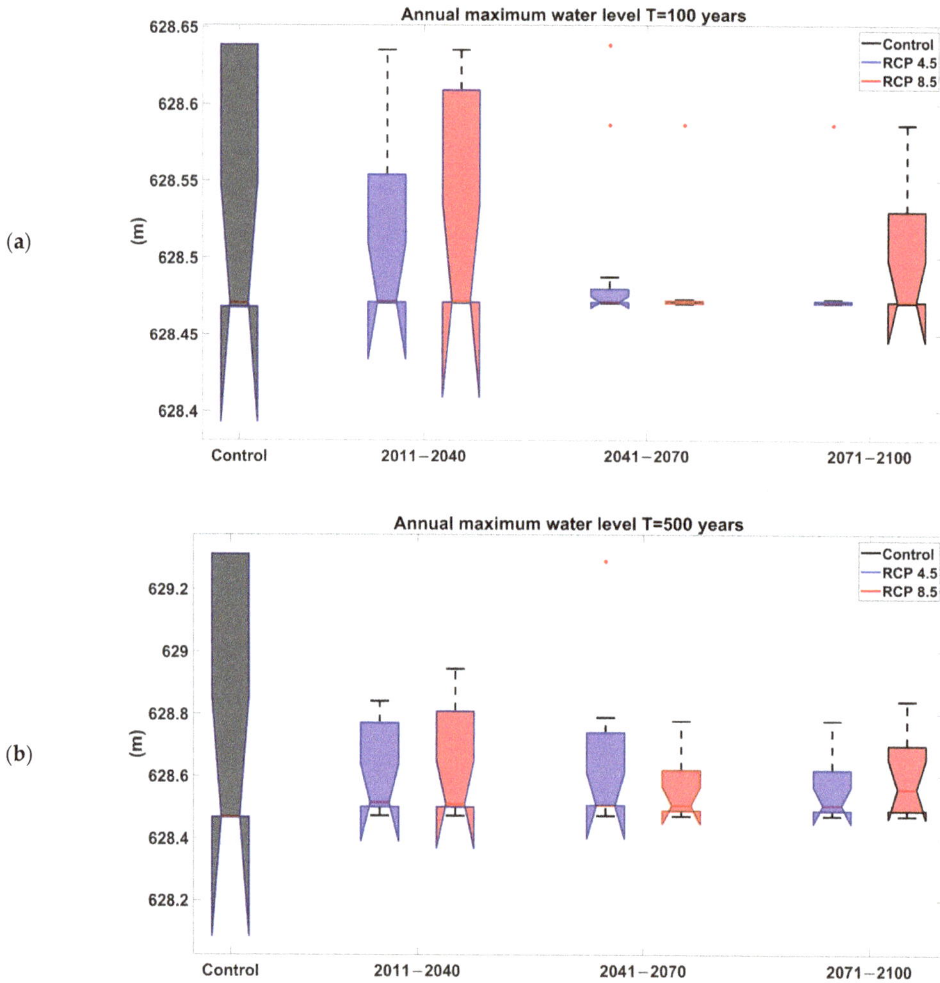

Figure 10. Annual maximum reservoir water level frequency box plotted by using the Gringorten formula. (**a**) Results for the 100-year return period, (**b**) results for the 500-year return period. Red dots represent outliers.

Similar results are obtained in the intermediate future period (2041–2070). Annual maximum peak outflows will be smaller than in the current conditions. The median frequency curves are similar to the previous case. However, a greater dispersion of results in this period is observed. In addition, the results are reversed according to the emission scenario (Figure S3), as in this case, the lowest annual maximum peak outflows are expected in the RCP 8.5 scenario. Nevertheless, the annual maximum peak outflows are small enough and they do not exceed the maximum outflow discharge that can be spilled by gates with a water level equal to the top of the conservation pool. Therefore, the initial reservoir water level is maintained.

In terms of annual maximum reservoir water levels, the median frequency curves are also similar. However, analyzing the results for each climate modeling chain, it can be observed that some models present significant increases in annual maximum reservoir water levels in the RCP 4.5 emission scenario compared with the RCP 8.5 scenario. In the RCP 4.5 scenario, the maximum reservoir water levels will increase for return periods of at least 25 years, whereas in the RCP 8.5 scenario, such increases will occur from the 50-year return period (Figure S4).

Finally, the last future period (2071–2100) presents similar results to the previous periods for the RCP 4.5 emission scenario. However, significantly lower annual maximum peak outflows are expected in the RCP 4.5 emission scenario (Figure S5). In addition, in the RCP 4.5 emission scenario, the 33rd and 67th percentiles are close to the median, while in the RCP 8.5 scenario, a higher dispersion is observed, as can be seen in the 33rd and 67th percentiles.

The reservoir levels in this period are similar to the previous cases. The median remains constant in both emission scenarios. A small variation is observed in the climate modeling chains, in the RCP 4.5 emission scenario, variations in the reservoir level can be observed for return periods of at least 50 years, while in the RCP 8.5 emission scenario, the reservoir level increases for some models from the 25-year return period.

Small changes in reservoir levels are observed regardless of climate variability, as the reservoir level fixed to the top of the conservation pool can be maintained while the outflow discharge is equal or smaller than the maximum spillway capacity. Spillway gate operation consists of opening the gates gradually to release an outflow discharge equal to the inflow discharge. Therefore, the initial reservoir level is maintained until the inflow discharge exceeds the maximum spillway capacity. Consequently, reservoir level increases are observed only when the inflow discharge is higher than the spillway capacity.

The results show a significant decrease in reservoir outflows due to climate change. This is in line with the results of other research for northern Spain.

6. Conclusions

A methodology to assess the influence of climate change on hydrological dam safety is presented, based on a stochastic rainfall generator, a continuous rainfall-runoff model and a method to identify the best operation of spillway gates in flood events. The methodology has been applied to the Eugui Dam in Spain. Climate projections of 12 climate modeling chains in three subperiods in the future and two emission scenarios supplied by the EURO-CORDEX program have been used.

In the Eugui Dam, future annual maximum peak outflows are expected to be equal to or lower than those of the current conditions in the reference period. Though rainfall in this catchment is expected to be similar to the current conditions, the number of wet days per year will be smaller in the future. In addition, the rainfall is expected to be more intense in the future. However, the soil moisture content is expected to be drier at the beginning of flood events in the future. Therefore, initial soil moisture condition is expected to be drier, and rainfall infiltration will be higher, absorbing a higher part of the initial precipitation in storms than in the current conditions. Consequently, runoff amounts are expected to be smaller in the future.

No significant variations in the annual maximum peak outflow frequency curves are observed in terms of the emission scenario, considering the median values of the 12 climate modeling chains. However, analyzing the climate modeling chains separately, a larger variation is observed depending on the period and emission scenario considered.

In the future scenarios, the median frequency curves of annual maximum reservoir water levels in the 12 climate modeling chains remain constant for all return periods in all the scenarios analyzed. As the dam spillway is operated by gates and the inflows are smaller than the maximum outflow discharge that the spillway gates can release with the reservoir water level at the top of the conservation pool, the maximum reservoir levels will not increase in most cases in a given flood event. Nevertheless, in all the cases, the maximum reservoir water levels reached in the future scenarios will be smaller than the water levels in the control scenario, considered as the current scenario. However, the dam crest level will be exceeded in some scenarios, as the conservation pool has been assumed to be full at the beginning of the flood events. Therefore, this methodology could point to the need to establish higher freeboards in the seasons when floods are expected to have a higher frequency. By modifying the freeboard, the effect of climate change can be dealt with in the areas or conditions where it is necessary. This stability is due, to a large extent, to the dam's regulation capacity and efficient management by means of floodgates, which allows floods to be regulated and hydrological security levels to be maintained. In agreement with other studies, which have shown that the inclusion of dams in global climate impact models can substantially reduce population exposure to floods [44].

Despite simulating a large amount of data (time series of 500 years with a time step of 15 min), computation times were acceptable, in the order of minutes. This computation speed has allowed simulations to be carried out using the projections of 12 climate modeling chains, three time periods in the future and two emission scenarios (72 simulations) in a short period of time. In addition, the sub-daily scale of rainfall and runoff time series made it possible to extract flood hydrographs that could be considered as input into the reservoir model.

The methodology proposed will allow a revision of the hydrological dam safety in dams with gated spillways, considering climate change and assessing its influence on hydrological dam safety.

Supplementary Materials: The following supporting information can be downloaded at: https://www.mdpi.com/article/10.3390/hydrology12060153/s1, Figure S1: Annual maximum peak outflow frequency curves plotted by using the Gringorten formula. (a) The first period (2011–2040) and RCP 4.5 emission scenario, (b) the first period (2011–2040) and RCP 8.5 emission scenario; Figure S2: Annual maximum reservoir water level frequency curves plotted by using the Gringorten formula. (a) The first period (2011–2040) and RCP 4.5 emission scenario, (b) the first period (2011–2040) and RCP 8.5 emission scenario; Figure S3: Annual maximum peak outflow frequency curves plotted by using the Gringorten formula. (a) The second period (2041–2070) and RCP 4.5 emission scenario, (b) the second period (2041–2070) and RCP 8.5 emission scenario; Figure S4: Annual maximum reservoir water level frequency curves plotted by using the Gringorten formula. (a) The second period (2041–2070) and RCP 4.5 emission scenario, (b) the second period (2041–2070) and RCP 8.5 emission scenario; Figure S5: Outflow frequency curves plotted by using the Gringorten formula. (a) The third period (2071–2100) and RCP 4.5 emission scenario, (b) the third period (2071–2100) and RCP 8.5 emission scenario; Figure S6: Annual maximum reservoir water level frequency curves plotted by using the Gringorten formula. (a) The third period (2071–2100) and RCP 4.5 emission scenario, (b) the third period (2071–2100) and RCP 8.5 emission scenario.

Author Contributions: Conceptualization, L.M. and A.P.; methodology, E.S., L.M., A.P., D.L.D.L., C.A. and S.G.; software, A.P., C.A. and S.G.; validation, E.S.; formal analysis, E.S.; investigation, E.S.; resources, L.M. and A.P.; data curation, E.S.; writing—original draft preparation, E.S., L.M. and A.P.; writing—review and editing, E.S., L.M., A.P. and D.L.D.L.; visualization, E.S.; supervision, L.M. and A.P.; project administration, L.M. and A.P.; funding acquisition, L.M., A.P. and S.G. All authors have read and agreed to the published version of the manuscript.

Funding: This research was funded by the project 'SAFERDAMS: Assessment of the impact of climate change on hydrological dam safety' (PID2019-107027RB-I00/AEI/10.13039/501100011033) and by the grant for pre-doctoral contracts to train PhD students PRE2020-091861, both funded by the Spanish Ministry of Science and Innovation. This research was also funded by the "Progetto ECS 0000024 Rome Technopole": CUP B83C22002820006, PNRR Missione 4 Componente 2 Investimento 1.5, finanziato dall'Unione europea—NextGenerationEU. The international collaboration with the Tuscia University for the realization of this research has been supported by the 'Programa Propio UPM 2021' funded by the Universidad Politécnica de Madrid with the collaboration of Santander bank.

Data Availability Statement: The original contributions presented in this study are included in the article. Further inquiries can be directed to the corresponding author.

Acknowledgments: The authors would like to acknowledge the Instituto Geográfico Nacional (IGN) for supplying the DTM, and the SAIH real-time system of the River Ebro Basin Authority for supplying the streamflow data.

Conflicts of Interest: The authors declare no conflict of interest.

Abbreviations

The following abbreviations are used in this manuscript:

AEMET	Agencia Estatal de METeorología
CN	Cuve Number
CN4GA	Curve Number for Green-Ampt
CORDEX	COoRdinated Reginal Climate Downscaling Experiment
COSMO4SUB	Continous Simlation Model for Small and Ungauged Basin)
DTM	Digitnal Terrain Model
NRSP	Neyman–Scott Rectangular Pulses
RCP	Representative Concentration Pathways
SRG	Stochastic Rainfall Generator
VEM	Volumetric Evaluation Method
WFIUH	Width Function-Based Instantaneous Unit Hydrograph

References

1. Lave, L.B.; Balvanyos, T. Risk analysis and management of dam safety. *Risk Anal.* **1998**, *18*, 455–462. [CrossRef] [PubMed]
2. Baecher, G.B.; Paté, M.E.; De Neufville, R. Risk of dam failure in benefit-cost analysis. *Water Resour. Res.* **1980**, *16*, 449–456. [CrossRef]
3. Apollonio, C.; Balacco, G.; Gioia, A.; Iacobellis, V.; Piccinni, A.F. Flood Hazard Assessment of the Fortore River Downstream the Occhito Dam, in Southern Italy. In *Computational Science and Its Applications—ICCSA 2017*; Lecture Notes in Computer Science; Gervasi, O., Murgante, B., Misra, S., Borruso, G., Torre, C.M., Rocha, A.M.A.C., Taniar, D., Apduhan, B.O., Stankova, E., Cuzzocrea, A., Eds.; Springer: Cham, Switzerland, 2017; Volume 10405. [CrossRef]
4. International Commission on Large Dams (ICOLD). *Lessons from Dam Incidents*, reduced ed.; ICOLD: Paris, France, 1973.
5. Sivapalan, M.; Takeuchi, K.; Franks, S.W.; Gupta, V.K.; Karambiri, H.; Lakshmi, V.; Liang, X.; McDonnell, J.J.; Mendiondo, E.M.; O'connell, P.E.; et al. IAHS decade on Predictions in Ungauged Basins (PUB), 2003–2012: Shaping an exciting future for the hydrological sciences. *Hydrol. Sci. J.* **2003**, *48*, 857–880. [CrossRef]
6. Młyński, D.; Wałęga, A.; Petroselli, A.; Tauro, F.; Cebulska, M. Estimating Maximum Daily Precipitation in the Upper Vistula Basin, Poland. *Atmosphere* **2019**, *10*, 43. [CrossRef]
7. Onof, C.; Chandler, R.E.; Kakou, A.; Northrop, P.; Wheater, H.S.; Isham, V. Rainfall modelling using Poisson-cluster processes: A review of developments. *Stoch. Environ. Res. Risk Assess.* **2000**, *14*, 384–411. [CrossRef]
8. Wheater, H.S.; Chandler, R.E.; Onof, C.J.; Isham, V.S.; Bellone, E.; Yang, C.; Lekkas, D.; Lourmas, G.; Segond, M.-L. Spatial-temporal rainfall modelling for flood risk estimation. *Stoch. Environ. Res. Risk Assess.* **2005**, *19*, 403–416. [CrossRef]
9. Ritschel, C.; Ulbrich, U.; Névir, P.; Rust, H.W. Precipitation extremes on multiple timescales—Bartlett–Lewis rectangular pulse model and intensity–duration–frequency curves. *Hydrol. Earth Syst. Sci.* **2017**, *21*, 6501–6517. [CrossRef]
10. De Luca, D.; Petroselli, A.; Galasso, L. A Transient Stochastic Rainfall Generator for Climate Changes Analysis at Hydrological Scales in Central Italy. *Atmosphere* **2020**, *11*, 1292. [CrossRef]

11. Willems, P.; Arnbjerg-Nielsen, K.; Olsson, J.; Nguyen, V. Climate change impact assessment on urban rainfall extremes and urban drainage: Methods and shortcomings. *Atmos. Res.* **2012**, *103*, 106–118. [CrossRef]
12. Maraun, D. Bias Correcting Climate Change Simulations—A Critical Review. *Curr. Clim. Change Rep.* **2016**, *2*, 211–220. [CrossRef]
13. Kendon, E.J.; Roberts, N.M.; Fowler, H.J.; Roberts, M.J.; Chan, S.C.; Senior, C.A. Heavier summer downpours with climate change revealed by weather forecast resolution model. *Nat. Clim. Change* **2014**, *4*, 570–576. [CrossRef]
14. Ban, N.; Schmidli, J.; Schär, C. Heavy precipitation in a changing climate: Does short-term summer precipitation increase faster? *Geophys. Res. Lett.* **2015**, *42*, 1165–1172. [CrossRef]
15. Viglione, A.; Blöschl, G. On the role of storm duration in the mapping of rainfall to flood return periods. *Hydrol. Earth Syst. Sci.* **2009**, *13*, 205–216. [CrossRef]
16. Blöschl, G.; Sivapalan, M.; Wagener, T.; Viglione, A.; Savenije, H. (Eds.) *Runoff Prediction in Ungauged Basins: Synthesis Across Processes, Places and Scales*; Cambridge University Press: Cambridge, UK, 2013.
17. Haberlandt, U.; Radtke, I. Hydrological model calibration for derived flood frequency analysis using stochastic rainfall and probability distributions of peak flows. *Hydrol. Earth Syst. Sci.* **2014**, *18*, 353–365. [CrossRef]
18. Grimaldi, S.; Volpi, E.; Langousis, A.; Papalexiou, S.M.; De Luca, D.L.; Piscopia, R.; Nerantzaki, S.D.; Papacharalampous, G.; Petroselli, A. Continuous hydrologic modelling for small and ungauged basins: A comparison of eight rainfall models for sub-daily runoff simulations. *J. Hydrol.* **2022**, *610*, 127866. [CrossRef]
19. Blöschl, G.; Hall, J.; Viglione, A.; Perdigão, R.A.; Parajka, J.; Merz, B.; Lun, D.; Arheimer, B.; Aronica, G.T.; Bilibashi, A.; et al. Changing climate both increases and decreases European river floods. *Nature* **2019**, *573*, 108–111. [CrossRef] [PubMed]
20. Kundzewicz, Z.W.; Pińskwar, I.; Brakenridge, G.R. Large floods in Europe, 1985–2009. *Hydrol. Sci. J.* **2013**, *58*, 1–7. [CrossRef]
21. IPCC. *Climate Change 2023: Synthesis Report. Contribution of Working Groups I, II and III to the Sixth Assessment Report of the Intergovernmental Panel on Climate Change*; Core Writing Team, Lee, H., Romero, J., Eds.; IPCC: Geneva, Switzerland, 2023; pp. 35–115. [CrossRef]
22. Arnell, N.W.; Reynard, N.S. The effects of climate change due to global warming on river flows in Great Britain. *J. Hydrol.* **1996**, *183*, 397–424. [CrossRef]
23. Giuntoli, I.; Vidal, J.P.; Prudhomme, C.; Hannah, D.M. Future hydrological extremes: The uncertainty from multiple global climate and global hydrological models. *Earth Syst. Dyn.* **2015**, *6*, 267–285. [CrossRef]
24. Dankers, R.; Arnell, N.W.; Clark, D.B.; Falloon, P.D.; Fekete, B.M.; Gosling, S.N.; Heinke, J.; Kim, H.; Masaki, Y.; Satoh, Y.; et al. First look at changes in flood hazard in the inter-sectoral impact model intercomparison project ensemble. *Proc. Natl. Acad. Sci. USA* **2014**, *111*, 3257–3261. [CrossRef]
25. Roudier, P.; Andersson, J.C.M.; Donnelly, C.; Feyen, L.; Greuell, W.; Ludwig, F. Projections of future floods and hydrological droughts in Europe under a +2 °C global warming. *Clim. Change* **2016**, *135*, 341–355. [CrossRef]
26. Alfieri, L.; Burek, P.; Feyen, L.; Forzieri, G. Global warming increases the frequency of river floods in Europe. *Hydrol. Earth Syst. Sci.* **2015**, *19*, 2247–2260. [CrossRef]
27. Lehner, B.; Döll, P.; Alcamo, J.; Henrichs, T.; Kaspar, F. Estimating the Impact of Global Change on Flood and Drought Risks in Europe: A Continental, Integrated Analysis. *Clim Change* **2006**, *75*, 273–299. [CrossRef]
28. Jacob, D.; Petersen, J.; Eggert, B.; Alias, A.; Christensen, O.B.; Bouwer, L.M.; Braun, A.; Colette, A.; Déqué, M.; Georgievski, G.; et al. EURO-CORDEX: New high-resolution climate change projections for European impact research. *Reg. Environ. Change* **2014**, *14*, 563–578. [CrossRef]
29. Garijo, C.; Mediero, L.; Garrote, L. Usefulness of AEMET generated climate projections for climate change impact studies on floods at national-scale (Spain). *Ing. Agua* **2018**, *22*, 153–166. [CrossRef]
30. Burton, A.; Fowler, H.J.; Blenkinsop, S.; Kilsby, C.G. Downscaling transient climate change using a Neyman–Scott Rectangular Pulses stochastic rainfall model. *J. Hydrol.* **2010**, *381*, 18–32. [CrossRef]
31. Wilks, D.S. Adapting stochastic weather generation algorithms for climate change studies. *Clim. Change* **1992**, *22*, 67–84. [CrossRef]
32. Fowler, H.J.; Kilsby, C.G.; O'connell, P.E.; Burton, A. A weather-type conditioned multi-site stochastic rainfall model for the generation of scenarios of climatic variability and change. *J. Hydrol.* **2005**, *308*, 50–66. [CrossRef]
33. Gabriel-Martin, I.; Sordo-Ward, A.; Garrote, L.; Castillo, L.G. Influence of initial reservoir level and gate failure in dam safety analysis. Stochastic approach. *J. Hydrol.* **2017**, *550*, 669–684. [CrossRef]
34. Sharafati, A.; Yaseen, Z.M.; Pezeshki, E. Strategic assessment of dam overtopping reliability using a stochastic process approach. *J. Hydrol. Eng.* **2020**, *25*, 04020029. [CrossRef]
35. King, L.M.; Simonovic, S.P. A deterministic Monte Carlo simulation framework for dam safety flow control assessment. *Water* **2020**, *12*, 505. [CrossRef]
36. De Luca, D.L.; Petroselli, A. STORAGE (STOchastic RAinfall GEnerator): A user-friendly software for generating long and high-resolution rainfall time series. *Hydrology* **2021**, *8*, 76. [CrossRef]

37. Girón, F.; Yagüe, J.; Martínez, R. Flood routing in reservoirs base on hydrological forecasting. In Proceedings of the Twentieth International Congress on Large Dams, Beijing, China, 19–22 September 2000; Rapports on Question 79, Response 25 (Q79-R25). International Commission on Large Dams: Paris, France, 2000; Volume 4, pp. 403–417.
38. Grimaldi, S.; Nardi, F.; Piscopia, R.; Petroselli, A.; Apollonio, C. Continuous hydrologic modelling for design simulation in small and ungauged basins: A step forward and some tests for its practical use. *J. Hydrol.* **2021**, *595*, 125664. [CrossRef]
39. Grimaldi, S.; Petroselli, A.; Romano, N. Curve-Number/Green-Ampt mixed procedure for streamflow predictions in ungauged basins: Parameter sensitivity analysis. *Hydrol. Process.* **2013**, *27*, 1265–1275. [CrossRef]
40. Natural Resources Conservation Service (NRCS). *Part 630 Hydrology, National Engineering Handbook*; U.S. Department of Agriculture: Washington, DC, USA, 2008.
41. Green, W.H.; Ampt, G.A. Studies on soil physics. *J. Agric. Sci.* **1911**, *4*, 1–24. [CrossRef]
42. EEA. *CLC2018 Technical Guidelines—European Environment Agency—European Topic Centre on Urban, Land and Soil Systems*; EEA: Wien, Austria, 2019.
43. Gringorten, I.I. A plotting rule for extreme probability paper. *J. Geophys. Res.* **1963**, *68*, 813–814. [CrossRef]
44. Boulange, J.; Hanasaki, N.; Yamazaki, D.; Pokhrel, Y. Role of dams in reducing global flood exposure under climate change. *Nat. Commun.* **2021**, *12*, 417. [CrossRef]

hydrology MDPI

Article

Regional Groundwater Flow and Advective Contaminant Transport Modeling in a Typical Hydrogeological Environment of Northern New Jersey

Toritseju Oyen * and Duke Ophori

Department of Earth and Environmental Studies, Montclair State University, 1 Normal Avenue, Montclair, NJ 07043, USA; ophorid@montclair.edu
* Correspondence: oyent1@montclair.edu

Abstract: This study develops a numerical model to simulate groundwater flow and contaminant transport in a "typical hydrogeological environment" of northern New Jersey, addressing freshwater decline. Focusing on the Lower Passaic water management area (WMA), we model chloride transport in a fractured-rock aquifer, where fracture networks control hydraulic conductivity and porosity. The urbanized setting—encompassing Montclair State University (MSU) and municipal wells—features heterogeneous groundwater systems and critical water resources, providing an ideal case study for worst-case contaminant transport scenarios. Using MODFLOW and MODPATH, we simulated flow and tracked particles over 20 years. Results show that chloride from MSU reached the Third River in 4 years and the Passaic River in 10 years in low-porosity fractures (0.2), with longer times (8 and 20 years) in high-porosity zones (0.4). The First Watchung Mountains were identified as the primary recharge area. Chloride was retained in immobile pores but transported rapidly through fractures, with local flow systems (MSU to Third River) faster than regional systems (MSU to Passaic River). These findings confirm chloride in groundwater, which may originate from road salt application, can reach discharge points in 4–20 years, emphasizing the need for recharge-area monitoring, salt-reduction policies, and site-specific assessments to protect fractured-rock aquifers.

Keywords: groundwater salinization; deicing salts; MODFLOW; MODPATH; fractured-rock aquifer; contaminant transport; northern New Jersey

1. Introduction

The rapid unchecked growth of urbanization and industrialization is significantly deteriorating water quality in New Jersey, as these anthropogenic activities continue to degrade the environment and compromise water resources, transforming land use and landcover patterns throughout the region [1]. This disturbing trend is documented by numerous studies indicating the decline in groundwater quality across the United States [2–5]. The complexities of groundwater pollution are multifaceted, encompassing both anthropogenic activities and uncontrollable natural events. Agricultural practices, while essential for food production, employ fertilizers that can compromise water resources. The application of deicing salts, another critical activity, has been heavily linked to salinization of groundwater resources [2,6]. Also, uncontrollable events such as heavy precipitation, flooding, and hurricanes can spread contaminants and worsen water resource degradation, particularly in urbanized regions like New Jersey. New Jersey's aging combined sewer systems are susceptible to overflow during flooding events, resulting in the deterioration of groundwater quality [2,6]. The impacts of events such as flooding and hurricanes could result in the degradation of water resources in urbanized regions like New Jersey. New Jersey has aging combined sewer systems that combine storm water and municipal sewage in the same sewer line. During heavy flooding events, there is overflow from the overwhelmed sewer system through manholes, and these sewer effluents flow into important water resources. The infiltration of this effluent into groundwater leads to a decline in groundwater quality. This occurs through point and non-point pollution sources. Deicing salt application is a major contributor to groundwater salinization [2,6,7]. Oyen and Ophori [8,9] revealed, in their studies on the spatiotemporal evaluation of groundwater, that the lower Passaic watershed experienced significant decline in freshwater quality when compared to many watersheds in northern New Jersey. Groundwater salinization driven by rising chloride concentrations may appear less urgent due to chloride's secondary regulatory status (maximum contaminant level [MCL] = 250 mg/L [10,11]). However, widespread exceedances of this threshold have been documented [2,12], and the potential of this phenomenon to compromise water quality warrants significant concern [6,7,10,13,14]. Also, increasing groundwater salinization has been shown to mobilize contaminants, including heavy metals [15,16], and poses a recurring challenge during remediation of organic compounds such as Trichloroethylene (TCE) [17,18].

Elevated salinity levels can facilitate the movement of contaminants from the soil matrix into groundwater. The presence of sodium ions in deicing salts can trigger cation exchange in soils, releasing heavy metals such as lead and zinc, which can then be transported to surface water and groundwater via runoff. A groundwater study done in Connecticut by McNaboe et al. [19] revealed that highly saline groundwater had a high concentration of Radon (Rn). It was observed that sodium ion from deicing salts mobilizes Rn to the subsurface. Numerous studies conducted across the United States have closely associated deicing salt use and expansive impervious surface areas in urban environments to rising levels of groundwater salinity [12,13,20]. Also, deicing salt applications have been linked to ecosystem degradation, as elevated salinity levels facilitate the increase of invasive species, thereby displacing native species from their habitats [2,20,21].

An additional concern associated with high-salinity groundwater is the phenomenon of "salting out", which involves a reduction in solubility of dissolved constituents in response to increased ionic strength within a solution [22]. Newell et al. [22] highlighted the complexities of remediating groundwater contaminated with per-and polyfluoroalkyl substances (PFAs), particularly in highly saline water resources. The salting-out effect process can lead to enhanced retention of PFAs in aquifer materials, rendering remediation efforts more challenging. This issue is more critical in regions characterized by fractured rock aquifers, such as the northern New Jersey hydrogeologic system. In fresh groundwater impacted with elevated salinity, PFAs become more persistent due to decreased solubility and increased sorption to aquifer materials. Following remediation of contaminated wells, contaminants trapped in aquifer pores are slowly re-introduced into the remediated wells, frustrating efforts of remediation [17,18].

The proliferation of anthropogenic activities has led to an increase in chloride availability in the environment, contributing to the salinization of freshwater resources [2,6]. The conservative properties of dissolved chlorides allow them to be resistant to degradation and sorption processes [23], making them an effective tracer in hydrologic investigations of metals and contaminated organic compounds. The profound influence of deicing salts on groundwater salinization warrants increased attention and awareness. Despite their perceived benefits, these salts pose a significant threat to the environment.

Groundwater modeling tools such as MODFLOW and MODPATH offer a valuable tool for assessing the transport dynamics of chlorides within the groundwater systems.

These groundwater models can quantify key parameters, such as travel distances and times, and identify potential contaminant point sources. By employing chloride as a surrogate for certain contaminants, the transport of these contaminants can be effectively modeled within the study area. By modeling the groundwater systems, these models facilitate informed management of subsurface resources [24–26]. The accuracy of groundwater models is rigorously evaluated through a calibration process, where hydraulic parameters such as hydraulic heads and recharge rates are manually adjusted to match calculated and observed hydraulic properties. This approach has proven effective in numerous studies [24,26]. Additionally, statistical summaries like root mean square error (RMSE) and mean standard error (MSE) are employed to validate model performance. Low MSE and RMSE values typically suggest that a model has an adequate fit. Nevertheless, researchers must exercise caution when interpreting model results, as calibration data should adequately represent actual field conditions [24].

Groundwater models are widely used to estimate aquifer properties across diverse geological settings and regions. Numerous groundwater modeling studies have been conducted in New Jersey [25,27–29], providing valuable insights into the understanding and effective management of groundwater systems. These studies are particularly relevant in northern New Jersey, where complex hydrogeological features prevail. Notably, Barry et al. [27] employed the MODFLOW model to simulate groundwater flow in the Central Passaic River Basin (CPRB) of New Jersey, generating a detailed groundwater recharge map. Also, Ophori et al. [12] developed a numerical model to simulate groundwater–surface water interaction during historical and recent pumping of groundwater at the CPRB. Serfes et al. [29] employed chemical equilibrium modeling to evaluate arsenic mobilization from Pyrite in the Passaic and Lockatong Formations of the Newark Basin, New Jersey, building their study on earlier studies that suggest that arsenic contaminations in groundwater are closely associated with pyrite oxidation. Lewis-Brown et al. [28] used groundwater modeling to assess groundwater contaminant transport in a fractured rock aquifer of New Jersey's Lockatong Formation, estimating travel times and flow paths.

This current study develops a comprehensive numerical model to investigate the transport dynamics of deicing constituents (chloride particles) within the sedimentary rock framework of the Brunswick aquifer, with a particular emphasis on determining the flow regime, particle tracking, and evaluating the sources and discharge points of chloride ions.

Urbanized fractured-rock aquifers face growing threats from deicing salts, yet predicting contaminant travel times remains fraught with uncertainty. This study tackles this gap by asking (1) How quickly can chloride infiltrating at First Watchung Mountain (Montclair State University) reach critical discharge points? and (2) How do porosity variations control these timelines? Using MODFLOW/MODPATH, we quantify transport dynamics within the sedimentary rock framework of northern New Jersey's Brunswick aquifer.

While previous studies have examined groundwater flow in the Passaic Basin [27,28], this study develops a numerical framework to quantify how fracture-network porosity controls contaminant transport timescales in a densely urbanized, salt-vulnerable setting. Our model uniquely integrates (1) site-specific hydraulic conductivity variations, (2) dual-porosity scenarios (0.2 vs. 0.4) to resolve discrepancies in prior travel-time estimates, and (3) backward particle tracking to identify municipal well recharge sources. These innovations provide a framework for predicting contaminant persistence in fractured-rock aquifers—a critical need for regions grappling with deicing salt impacts and long-term remediation challenges.

1.1. Study Area

The study area covers about 105 km^2, underlain by the Brunswick aquifer of the Newark Basin, within the Lower Passaic watershed in New Jersey. It is bounded by the Peckman River on the west, the Passaic River on the north and east, and the Upper Montclair community on the south. The study area (Figure 1), which is mostly urbanized, has a commuter university community located in Montclair Township. The study area also covers parts of Clifton, Paterson, Cedar Grove, Little Falls, Glen Ridge, Bloomfield, Belleville, and Nutley (comprising both Essex and Passaic counties) of New Jersey. Montclair State University (MSU), a commuter university, is where over 20,000 students attend. Three municipal wells (Glenfield, Lorraine, and Rand wells) are located in Montclair Township and supply water to Montclair Township and nearby communities.

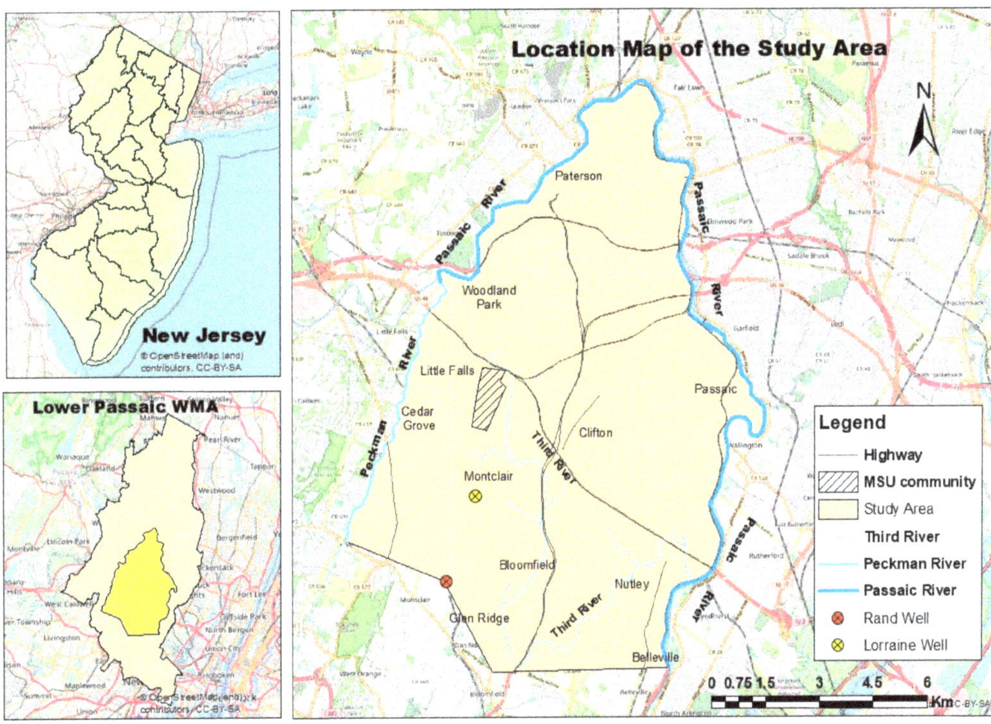

Figure 1. Location of the study area.

Of the three wells, two (Lorraine well and Rand well) are located within the study area. The study area serves as a microcosm of northern New Jersey, reflecting its "typical hydrogeological environment". Within this area, the Montclair State University region, located on parts of the First Watchung Mountain, is identified as critically important, not only due to its geographic position on high topographic relief and its role as a primary recharge zone, but also because of the daily gathering of large groups of individuals within the university community, which can lead to heightened environmental impacts. While large populations are often associated with increased environmental contamination, significant measures are implemented to ensure hazard-free conditions as individuals gather in the university community. At times, these measures may have unintended negative consequences for the environment. One example is the application of deicing salts during snowy and icy conditions, which, while ensuring safety and normalcy, can pose environmental risks. These deicing salts (mostly chloride constituents) infiltrate into groundwater and thereby increase groundwater salinity. These attributes described in the university community and other considerations guided the selection of the study area. Another important consideration is the location of three public supply wells around the study area, of which one is indicated to be contaminated. In a 2022 water report, the three wells were rated to be highly susceptible to volatile organic compounds (VOCs), radionuclides, and inorganic contaminants, among others [30]. Both the Glenfield and Lorraine wells currently run with attached carbon absorbers, while the Rand well does not run currently. According to Staff [31], Montclair Township has secured federal government funds to rehabilitate the Rand municipal supply well, and to address concerns associated with elevated levels of perfluorooctane sulfonate (PFOS) and perchlorate contaminants. The supposed Rand well rehabilitation could be one reason why water reports from Montclair Water Bureau [30] indicate that Rand well is not running.

1.2. Geologic Setting

The area is underlain by the Brunswick aquifer, an Early Mesozoic-age aquifer that forms part of the Newark Basin in New Jersey. Composed primarily of consolidated sedimentary and igneous rock units from the late Triassic and early Jurassic periods, the Brunswick aquifer serves as a true reflection and microcosm of northern New Jersey's geology. Its complex geological history includes a variety of formations, such as the Passaic Formation, Orange Mountain Basalt, and Feltville Formation. The Passaic Formation primarily consists of red mudstone and siltstone, with lesser amounts of gray and black shale. It is notable for its fine-grained facies, which transition laterally into sandstone and conglomerate in the basin's margins. This formation can reach a maximum thickness of about 11,800 feet, making it one of the most substantial units within the region. The Orange Mt. Basalt is mostly dark-green-gray to greenish-black basalt which is part of the series of "interlayered basalt and clastic sedimentary" rocks of early Jurassic age, with a layer thickness of up to 597 feet. It consists of several basalt flows that were fed by diabase sheets and dikes, which intruded and thermally metamorphosed older sedimentary deposits. The Feltville Formation is included in the sequence of sedimentary rocks that were deposited in a variety of environments, including fluvial-deltaic settings. It is interlayered with basalt flows, indicating a dynamic geological history. The Feltville Formation consists of interbedded brownish red to light-grayish-red, fine-to-coarse-grained stone, gray and black coarse siltstone, and silty mudstone. This formation reaches a layer thickness of about 510 feet [32–34].

1.3. Hydrogeological Setting

The Brunswick aquifer is part of the fractured rock aquifer of Newark Basin; it consists of sandstone, conglomerate, siltstone, and shale. Fractures act as storage and conduits for groundwater in this aquifer, and water is usually fresh, non-corrosive, and calcium-bicarbonate dominated. The aquifer yields a median value of over 100 to 250 gallons per minute (gpm) [35]. The aquifer displays heterogeneous characteristics, marked by spatial variations in permeability and transmissivity across different strata. Notably, deep formations exhibit low permeability (transmissivity less than 1 gpd/ft), whereas shallow formations can have high transmissivity (up to 30,000 gpd/ft). Groundwater flow within the aquifer is influenced by local topographic gradients and the stratigraphic arrangement of the units [33]. The topography of the study area ranges from about 50 ft to 600 ft (15 m to 183 m). In the Brunswick aquifer, groundwater flows across various strata through fractures [29]. Groundwater storage primarily occurs within extensive fracture networks, while long-distance transmission is facilitated by bedding parts and highly permeable units between confining beds. The Brunswick aquifer is characterized by a complex network of fractures that significantly influence its hydrogeological properties [33]. These hydrogeologic features of the Brunswick aquifer present a system with varied hydrogeological properties, such as varied hydraulic conductivity and porosity for the aquifer formation and the water-bearing fractures (WBFs).

1.4. The Typical Hydrogeological Environment of Northern New Jersey

The characteristic features of the microcosm of the hydrogeologic environment of northern New Jersey may be summarized by its topography, climate, vegetation, geology, and land use. The topographic environment varies from the rolling hills of the piedmont region to the highlands and the Ridge and Valley provinces [36]. There are regularly alternating topographic peaks and troughs that can be simplified to a local surface appearing sinusoidal. This surface could control the water table that determines the groundwater flow. Regional slopes measured from topographic maps having a scale of 1:250,000 average 0.02. The climate is humid continental with moderate cold, snowy winters and warm humid summers. The average annual precipitation is 40 inches, and freeze-free days range from 163 to 217 [37]. Northern New Jersey is characterized by hardwood forest, with dominant trees like sugar maple, red maple, beech, yellow birch, sweet birch, and red oak, which play a role in the detention of surface runoff to recharge groundwater [38]. Most of northern New Jersey is covered by the Newark Basin, which includes the Brunswick aquifer. The Newark Basin consists of conglomerates, siltstone, and shale. Prominent fractures are known to store and conduct groundwater in the basin. The Brunswick aquifer is of early Mesozoic age [33]. It is a highly fractured aquifer of mainly sedimentary and igneous rock of the late Triassic and early Jurassic periods. It is a prolific aquifer with yields of up to 100 to 250 gallons per minute [35]. The northeast part of northern New Jersey is highly urbanized. This urbanization has led to pollution due to the application of deicing salts on the urban roads during snowy and icy conditions [12]. For the aforementioned reasons, the Montclair State University region and its environs is identified as a "typical environment" of northern New Jersey. It is an area in which the daily gathering of large groups of people can lead to significant environmental problems.

2. Conceptual Model Design and Calibration

This study applies the techniques of groundwater flow models to describe the groundwater flow conditions of the study area. The study employed the use of field observations and well data to develop a 3-D numerical groundwater flow model to simulate the groundwater flow conditions. A finite difference code, MODFLOW [39] which is integrated in the Groundwater Modeling System (GMS), was used to execute the simulation of steady-state groundwater flow. Details on the application and concepts with regard to the theory of groundwater flow and the use of MODFLOW are found in Harbaugh et al. [39]. This code is used to provide a conceptual model framework of the groundwater flow. The geologic map of the study area was imported using GIS tools within the GMS software and indicated as the base map used to develop a polygon layer for the model simulation. The model domain was built around this polygon layer. By the sides of the polygon are the rivers bounding the study area which were assigned general head boundary. The topographic map of the area (Figure 2a) was integrated to provide spatial elevation data for the ground surface at locations corresponding to the wells used in the study.

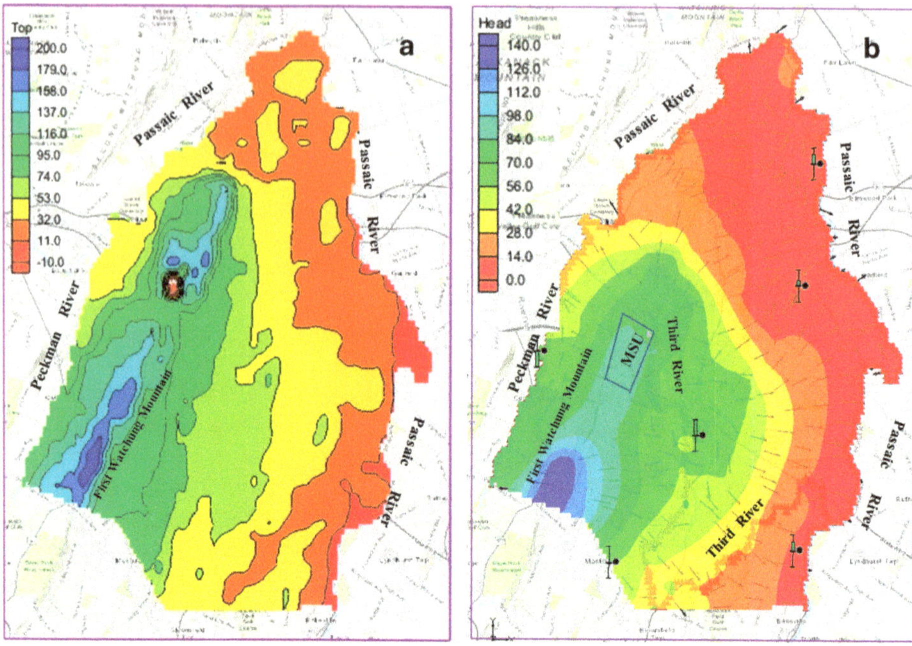

Figure 2. Map showing (**a**) topography, (**b**) calibrated head of the study area.

Within the model coverage, the study area was subdivided into zones based on hydraulic conductivity and recharge patterns, informed by land use, and the geology and topography of the area. The hydraulic conductivity values were assigned using the available observed data. If such data were unavailable, the standard literature ranges relevant to the study area's geology informed the determination of the hydraulic conductivity values. The recharge was ascertained from precipitation in the study area, and the recharge values were varied based on the land use and land cover in the area. Anderson et al. [24] indicated that spatial variations in recharge are highly influenced by not just precipitation but local topography, permeability, the water table, and several other hydrogeological properties and conditions. The model layers consisted of, from top to bottom, an unconsolidated water-bearing formation on the top surface to a no-flow boundary at the bottom bedrock layer. Since permeability in rocks has been determined to decrease with depth, the model assumes that the hydraulic conductivities of the rocks decrease with depth. The area is bounded by a small fracture filled with permeable materials on the west part of the study area. Well and water level data were integrated into coverage layers to serve as observational tools. The observed water level and elevation data, as well as the recharge and hydraulic conductivity values, were used to calibrate the hydraulic heads in the study area. The top elevation values were assigned as the starting heads for the top layer. The geology of the Brunswick aquifer has been extensively studied and documented [5,32–34]. Information from these studies guided the determination of the number of layers used in the simulation of the groundwater flow. The model consists of ten layers representing the Brunswick aquifer. The observation wells in the model are developed in varied geologic units, claystone, shale, siltstone, and sandstone.

After adequate representation of the groundwater system on the model coverages, a grid frame was constructed to facilitate the groundwater model. The MODFLOW model was set up with a 12,500 m by 17,500 m grid dimension, and the simulation was run under steady state conditions. The study area was divided into zones for hydraulic conductivities and recharge on the geological map (Supplementary Materials, Figures S1 and S2). The divisions were advised by the land use, geology, and topography of the area. The groundwater flow across these boundaries was presumed infinitesimal. The river and stream networks were captured under the river coverage, and the stage of rivers was assigned based on the knowledge of the stream patterns in the area.

Source terms included recharge from precipitation, spatially distributed at 5–25% of annual rainfall [40], with higher rates assigned to the First Watchung Mountain recharge zone. Contaminant sources (chloride particles) were introduced at MSU as point sources simulating deicing salt infiltration. Sinks comprised pumping municipal wells (Lorraine and Rand wells) and discharge areas—rivers (Peckman, Third, Passaic).

The calibrated groundwater flow simulation model indicates that groundwater heads in the study area range from about 140 m at high topographic reliefs, such as areas around the First Watchung mountain, to below 5 m at the Passaic River, east of the area (Figure 2). The model was calibrated by adjusting hydraulic parameters to achieve a close match between simulated and observed hydraulic head values. Initial recharge values were modified using the local annual precipitation values [37] to capture the spatial variability more effectively. Recharge values are typically 5–25% of the annual precipitation, and it is mostly dependent on the soil type and landcover [40]. Initial recharge values in some of the portioned zones were reduced by an order of magnitude, as further adjustment resulted in spatially varied coverage across the study area. The calibrated model assumed a uniform recharge rate through the simulation period, with precipitation serving as the primary recharge mechanism. Hydrogeological properties reported in literature [24,27] guided the initial hydraulic conductivity values applied in the model, and the hydraulic conductivity values were manually adjusted during the calibration process. The hydraulic conductivity range obtained from the calibrated model in the study provides a good representation of the Brunswick aquifer and accurately models the study area. The calibrated hydraulic conductivity values closely match field-measured values of the aquifer [41] and fall within the expected range for similar rock types reported in literature and studies in the region [27]. The calibrated hydraulic conductivity values were within the range of 5.59×10^{-6} to 1.96×10^{-3} m/s.

Well data [42] for six observation wells within the study area were used in the calibration of the model. A calibration target bar by the observational wells indicates the amount of error between the observed head and the calculated head. The calibration target bar is green when the error is within acceptable tolerance, while yellow- and red-colored bars show the error associated with the model calculated head is not within the acceptable tolerance. The trial-and-error method is used to adjust the hydraulic parameters targeting an adequately calibrated model until the measured head values closely match the observed heads, with minimal error, and effectively represent the study area. A plot (Figure 3) of observed vs. calculated head provides a deeper visual understanding of the calibrated model.

A sensitivity analysis was performed to evaluate the robustness of the groundwater flow model in the face of minor changes in the aquifer's hydraulic parameters. This investigation focused on the model's sensitivity to certain changes in hydraulic conductivity and recharge values. A systematic approach was employed, wherein the data for each parameter were steadily varied, and the simulation repeated to quantify the resulting deviations in hydraulic heads. The RMSE was monitored after each sensitivity analysis run, to assess the model's stability and deviation from the calibrated state.

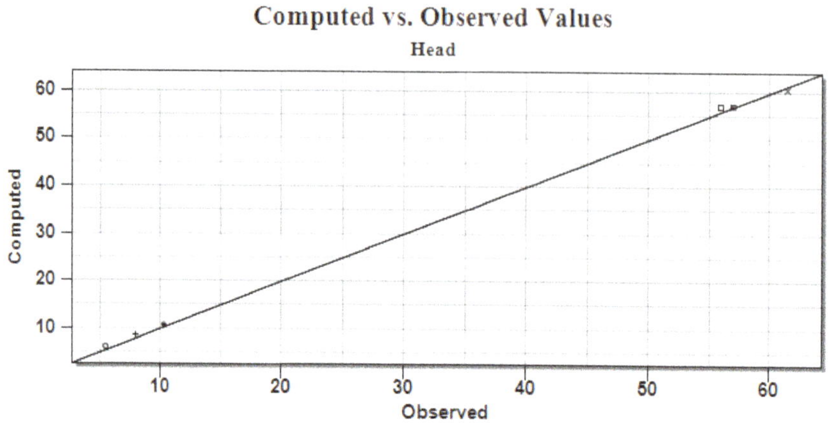

Figure 3. A plot of observed vs. calculated head.

The model's sensitivity to hydraulic conductivity and recharge was tested by varying each parameter incrementally from 5% to 50% while keeping other parameters constant. Results showed that changes in hydraulic conductivity up to 10% indicate minimal deviation in heads (±1 m). At 20% variation, heads shifted moderately, falling within the yellow range (±2 m). Beyond 35% variation, there was noticeable deviation, with heads falling within the red range, and model errors increased sharply, indicating model instability. For the recharge parameter, variations up to 10% had little impact (±1 m). At 15% variation, some heads deviated, falling within the yellow range (±2 m). Beyond 30% variation, the calculated heads fell within the red region, showing that the model is not stable for 30% variations in the groundwater system's recharge. The model proved more sensitive to hydraulic conductivity than to recharge when partitions were adjusted individually. However, simultaneous recharge adjustments across all partitions increased sensitivity beyond conductivity's influence. The analysis demonstrated that these hydrogeological parameters were sensitive to variations beyond 15% (Supplementary Materials, Table S1).

The calibrated model is presented in Figure 2b. It reveals a close match between the observed and modeled groundwater heads. All the simulated heads in the wells aligned closely with field-observed values, indicating accurate model performance. The flow simulation reveals that all six monitoring wells were efficiently simulated to match the field-observed values (the wells had calibration target bars that indicated green colors). A plot of the observed hydraulic heads versus modeled hydraulic heads is shown in Figure 3. This visual representation reveals a strong correlation between the observed and modeled heads, indicating a reasonably accurate representation of the groundwater flow system in the area. The model reveals that the highest hydraulic heads were in the western part of the study area, and the reverse was observed in the east of the study area, which has lower hydraulic heads. The model shows a diverging flow from the recharge area at high topographic reliefs to other parts of the study area (Figure 2b).

In general, groundwater moves from the peak of the First Watchung Mountain in the southwest towards the west, north, and east of the study area. The velocity vectors reveal that groundwater moves from the southwest part of the study area westward towards the Peckman River, and from the southwest to the east and northeast to the Passaic River. The calibrated groundwater flow model indicates that a significant portion of the groundwater flow in the study area originates or passes through the First Watchung mountain area (Figure 2b). This area serves as the primary recharge area for a significant part of the study area, including the two municipal wells that were investigated in this study. The boundary conditions used in the modeling indicate that inflow into the groundwater system enters as precipitation, by means of rainfall and/or snowfall, and outflow occurs mostly as streamflow through the Peckman River, the Third River, and the Passaic River.

Groundwater Flow Regime

The study area is unique, characterized by multiple groundwater flow systems (Figures 4–6). The primary recharge area plays a vital role in sustaining the region's water resources, with the highest hydraulic head values located around the First Watchung Mountain, aligning with its elevated topographic features. Tóth [43] established a foundational framework for understanding groundwater flow systems, demonstrating that multiple flow systems can develop in regions with adequate recharge. Tóth [43] categorized these systems by linking recharge and discharge areas, identifying three distinct types: local, intermediate, and regional. Local flow systems develop between nearby recharge and discharge zones, typically involving short flow paths. Intermediate flow systems originate in recharge areas but extend farther downstream, discharging at locations beyond the nearest discharge points. Regional flow systems, on the other hand, originate from large-scale recharge areas and extend to distant discharge locations, featuring significantly longer flow paths. This classification provides a clear structure for analyzing groundwater movement in various hydrological settings.

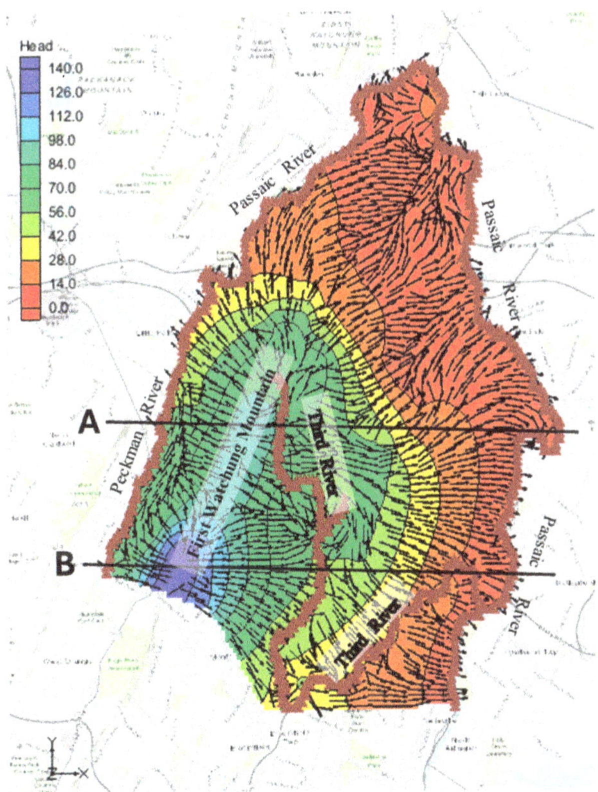

Figure 4. Groundwater flow model velocity vectors.

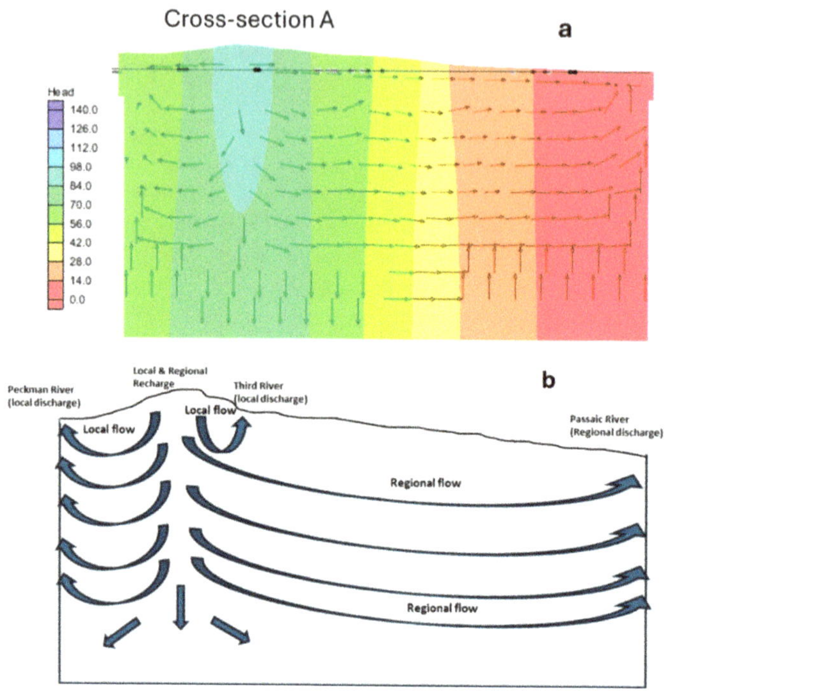

Figure 5. (**a**) Model Cross-section A (velocity vectors). (**b**) Schematic of groundwater flow for Cross-section A.

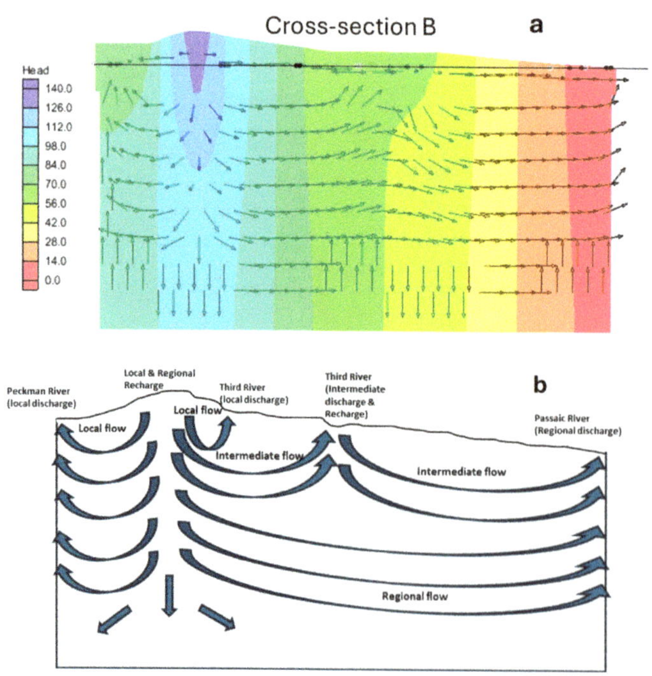

Figure 6. (**a**) Model Cross-section B (velocity vectors). (**b**) Schematic of groundwater flow for Cross-section B.

To better understand the diverse groundwater flow systems in the study area, two cross-sections (A and B, shown in Figures 4–6) were assessed. Cross-section A reveals both local and regional flow systems (Figure 5). In the local flow system, groundwater moves from the recharge area near the First Watchung Mountain westward to the Peckman River and eastward to the Third River. The regional flow system, originating from the same recharge area, extends to the distant Passaic River. Cross-section B includes local, intermediate, and regional flow systems (Figure 6). The local flow system similarly transports groundwater from the First Watchung Mountain recharge area to the nearby Peckman and Third Rivers. Intermediate flow systems, however, originate from the same recharge area but extend beyond local discharge points, reaching the midstream of the Third River but not the distant Passaic River. Additionally, intermediate systems recharge at the Third River and discharge at the Passaic River. The regional flow system, like in Cross-section A, flows from the First Watchung Mountain recharge area to the Passaic River. Schematics of the groundwater flow systems in Cross-sections A and B (Figures 5b and 6b) further illustrate the multiple flow systems present in the study area, enhancing understanding of their dynamics.

3. Particle Tracking

Particle tracking analyzes groundwater flow paths and travel times and distances of advective contaminants. It uses hypothetical particles situated in certain locations to simulate solute movement, using the velocities from the groundwater flow models [24]. A particle tracking model enables forward and backward tracking of particles in time, facilitating identification of water sources and contaminant sources. Particle tracking analysis also offers a valuable assessment of groundwater flow systems, as the model generates flow lines that clearly depict the types of flow systems present in a region. The model focuses solely on advective transport and does not consider the contaminants' diffusion, dispersion, reactions, and degradation processes. The MODPATH code [44] is used to execute particle tracking, delineating areas contributing water to wells and calculating the travel times and paths. According to Barry et al. [27], particle tracking analysis is influenced by model boundary conditions, the hydraulic properties of the aquifer (hydraulic conductivity, porosity, etc.), and pumping well attributes (pumping rates and well depth). This implies that a pumping well can change the flow paths of particles. This study uses particle tracking to describe the aquifer flow patterns.

Particle tracking analysis was employed to evaluate groundwater flow systems and identify the types of flow systems in the study area. Particles were tracked in order to determine the transport characteristics of a conservative contaminant (chloride)'s movement through a "typical hydrogeologic environment" of northern New Jersey. Hypothetical particles were introduced into the recharge areas near the First Watchung Mountain. The analysis illustrates flow paths as particles travel from the recharge areas to discharge areas in the Peckman River, Third River, and Passaic River (Figure 7). The cross-section of the particle tracking analysis reveals local flow systems, with particles moving toward the nearby Peckman and Third Rivers (Figure 7b). Additionally, the analysis demonstrates regional flow systems, as particles move from the recharge areas to discharge points at the distant Passaic River. In the local flow systems, recharge and discharge zones are distributed intermittently across the basin. This indicates that drinking water sources that are closely located may have entirely different origins. Consequently, significant variations in chemical quality can be anticipated [43]. As a result, contaminants in the Third River and Peckman River may originate from sources unrelated to the First Watchung Mountain.

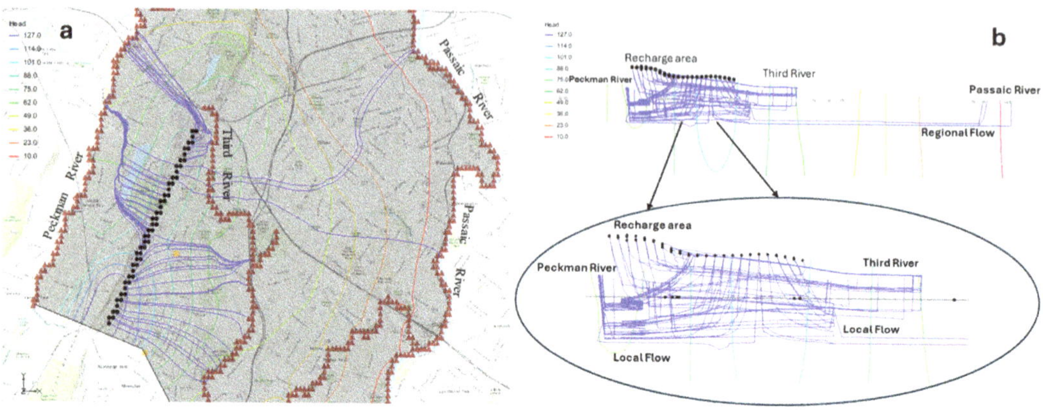

Figure 7. Some flow paths of the groundwater flow system: (**a**) aerial view; (**b**) cross-sectional view.

The conservative nature of chloride makes it an ideal tracer in hydrologic investigations, facilitating accurate modeling of advective groundwater flow. This attribute enables chloride to serve as a reliable proxy for delineating contaminant pathways in groundwater systems. Due to its conservative nature, chloride ions typically do not undergo significant sorption in the aquifer. Instead, they may be retained in the immobile solution within the pores of the aquifer material, while solute transport continues in the mobile media of the fractures. Therefore, variability in aquifer porosity in a fractured rock system could result in varied impacts on particle paths, necessitating the need to use varied porosity values during the groundwater modeling process. For this reason, the particle tracking simulation employs two different porosity values to represent the Brunswick formation and the Brunswick fractured rock system.

Travel paths, times, and distances of chloride particles from points of particle introduction to discharge points were determined using the MODPATH code. For this study, chloride particles were introduced at the Montclair State University community (MSU, a region identified as critically important) to conduct forward particle tracking from source to exit points. This is to simulate how chloride (possibly from deicing salts) may travel through a "typical hydrogeologic environment" of northern New Jersey. Particles were also located at municipal well locations for backward chloride tracking to determine possible sources of contaminants (e.g., deicing salts). Particles were placed strategically at MSU, and around the municipal wells, then respectively tracked forward for 1, 4, 8, 10, through 20 years of travel times from the university community. For the municipal wells, particles were tracked backward in time to their origin. About 59–200 particles were introduced to provide adequate capture of the hypothetical contaminant pathways. The forward and backward particle tracking models reveal the extent of particles' travel and their associated travel times, paths, and distances. Porosity values for the groundwater system were set at 0.4 and 0.2, while the municipal wells were set to pump at 200 m^3/s. The impact of variability of porosity on particle paths was observed. The particle tracking analysis revealed that particles introduced at MSU exited at surface water bodies, as discharge points, after a period of time. The porosity values of 0.4 and 0.2 resulted in distinctly varied particle movement.

The forward tracking analysis, utilizing a porosity value of 0.2, revealed that chloride particles from MSU reached the Third River, Cedar Grove reservoir, and the Peckman River within a 4-year travel time (Figure 8). The particles traveling from MSU covered travel distances in local flow systems, ranging from 314–317 m (Third River), 815–822 m (Cedar Grove reservoir), and 2035 m (Peckman River). The forward analysis, based on a porosity value of 0.4 (Figure 9), had particles that traveled shorter distances, of 141–150 m (Third River), 273–290 m (Cedar Grove reservoir), and 600 m (Peckman River) during the same time. The variation in travel distances signified potential travel of contaminants (possibly chloride from deicing salts) in different local flow systems of a typical hydrogeological environment in northern New Jersey.

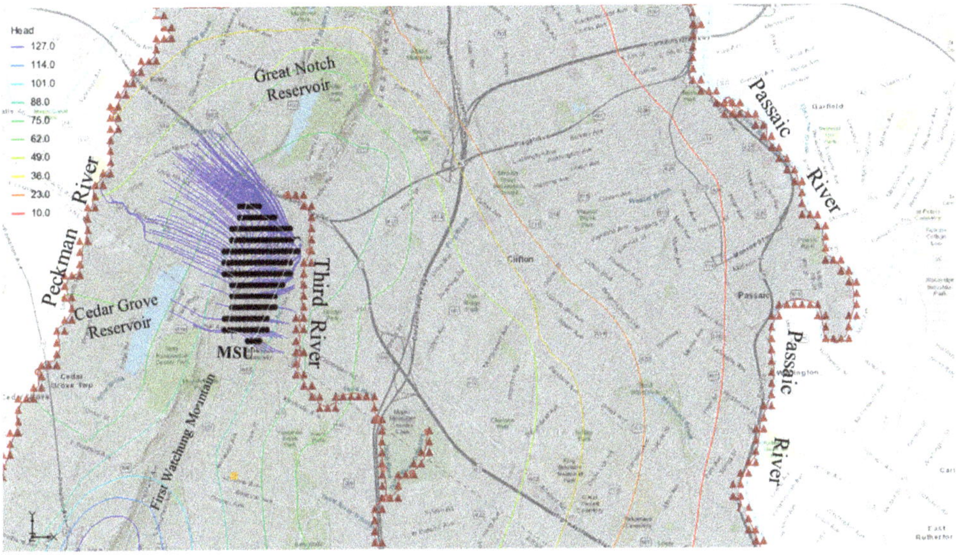

Figure 8. Hypothetical chloride particle transport during a 4-year period—porosity of 0.2.

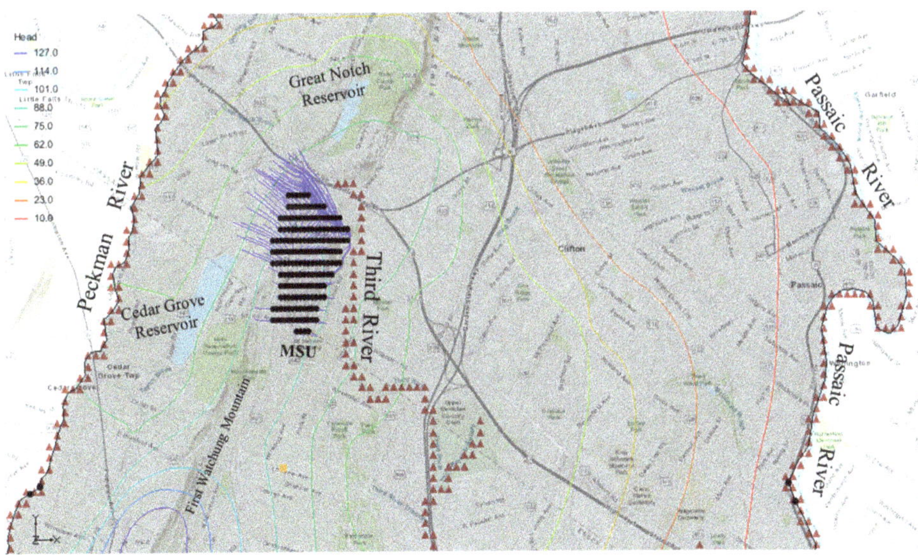

Figure 9. Hypothetical chloride particle transport during a 4-year period—porosity of 0.4.

A forward particle tracking analysis was conducted to simulate the movement of chloride particles from MSU to the Passaic River. Using a porosity value of 0.2, the results indicated that the particles reached the river within a 10-year travel time, covering a distance of 7624 m (Figure 10). On the contrary, the same analysis employing a porosity value of 0.4 revealed that the particles traveled 1040 m (Figure 11), which is less than halfway to the Passaic River.

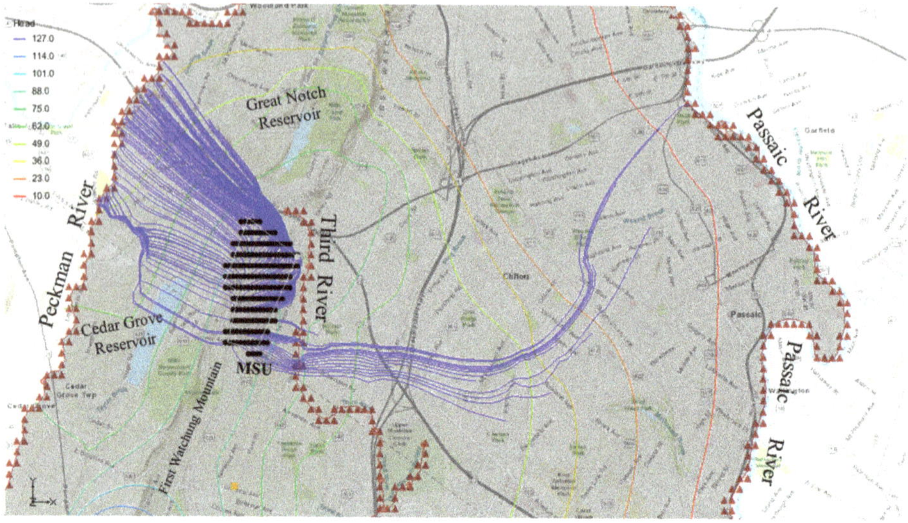

Figure 10. Hypothetical chloride particle transport during a 10-year period—porosity of 0.2.

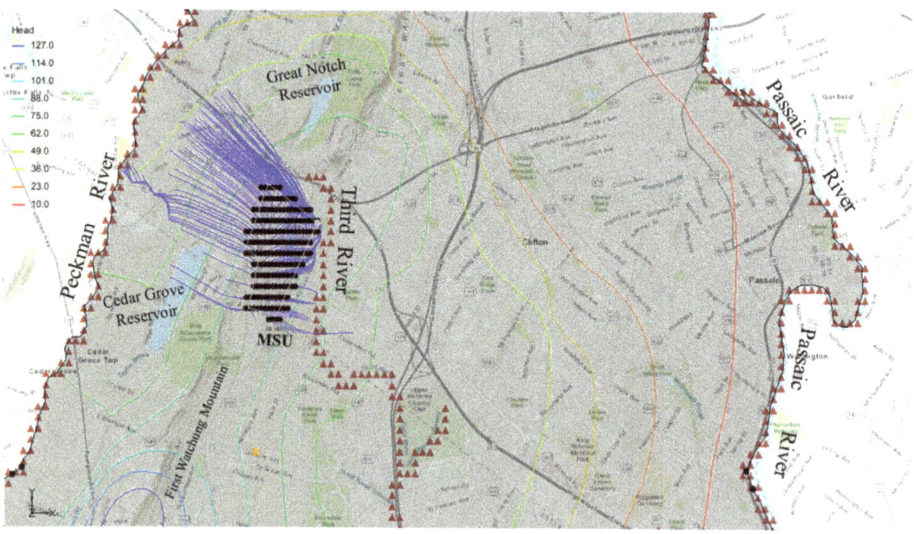

Figure 11. Hypothetical chloride particle transport during a 10-year period—porosity of 0.4.

Utilizing a porosity value of 0.4, a forward particle tracking analysis demonstrated that chloride particles migrated to the areas around the Third River, Cedar Grove reservoir, and the Peckman River within an 8-year timeframe (Figure 12). The particles from MSU traveled within the local flow system distances ranging from 315–318 m (Third River), 817–824 m (Cedar Grove reservoir), and 2038 m (Peckman River). After 20 years of travel, the forward particle tracking analysis revealed that chloride particles traveling within the regional flow system, through an aquifer of 0.4 porosity, arrived at the Passaic River (Figure 13), with migrating distances of 7634 m from MSU.

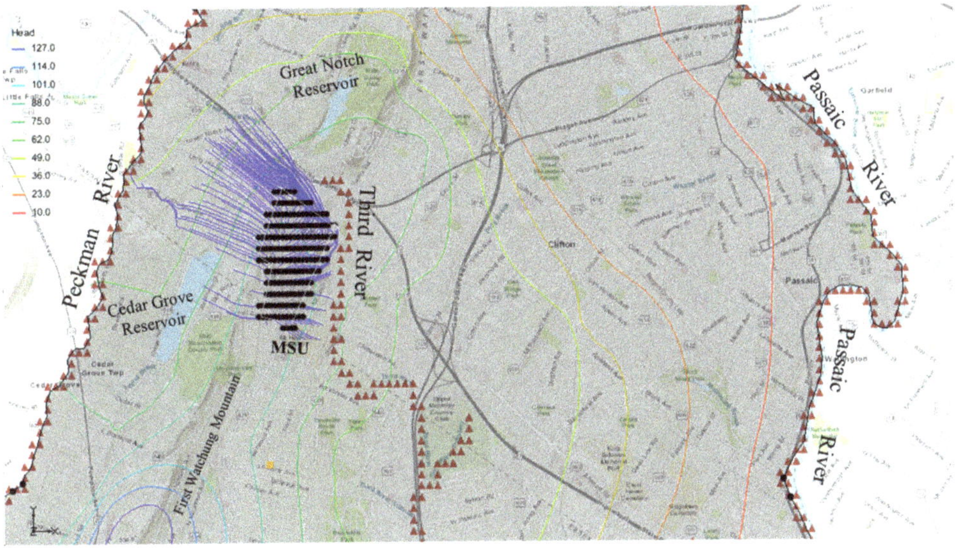

Figure 12. Hypothetical chloride particle transport during an 8-year period—porosity of 0.4.

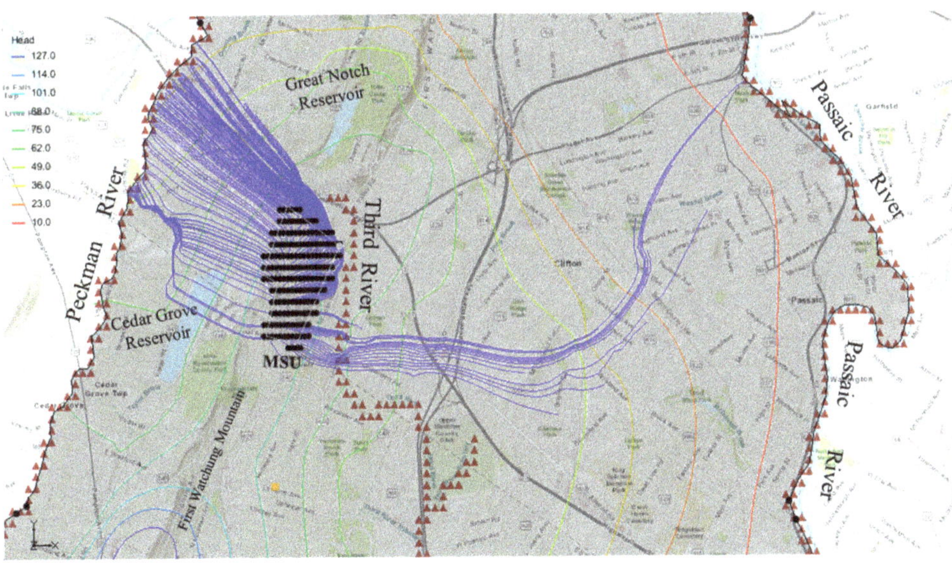

Figure 13. Hypothetical chloride particle transport during a 20-year period—porosity of 0.4.

A backward particle tracking analysis was employed to investigate the movement of chloride particles from two Montclair Township municipal wells in order to determine areas contributing recharge waters to these wells, and also to identify potential contamination sources to the wells (Figure 14). The results of this analysis, regardless of the porosity values used, revealed that the wells are recharged by water from the First Watchung Mountain. This water would possibly carry potential harmful contaminants to the municipal wells. The analysis revealed that water from MSU would not recharge the municipal wells, nor would it contribute contaminants through the groundwater transport process.

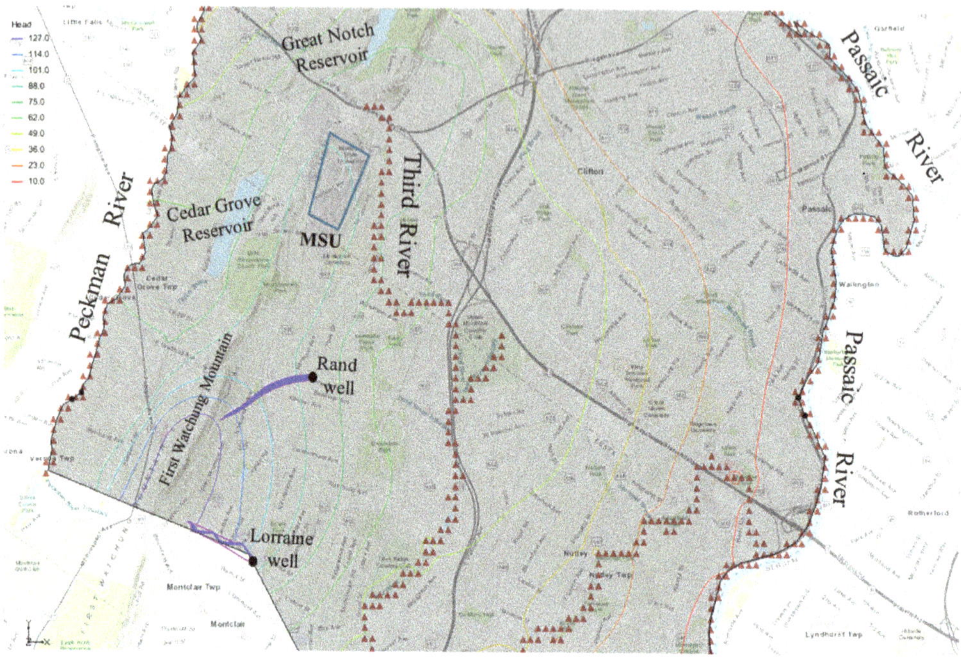

Figure 14. Hypothetical backward chloride particle tracking analysis from supply wells.

The arrival times of the chloride particles at reservoirs and surface water bodies varied with different porosity values. The tracking analysis revealed that chloride particles would arrive faster at the various discharge points with lower porosity values relative to higher values, corroborating studies from Lewis-Brown et al. [28]. This analysis suggests that in fractured groundwater systems with geology like the Brunswick aquifer, contaminant travel would vary in distance and time as a result of the variation in aquifer properties (such as porosity). This could pose a threat to remediation processes, as some of the rock may hold back contaminants in their pores long after the bulk of the contaminants has been flushed in the mobile phase of transport.

The particle tracking analysis indicates that MSU will contribute neither recharge water nor contaminants to the municipal wells. However, contaminants could be carried in runoff to areas that directly recharge the municipal wells. There is a chance that contaminants that arrive at the Third River may interact with areas that directly recharge the municipal wells. Particles arriving at the Third River may be recharged to reach the Passaic River earlier than the 10-year and 20-year travel times that were predicted with aquifer porosities of 0.2 and 0.4, respectively. Since the Third River is a tributary of the Passaic River, it is plausible that chloride particles could migrate to the Passaic River within a 4-year timeframe via surface-water pathways. This tracking analysis does not account for processes such as runoff, which could potentially facilitate even shorter travel times to the Passaic River than the 10–20 years indicated by the tracking. This simulation also does not account for solute dispersion. Consequently, the travel times computed through particle tracking analysis do not represent the initial arrival of reactive solutes at rivers or wells. Rather, the analysis provides an approximation of how long it will take for the main cluster of particles (simulating conservative contaminants like chlorides from deicing salts) to reach discharge locations (rivers or wells). As indicated by Nicholson and Watt [45], it is crucial to recognize that when a well is recharged from underlying water-bearing zones separated by a confining layer, the recharge area contributing to the well may not necessarily include the wellhead site. Nevertheless, this hypothetical groundwater transport study provides an elaborate understanding of the travel pathways, times, and distances of potential contaminants (like salts from deicing) in typical regional groundwater flow systems in northern New Jersey.

3.1. Implications for Contaminant Transport and Remediation

The heterogeneous nature of aquifer porosity values in fractured rock aquifers, particularly in systems like the Brunswick aquifer, creates distinct challenges for contaminant management. This heterogeneity leads to contaminant accumulation within aquifer material pores, where migration into the rock matrix significantly reduces particle mobility. Even after remediation efforts like well cleanup and flushing with clean water, these retained contaminants can gradually remobilize, undermining treatment effectiveness [17]. Such persistence presents substantial groundwater management challenges, especially when compounded by salinization risks from deicing salt constituents like chloride [2,6,12].

Particle tracking analyses provide critical insights into contaminant behavior within northern New Jersey's characteristic "typical hydrogeological environment". These studies reveal how chloride and similar contaminants move through local, intermediate, and regional groundwater systems, with transport characteristics that are highly varied between different flow regimes [21,28,43]. Three key findings emerge from this study:

- Contaminant travel paths, times, and distances show substantial variation across flow systems.
- Aquifer porosity exerts significant control on transport dynamics, with lower porosity values (0.2) enabling faster movement compared to higher values (0.4) [24,28].
- These transport patterns directly inform monitoring well placement and management strategy design [24].

These results enable more targeted protection of critical water resources, including the Third River and Peckman River, through enhanced early detection capabilities and optimized intervention approaches [12]. The demonstrated relationship between porosity and contaminant mobility particularly emphasizes the need for precise hydrogeological characterization in remediation planning.

3.2. Management Implications for Fractured-Rock Aquifer Protection

This study highlights two pressing management challenges with direct consequences for water protection. First, contaminants like chloride do not simply flow through fractured aquifers—they linger. Trapped in rock pores long after initial infiltration, they gradually release back into groundwater, undermining remediation efforts. Second, the First Watchung Mountain's critical role as the region's primary recharge area makes it disproportionately vulnerable; every deicing salt application here ripples through the entire groundwater system.

These insights demand targeted action:

- Strategic Monitoring—Focus surveillance on the First Watchung recharge zone and key discharge points like the Third River, where our models show that contaminants arrive first (within 4 years). Prioritizing these early-warning areas allows faster response to salinity spikes.
- Adaptive Remediation—Cleanup designs must account for the dual behavior of fractured rock. While pumps can flush contaminants from fractures, the surrounding matrix slowly releases stored pollutants—requiring longer treatment timelines and technologies tailored to porosity differences.
- Recharge-Zone Safeguards—Reduce salt infiltration at its source by implementing protective buffers around high-recharge areas. Alternative deicers, application limits,

and permeable surfaces can significantly cut chloride loading while maintaining winter safety.

- Infrastructure Alignment—Groundwater flow maps should guide development. Avoiding disruption to natural flow paths—like siting new roads away from recharge zones—preserves the system's natural resilience.

By matching management strategies to the aquifer's unique system, we can better protect water supplies from both immediate contamination and long-term salinity buildup.

4. Conclusions

This study provides a comprehensive analysis of groundwater flow and contaminant transport dynamics in a "typical hydrogeological environment" of northern New Jersey, characterized by fractured-rock aquifers with complex porosity heterogeneity. Our integrated modeling approach, combining MODFLOW and MODPATH simulations, reveals three distinct but interconnected groundwater flow systems: (1) local systems discharging to the Peckman and Third Rivers within 4–8 years, (2) intermediate systems that terminate before reaching the regional discharge zone, and (3) regional systems requiring 10–20 years to reach the Passaic River. The First Watchung Mountain was identified as the dominant recharge source for both municipal wells and regional flow systems, highlighting its critical role in the area's groundwater sustainability. Our findings demonstrate that chloride contaminants from urban sources like deicing salts can traverse these fractured systems surprisingly quickly, reaching sensitive discharge points in as little as four years under certain porosity conditions.

The stark contrast in transport times between low-porosity fractures (0.2) and higher-porosity zones (0.4) demonstrates the multifaceted behavior of contaminant movement through these geological formations. While fractures serve as rapid conduits, the surrounding rock matrix acts as both a temporary sink and potential long-term source of contamination through gradual remobilization. This dual behavior presents unique challenges for remediation efforts, particularly in urban areas where road salt application is a persistent pressure on groundwater quality. The particle tracking analysis provides particularly valuable insights, showing clear separation between pathways serving municipal wells and those discharging at surface waters—a distinction with important implications for targeted protection strategies.

Looking ahead, these findings open several promising avenues for further research. The clear influence of porosity on transport dynamics suggests that more detailed characterization of fracture networks could significantly improve predictive models. There is also compelling evidence that transient hydrological conditions and geochemical interactions deserve greater attention in future modeling efforts. Perhaps most importantly, this study establishes a methodological framework that can be adapted to other urbanized areas underlain by fractured-rock aquifers, where balancing water supply needs with quality protection remains an ongoing challenge. The lessons learned here in northern New Jersey's complex hydrogeological environment may well inform groundwater management strategies in similar settings across the globe.

Supplementary Materials: The following supporting information can be downloaded at: https://www.mdpi.com/article/10.3390/hydrology12070167/s1.

Author Contributions: Conceptualization, T.O.; methodology, T.O.; software, T.O. and D.O.; validation, T.O.; formal analysis, T.O.; writing—original draft preparation, T.O.; data curation, T.O.; writing—review and editing, T.O. and D.O.; supervision, D.O. All authors have read and agreed to the published version of the manuscript.

Funding: This research received no external funding.

Data Availability Statement: All data are publicly available, and sources are cited in the References section.

Conflicts of Interest: The authors declare no conflicts of interest. The funders had no role in the design of the study; in the collection, analyses, or interpretation of data; in the writing of the manuscript; or in the decision to publish the results.

References

1. Lathrop, R.G. Measuring Land Use Change in New Jersey: Land Use Update to Year 2000. A Report on Recent Development Patterns 1995 to 2000. 2004. Available online: https://crssa.rutgers.edu/projects/lc/docs/landuse_upd.pdf (accessed on 11 May 2022).
2. Kaushal, S.S.; Likens, G.E.; Pace, M.L.; Reimer, J.E.; Maas, C.M.; Galella, J.G.; Utz, R.M.; Duan, S.; Kryger, J.R.; Woglo, S.A.; et al. Freshwater salinization syndrome: From emerging global problem to managing risks. *Biogeochemistry* **2021**, *154*, 255–292. [CrossRef]
3. Reilly, T.; Reilly, T.E.; Dennehy, K.F.; Alley, W.M.; Cunningham, W.L. Ground-Water Availability in the United States. In *U.S. Geological Survey Circular 1323*; U.S. Geological Survey: Reston, VA, USA, 2008; Volume 70. Available online: https://pubs.usgs.gov/circ/1323/pdf/Circular1323_book_508.pdf (accessed on 31 May 2023).
4. Sathe, S.S.; Mahanta, C. Groundwater flow and arsenic contamination transport modeling for a multi aquifer terrain: Assessment and mitigation strategies. *J. Environ. Manag.* **2019**, *231*, 166–181. [CrossRef] [PubMed]
5. Serfes, M.; Bousenberry, R.; Gibs, J. *New Jersey Ambient Ground Water Quality Monitoring Network: Status of Shallow Ground-Water Quality, 1999–2004*; New Jersey Geological Survey: Trenton, NJ, USA, Information Circular; 2007. Available online: https://www.nrc.gov/docs/ML1408/ML14086A280.pdf (accessed on 11 December 2019).
6. Rossi, M.L.; Kremer, P.; Cravotta, C.A.; Seng, K.E.; Goldsmith, S.T. Land development and road salt usage drive long-term changes in major-ion chemistry of streamwater in six exurban and suburban watersheds, southeastern Pennsylvania, 1999–2019. *Front. Environ. Sci.* **2023**, *11*, 1153133. [CrossRef]
7. Perera, N.; Gharabaghi, B.; Howard, K. Groundwater chloride response in the Highland Creek watershed due to road salt application: A re-assessment after 20 years. *J. Hydrol.* **2013**, *479*, 159–168. [CrossRef]
8. Oyen, T.; Ophori, D. Estimating the spatial distribution of groundwater Salinity in the New Jersey Northeast Water region using Geostatistical methods. *Geol. Soc. Am. Abstr.* **2023**, *55*, 395885. [CrossRef]
9. Oyen, T.; Ophori, D. Spatiotemporal Analysis of Available online Freshwater Resources in Watersheds Across Northern New Jersey. *Hydrology* **2025**, *12*, 149. [CrossRef]
10. US EPA, Drinking Water Regulations and Contaminants | US EPA. Available online: https://www.epa.gov/sdwa/drinking-water-regulations-and-contaminants (accessed on 11 May 2022).
11. New Jersey Ground Water Quality Standards. N.J.A.C. 7:9C. Title 7 of the New Jersey administrative code, in pursuant to the Water Pollution Control Act (N.J.S.A. 58:10A-1 et seq.) and the Water Quality Planning Act (N.J.S.A. 58:11A-1 et seq.), Last Amended: 1 June 2020. NJDEP 2020. Available online: https://dep.nj.gov/wms/bears/ground-water-quality-standards-gwqs/ (accessed on 11 May 2022).
12. Ophori, D.; Firor, C.; Soriano, P. Impact of road deicing salts on the Upper Passaic River Basin, New Jersey: A geochemical analysis of the major ions in groundwater. *Environ. Earth Sci.* **2019**, *78*, 1–13. [CrossRef]
13. Kaushal, S.S.; Groffman, P.M.; Likens, G.E.; Belt, K.T.; Stack, W.P.; Kelly, V.R.; Band, L.E.; Fisher, G.T. Increased salinization of fresh water in the Northeastern United States. *Proc. Natl. Acad. Sci. USA* **2005**, *102*, 13517–13520. [CrossRef]
14. McCleskey, R.B. New method for electrical conductivity temperature compensation. *Environ. Sci. Technol.* **2013**, *47*, 9874–9881. [CrossRef]
15. Novotny, D.; Muehring, D.; Zitomer, D.H.; Smith, D.W.; Facey, R. Cyanide and metal pollution by urban snowmelt: Impact of deicing compounds. *Water Sci. Technol.* **1998**, *38*, 223–230. [CrossRef]
16. Sarkar, D. Preliminary studies on mercury solubility in the presence of iron oxide phases using static headspace analysis. *Environ. Geosci.* **2003**, *10*, 151–155. [CrossRef]
17. Gefell, M.J.; Lipson, D.S. Matrix diffusion-derived plume attenuation and implications for remediation in crystalline bedrock. In Proceedings of the 13th Annual Battelle Conference on Chlorinated and Recalcitrant Compounds, Denver, CO, USA, 2–6 June 2024; Available online: https://proceedings.battelle.org/index.cfm/conference/view/ID/64?sessions=84 (accessed on 6 June 2024).
18. Hort, H.M.; Robinson, C.E.; Sawyer, A.H.; Li, Y.; Cardoso, R.; Lee, S.A.; Roff, D.; Adamson, D.T.; Newell, C.J. Conceptualizing Controlling Factors for PFAS Salting Out in Groundwater Discharge Zones Along Sandy Beaches. *Groundwater* **2024**, *62*, 860–875. [CrossRef] [PubMed]
19. McNaboe, L.A.; Robbins, G.A.; Dietz, M.E. Mobilization of Radium and Radon by Deicing Salt Contamination of Groundwater. *Water Air Soil. Pollut.* **2017**, *228*, 94. [CrossRef]

20. Novotny, E.V.; Sander, A.R.; Mohseni, O.; Stefan, H.G. Chloride ion transport and mass balance in a metropolitan area using road salt. *Water Resour. Res.* **2009**, *45*, 12. [CrossRef]
21. Kilgour, B.; Gharabaghi, B.; Trudel, L.; Jarvie, S.; Perera, N. Ecological benefit of the road salt code of practice. *Water Qual. Res. J.* **2014**, *49*, 43–52. [CrossRef]
22. Newell, C.J.; Javed, H.; Li, Y.; Johnson, N.W.; Richardson, S.D.; Connor, J.A.; Adamson, D.T. Enhanced attenuation (EA) to manage PFAS plumes in groundwater. *Remediat. J.* **2022**, *32*, 239–257. [CrossRef]
23. Trowbridge, P.R.; Kaul, J.S.; Sassan, D.A.; Heath, D.L.; Walsh, E.M. Relating Road salt to exceedances of the water quality standard for chloride in New Hampshire streams. *Environ. Sci. Technol.* **2010**, *44*, 4903–4909. [CrossRef]
24. Anderson, M.P.; Woessner, W.W.; Hunt, R.J. *Applied Groundwater Modeling: Simulation of Flow and Advective Transport*, 2nd ed.; Academic Press: Cambridge, MA, USA, 2015.
25. Ophori, D.; Barry, F.; Hoffman, J.; Canace, R. A three-dimensional (3-D) model of groundwater and surface water interaction in the central Passaic River basin, New Jersey. *J. Environ. Hydrol.* **2021**, *29*, 2.
26. Yidana, S.M.; Ganyaglo, S.; Banoeng-Yakubo, B.; Akabzaa, T. A conceptual framework of groundwater flow in some crystalline aquifers in Southeastern Ghana. *J. Afr. Earth Sci.* **2011**, *59*, 185–194. [CrossRef]
27. Barry, F.; Ophori, D.; Hoffman, J.; Canace, R. Groundwater flow and capture zone analysis of the Central Passaic River Basin, New Jersey. *Environ. Geol.* **2009**, *56*, 1593–1603. [CrossRef]
28. Lewis-Brown, J.C.; Carleton, G.B.; Imbrigiotta, T.E. *Hydraulic and Solute-Transport Properties and Simulated Advective Transport of Contaminated Ground Water in a Fractured-Rock Aquifer at the Naval Air Warfare Center, West Trenton, New Jersey, 2003*; USGS Scientific Investigations Report; U.S. Geological Survey (USGS): Reston, VA, USA, 2006. [CrossRef]
29. Serfes, M.E.; Herman, G.C.; Spayd, S.E.; Reinfelder, J. Sources, mobilization and transport of arsenic in groundwater in the Passaic and Lockatong Formations of the Newark Basin, New Jersey. In *Contributions to the Geology and Hydrogeology of the Newark Basin, Bulletin 77*; Charles, H.G., Serfes Michael, E., Eds.; NJ Geological Survey: Trenton, NJ, USA, 2010; Chapter E; pp. E1–E40. Available online: https://www.researchgate.net/publication/270273467_Sources_mobilization_and_transport_of_arsenic_in_groundwater_in_the_Passaic_and_Lockatong_Formations_of_the_Newark_Basin_New_Jersey (accessed on 11 May 2022).
30. Montclair Water Bureau. Montclair & Glen Ridge Year 2022 Annual Drinking Water Quality Report. 2022. Available online: https://www.montclairnjusa.org/files/sharedassets/public/v/1/government/departments/water-bureau/documents/water-quality-2022.pdf (accessed on 21 December 2023).
31. Staff, B. Rep. Sherrill Secures $2,056,000 in Funding to Fix Montclair's Rand Well. Montclair Local, 24 December 2022. Available online: https://montclairlocal.news/2022/12/rep-sherrill-secures-2056000-in-funding-to-fix-montclairs-rand-well/ (accessed on 21 December 2023).
32. Drake, A.A., Jr.; Volkert, R.A.; Monteverde, D.H.; Herman, G.C.; Houghton, H.F.; Parker, R.A.; Dalton, R.F. *Bedrock Geologic Map of Northern New Jersey*; U.S. Geological Survey: Trenton, NJ, USA, 1997. [CrossRef]
33. Herman, G.C. Hydrogeological framework of bedrock aquifers in the Newark Basin, New Jersey. In Proceedings of the 18th Annual Meeting of the Geological Association of New Jersey, South Brunswick, NJ, USA, October 2010; Available online: https://www.researchgate.net/publication/270217175_Hydrogeological_framework_of_bedrock_aquifers_in_the_Newark_Basin_New_Jersey (accessed on 11 May 2022).
34. Olsen, P.E. The latest Triassic and early Jurassic formations of the Newark basin (eastern North America, Newark Supergroup): Stratigraphy, structure, and correlation. *N. J. Acad. Sci.* **1980**, *25*, 25–51.
35. Herman, G.C.; Canace, R.J.; Stanford, S.D.; Pristas, R.S.; Sugarman, P.J.; French, M.A.; Hoffman, L.J.; Serfes, M.S.; Mennel, W.J. Aquifers of New Jersey. NJDEP New Jersey Geological Survey Open-File Map OFM 24, 1:500,000 Scale. 1998. Available online: https://www.nj.gov/dep/njgs/pricelst/ofmap/ofm24.pdf (accessed on 11 May 2022).
36. Watt, M.K. *A Hydrologic Primer for New Jersey Watershed Management*; USGS: West Trenton, NJ, USA, 2000. Available online: https://pubs.usgs.gov/wri/2000/4140/report.pdf (accessed on 11 December 2019).
37. ONJSC (Office of the New Jersey State Climatologist). New Jersey Climate Overview. Available online: https://climate.rutgers.edu/stateclim_v1/njclimoverview.html (accessed on 11 May 2022).
38. New Jersey Department of Environmental Protection. *New Jersey Forest Action Plan*; New Jersey Department of Environmental Protection: Trenton, NJ, USA, Forest Service; December 2020. Available online: https://www.nj.gov/dep/parksandforests/forest/njsfap/ (accessed on 11 December 2019).
39. Harbaugh, A.W.; Banta, E.R.; Hill, M.C.; McDonald, M.G. *MODFLOW-2000, The U.S. Geological Survey Modular Ground-Water Model: User Guide to Modularization Concepts and the Ground-Water Flow Process*; USGS Open-File Report; U.S. Geological Survey (USGS): Trenton, NJ, USA, 2000. [CrossRef]
40. Kanmani, S.; Gandhimathi, R.; Shanmuharajan, M.B.; Ponraj, M.; Srinivasan, N.P.; Sharmiladevi, K.; Vivek, S. Leachate transport phenomenon on groundwater quality: Modeling using modflow and MT3DMS tools. *Glob. Nest J.* **2023**, *25*, 44–55. [CrossRef]

41. New Jersey Department of Environmental Protection. New Jersey Geographic Information Network: DGS02-1 New Jersey Geological Survey Hydro Database, NJDEP Bureau of GIS. Available online: https://njogis-newjersey.opendata.arcgis.com/documents/b62f98338c1841ac9bf0196e8de2489f/about (accessed on 11 December 2019).
42. USGS Water Resources of the United States—National Water Information System (NWIS) Mapper, USGS, National Water Information System: Mapper. Available online: https://maps.waterdata.usgs.gov/mapper/index.html? (accessed on 11 May 2022).
43. Tóth, J. A theoretical analysis of groundwater flow in small drainage basins. *J. Geophys. Res.* **1963**, *68*, 4795–4812. [CrossRef]
44. Pollock, D.W. *User's Guide for MODPATH/MODPATH-PLOT, Version 3; A Particle Tracking Post-Processing Package for MODFLOW, The U.S. Geological Survey Finite-Difference Ground-Water Flow Model*; USGS Open-File Report; U.S. Geological Survey (USGS): Reston, VA, USA, 1994. [CrossRef]
45. Nicholson, R.S.; Watt, M.K. Simulation of Ground-Water-Flow Patterns and Areas Contributing Recharge to Streams and Water-Supply Wells in a Valley-Fill and Carbonate-Rock Aquifer System, Southwestern Morris County, New Jersey. USGS Water-Resources Investigations Report 97-4216. 1998. Available online: https://pubs.usgs.gov/wri/1997/4216/report.pdf (accessed on 21 December 2023).

hydrology

MDPI

Article

A Comparative Analysis of Remotely Sensed and High-Fidelity ArcSWAT Evapotranspiration Estimates Across Various Timescales in the Upper Anthemountas Basin, Greece

Stefanos Sevastas [1], Ilias Siarkos [2,*] and Zisis Mallios [1]

[1] School of Civil Engineering, Aristotle University of Thessaloniki, 54124 Thessaloniki, Greece; sevastas@civil.auth.gr (S.S.); zmallios@civil.auth.gr (Z.M.)
[2] Department of Civil Engineering, Democritus University of Thrace, 67100 Xanthi, Greece
* Correspondence: isiarkos@civil.duth.gr; Tel.: +30-6986-505775

Abstract: In data-scarce regions and ungauged basins, remotely sensed evapotranspiration (ET) products are increasingly employed to support hydrological model calibration. In this study, a high-resolution hydrological model was developed for the Upper Anthemountas Basin using ArcSWAT, with a focus on comparing simulated ET outputs to three freely available remote sensing-based ET products: the MODIS MOD16 Collection 5, the updated MODIS MOD16A2GF Collection 6.1, and the SSEBop Version 5 dataset. ET estimates derived from the calibrated SWAT model were compared to all remote sensing products at the basin scale, across various temporal scales over the 2002–2014 simulation period. Results indicate that the MOD16 Collection 5 product achieved the closest correspondence with SWAT-simulated ET across all temporal scales. The MOD16A2GF Collection 6.1 product exhibited moderate overall agreement, with improved performance during early summer. The SSEBop Version 5 dataset generally displayed weaker correlation, but demonstrated enhanced alignment during the driest years of the record. Strong correspondence is observed when averaging the ET values from all satellite products. These findings underscore the importance of exercising caution when utilizing remotely sensed ET products as the sole basis for hydrological model calibration, particularly given the variability in performance among different datasets.

Keywords: watershed management; hydrological modeling; evapotranspiration; GIS spatial techniques; ArcSWAT; remote sensing data analysis

1. Introduction

Evapotranspiration (ET) constitutes a principal component of the terrestrial water balance, governing not only the partitioning of incoming precipitation into runoff and subsurface flow but also exerting a fundamental control on soil moisture dynamics and groundwater recharge. An accurate representation of ET is therefore essential in hydrological modeling, as it serves as a critical link among the water, energy, and carbon cycles and regulates surface–atmosphere exchanges at multiple scales [1,2].

In recent years, freely available satellite-derived ET products (e.g., MODIS ET products, GLEAM, WaPOR, ALEXI, TerraClimate, SSEBop) have emerged as vital tools for enhancing the calibration of hydrological models, such as the SWAT (Soil and Water Assessment Tool) model. These products are employed either as supplementary information alongside observed streamflow data or, in the context of fully ungauged basins, as the principal calibration targets. With regard to the former case, Rientjes et al. [3] found that supplementing streamflow data with MODIS ET improved both runoff and soil moisture simulations in a humid tropical catchment. Tobin and Bennett [4] showed that regulating the hydrological model with MOD16 ET tightened parameter ranges and mitigated systematic biases, relative to flow-only calibration. Rajib et al. [5] reported that spatially explicit calibration using remotely sensed ET and biophysical parameters significantly elevated model predictability within a large watershed in the Prairie Pothole Region of the United States. Parajuli et al. [6] demonstrated that SEBAL and MODIS-based ET data effectively captured seasonal ET peaks in semi-arid watersheds when used in calibration, while Herman et al. [7] similarly illustrated that blending multiple ET products (MODIS, SSEBop, GLEAM) lowered forecast errors during high evaporative demand periods. Sirisena et al. [8] applied joint calibration with streamflow and remote sensing ET in a data-poor Australian basin, achieving more realistic hydrograph recession and baseflow dynamics. Dile et al. [9] evaluated SSEBop, MOD16, and GLEAM ET and showed that, after bias correction, SSEBop ET estimates could reliably support water balance closure in data-scarce Ethiopian watersheds.

Moreover, Shah et al. [10] demonstrated that multivariate calibration with both MODIS and ALEXI ET reshaped optimal parameter sets and improved seasonal runoff and ET dynamics, while Sun et al. [11] introduced a two-phase approach, in which calibrating vegetation parameters with ET before adjusting against discharge enhanced parameter identifiability and overall predictive skill. More recently, Bennour et al. [12] achieved robust seasonal water balance simulations in the Lake Chad Basin by assimilating ET and soil moisture observables, while Alemayehu et al. [13] combined streamflow signatures with ET to reduce uncertainty and enhance spatial realism in poorly gauged Ethiopian basins. Parajuli et al. [14] found that, while the model developed using SWAT reproduced seasonal ET variability well, satellite products resolved finer intra-monthly fluctuations, whereas Koltsida and Kallioras [15] confirmed that a multi-variable calibration incorporating ET and discharge improves the model performance in Mediterranean catchments. Finally, Dangol et al. [16] demonstrated that a multivariate framework integrating ET, leaf area index, and surface temperature further tightens parameter uncertainty and elevates model fidelity across diverse land use conditions.

On the contrary, only a limited number of studies utilized satellite-derived ET as the sole criterion for calibrating hydrological models. For example, Immerzeel and Droogers [17] showed that SWAT can be effectively calibrated in data-scarce regions using a time series of MODIS-derived ET. Optimal model performance was achieved by jointly tuning soil, meteorological, and groundwater parameters over an eight-month subbasin ET record. Verma [18] applied SWAT calibration using MODIS ET in the Sirsa River Basin, demonstrating robust reproduction of seasonal hydrographs under Western Himalayan conditions. Odusanya et al. [19] successfully employed ET time series from three satellite products to perform multisite SWAT calibration and validation in southwestern Nigeria, despite sparse gauging networks. Collectively, these studies emphasize the capability of remotely sensed ET products to serve as effective supplements or, in some cases, substitutes for traditional streamflow data in the calibration of hydrological models.

Despite these advances, questions remain regarding the temporal consistency and relative performance of different ET products when aligned against a well-calibrated hydrological model. In the literature, two primary concerns are identified. First, hydrological model calibration often depends on a single streamflow gauge, which may fail to capture spatial variability in ET across heterogeneous land cover and microclimates [8,20,21]. Second, substantial discrepancies among ET products can lead to divergent water balance outcomes, with implications for water allocation, runoff, and drought forecasting and ecosystem management in Mediterranean catchments where evaporative demand is high [1,22].

In this context, the present study aims to quantify the temporal concordance of remotely sensed and high-fidelity ArcSWAT-simulated actual evapotranspiration (AET) (henceforth referred to as evapotranspiration, ET, for simplicity) time series in the Upper Anthemountas Basin, Greece. Specifically, it examines monthly averages, seasonal totals, and annual sums over an 11-year (2004–2014) simulation period. Furthermore, the reliability of this comparison hinges on the use of a detailed, high-resolution SWAT model, which integrates a fine-grained Digital Elevation Model (DEM), a finely stratified soil texture map, and precise land cover classifications. High-precision and fine-resolution input data enhance the representation of watershed characteristics and facilitate a more accurate simulation of local hydrologic responses to topographic complexity, soil heterogeneity, and vegetation dynamics, thus improving the modeling of spatially distributed processes (i.e., runoff generation, nutrient transport) [23–25]. Using such data also increases the likelihood of accurate model calibration and reduces predictive uncertainty [23,26,27], thereby providing a robust basis for the comparative assessment of satellite-derived ET estimates. Ultimately, this study seeks to clarify the strengths and limitations of current remote sensing ET products in representing modeled ET dynamics, providing insights for incorporating satellite data into robust hydrological model calibration strategies.

2. Materials and Methods

2.1. Study Area

The Upper Anthemountas Basin is part of the entire Anthemountas Basin, located in Chalkidiki Peninsula, Greece, to the southeast of the city of Thessaloniki (Figure 1). Encompassing an area of approximately 110 km^2, it is a semi-mountainous basin that borders to the west with the Lower Anthemountas Basin and to the south with the Nea Moudania Basin. A few notable geomorphological properties of the basin are provided in Table 1.

The study area is a rural region with a varied landscape, primarily composed of mixed forests and cultivated land, which together make up 93% of the total area [25,28]. Agriculture is the dominant economic activity, while livestock activities partially contribute to the economic development of the area as well. Agricultural zones, covering 46% of the total area, exhibit complex patterns of wheat, corn, cotton, alfalfa, and olive cultivation, along with a scarce number of orchard and vegetable parcels [29]. On the other hand, forests largely occupy the north and south side of the basin and consist of various evergreen scrub oak species, such as Kermes oak (*Quercus coccifera*) and holm oak (*Quercus ilex*), combined with woody dense shrublands of Mediterranean woundwort (*Stachys cretica*) and conehead thyme (*Thymus capitatus*). This particular type of mixed shrubland biome, commonly known as "Maquis", "Macchia", or "Macchie", is distinctively unique and predominantly found in the Mediterranean region. Deciduous trees of Hungarian oak (*Quercus frainetto*) dominate the western part of the basin. Plane trees (*Platanus*) are encountered alongside the main channel river banks, while a few chestnut and pine trees are also found scattered within the basin [30]. In total, woody vegetation (forest trees and shrub ecosystems) accounts for the largest proportion of land cover in the study area (47%). Various types of grasslands, comprising only 3% of the total area, are dispersed throughout the watershed, mainly consisting of "*Bromus squarrosus*", a species of brome grass. Lastly, a few limited areas (1%) of abandoned magnesite mines are sprinkled within the catchment.

In terms of geology, the region is intricate, featuring a wide array of sediments (clay and marl with intercalated hard-grained materials, such as sands, conglomerates, and sandstones), intermixed with fractured igneous and metamorphic rocks that include epigneisses, greenschists, dunites, and peridotites [25,31]. Soil formation in the study area mirrors the spatial variability of parent materials, reflecting the basin's complex geology. According to Misopolinos et al. [32], Inceptisols are the predominant soil order, followed by Vertisols, with lesser contributions from Alfisols and Entisols. Regarding soil texture, medium texture soils (loams and sandy clay loams), along with heavier soils (clay loams) dominate the central and southern part of the watershed, while coarser soils (sandy loams) cover, for the most part, the northern part of the basin, but are also found scattered all over the area [30].

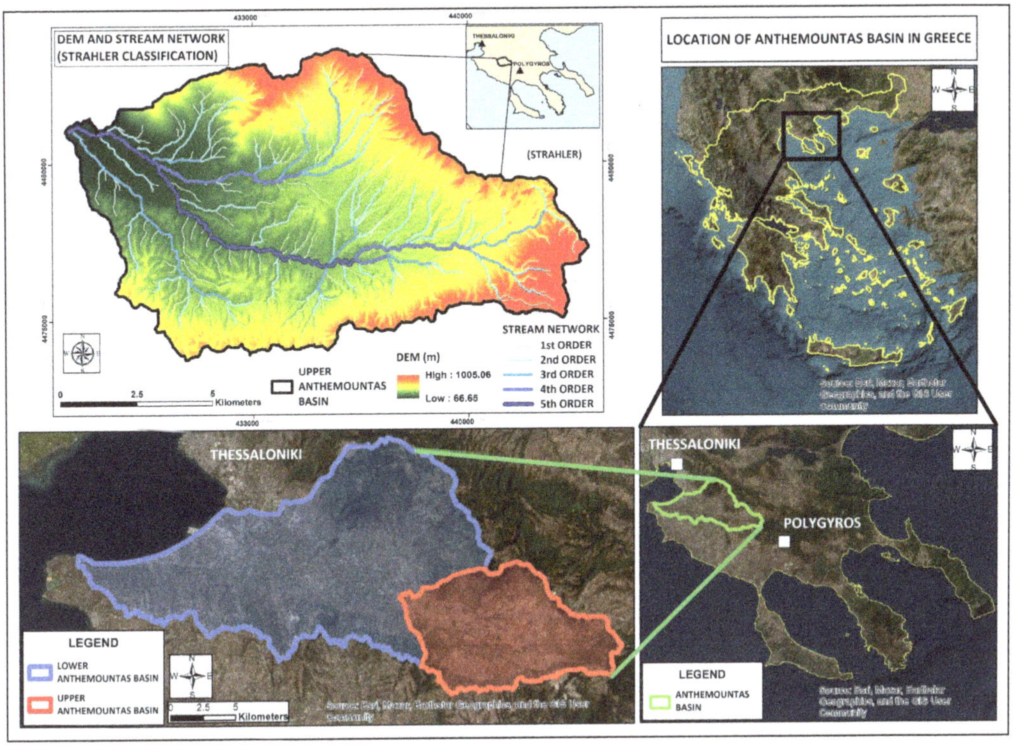

Figure 1. The geographical location, boundaries, and segmentation (Lower and Upper) of the Anthemountas Basin, along with boundaries, elevation data, and stream network of the Upper Anthemountas Basin, which serves as the study area.

Table 1. Basic geomorphological features of the Upper Anthemountas Basin.

Mean Elevation (m)	Min Elevation (m)	Max Elevation (m)	Mean Slope (%)	Perimeter (km)	Area (km^2)
352.3	66.7	1005.1	20.8	65.9	106.5

The climate of the study area is typically Mediterranean, and the weather pattern is characterized by high temperatures during summer and relatively low annual precipitation (470 mm) [25,31]. Based on various climate indexes that were applied on the reference area (e.g., Pinna, De Martonne, Emberger, Daget), the Upper Anthemountas Basin can be classified, on average, as "Semi-Dry Mediterranean" [30]. From a hydrological perspective, the catchment features a dense and well-developed stream network [25,29], culminating in a fifth-order stream as classified by the Strahler method (Figure 1) [30]. Nevertheless, throughout most of the year, the stream's flow is minimal, which can be attributed to low precipitation and to the semi-permeable nature of certain upper geological and soil layers. Consequently, noticeable surface outflows on the stream occur only briefly, mainly following periods of intense rainfall [25,29].

2.2. General Methodological Framework

The methodology adopted in this study comprises a sequence of four (4) well-defined and logically structured steps. These steps are designed to support the development and calibration of a robust and transparent hydrological model, as well as to facilitate a systematic comparison between the simulated ET outputs and remotely sensed ET estimates. The complete approach is illustrated in Figure 2 and outlined below.

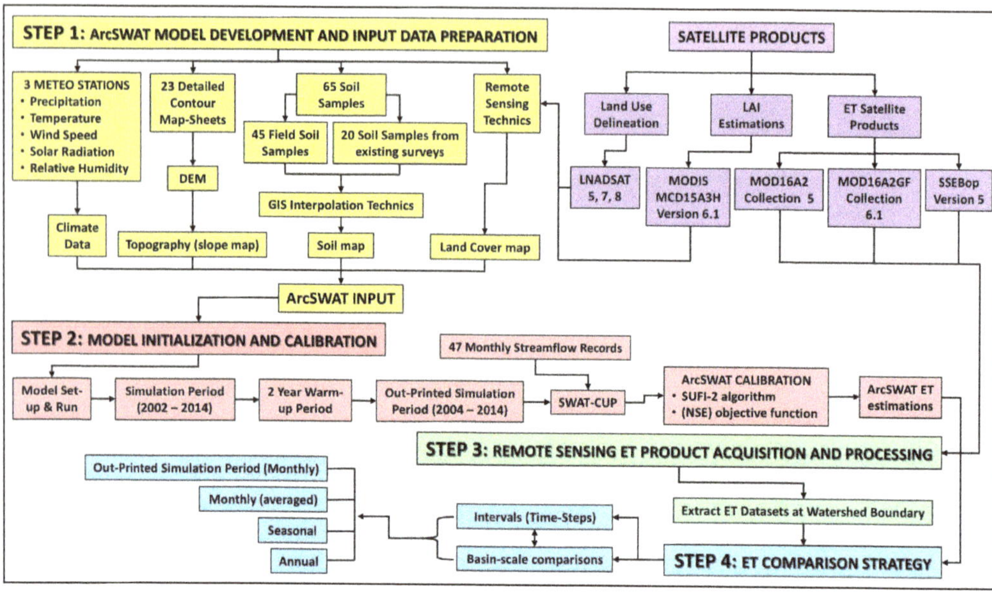

Figure 2. Schematic representation of the methodological framework developed in the present study.

Step 1: Hydrological Model Development and Input Data Preparation.

The hydrological model for the study area was constructed using the SWAT framework and refined through the integration of high-resolution spatial datasets, alongside comprehensive field and laboratory measurements. Specifically, the topographic input includes a DEM derived from detailed contour maps, while climate data were sourced from three (3) local meteorological stations. Soil information was compiled through extensive field sampling and supplemented by existing surveys, with key physical properties measured and interpolated to generate a spatially detailed distributed soil map. Land use classification was achieved through the integration of high-resolution satellite imagery, official forest maps, and remote sensing techniques to refine vegetation-related parameters within the model.

Step 2: Model Initialization and Calibration.

A two-year warm-up run established equilibrium conditions for soil moisture and groundwater storage, while the overall simulation period expanded from 2002 to 2014. The calibration process employed forty-seven (47) monthly streamflow records and focused on optimizing key hydrological parameters. Due to the limited number of observations, all available data were used for calibration, and validation was omitted. The SUFI-2 algorithm within SWAT-CUP was utilized, and the Nash–Sutcliffe Efficiency (NSE) criterion was selected as the objective function, thereby ensuring robust capture of the hydrological component.

Step 3: Remote Sensing ET Product Acquisition and Processing.

Three (3) gridded freely available ET datasets were selected to be compared against the model outputs: (i) the MOD16A2 Collection 5 product at 1 km resolution; (ii) the MOD16A2GF Collection 6.1 product at 500 m resolution; and (iii) the SSEBop Version 5 product at 1 km resolution. Each product was clipped to the watershed boundary and spatially averaged to derive basin-scale mean ET time series, ensuring consistency with the SWAT aggregation scale. This basin-wide comparison approach was selected for two main reasons: (a) the relatively small watershed size (~110 km^2); and (b) the goal of ensuring uniformity and simplifying the analysis, as well as to enhance computational efficiency. By focusing on the watershed boundaries, the analysis was streamlined, ensuring manageable processing times while maintaining the spatial relevance of the results.

Step 4: ET Comparison Strategy.

Comparisons between SWAT-simulated ET and satellite-derived products were conducted at monthly (averaged), seasonal, and annual intervals, as well as over the entire simulation period at a monthly time step, in order to capture the overall temporal dynamics of ET.

2.3. *Performance Evaluation Metrics*

Various statistical efficiency metrics were implemented in this research, including RMSE, R^2, bR^2, NSE, MNS, SRS, KGE, and PBIAS (Table 2) [33–36]. Of the aforementioned indices, RMSE was adopted for evaluating the DEM created as an input for the hydrological model, whereas the rest were utilized during the model calibration procedure for assessing its performance.

Table 2. Performance evaluation metrics utilized in the study.

Statistical Metrics (Indices)	Formula *	Value Range	Optimal Value
Root Mean Square Error (RMSE)	$RMSE = \sqrt{\frac{\sum_{i=1}^{n}(x_i-y_i)^2}{n}}$	[0,+∞)	0 (Lower is better)
Coefficient of Determination (R^2)	$R^2 = \left[\frac{\sum_{i=1}^{n}(x_i-\overline{x})(y_i-\overline{y})}{\sqrt{\sum_{i=1}^{n}(x_i-\overline{x})^2}\sqrt{\sum_{i=1}^{n}(y_i-\overline{y})^2}}\right]^2$	[0,1]	1 (Higher is better)
Modified Coefficient of Determination (bR^2)	$bR^2 = \begin{cases}\left(\frac{\overline{y}}{\overline{x}}\right) * R^2, \ if\ \overline{y} \le \overline{x} \\ \left(\frac{\overline{x}}{\overline{y}}\right) * R^2, \ if\ \overline{y} > \overline{x}\end{cases}$	[0,1]	1 (Higher is better)
Nash–Sutcliffe Efficiency (NSE)	$NSE = 1 - \frac{\sum_{i=1}^{n}(x_i-y_i)^2}{\sum_{i=1}^{n}(x_i-\overline{x})^2}$	(−∞,1]	1 (Higher is better)
Modified Nash–Sutcliffe Efficiency (MNS)	$MNS = 1 - \frac{\sum_{i=1}^{n}\lvert x_i-y_i\rvert^j}{\sum_{i=1}^{n}\lvert x_i-\overline{x}\rvert^j}\ (where\ j \in N)$	(−∞,1]	1 (Higher is better)
Ratio of the Standard Deviation of Observations to the Root Mean Square Error (RSR)	$RSR = \sqrt{\frac{\sum_{i=1}^{n}(x_i-y_i)^2}{\sum_{i=1}^{n}(x_i-\overline{x})^2}}$	[0,+∞)	0 (Lower is better)
Kling–Gupta Efficiency (KGE)	$KGE = 1 - \sqrt{(r-1)^2 + \left(\frac{SDy}{SDx} - 1\right)^2 + \left(\frac{\overline{y}}{\overline{x}} - 1\right)^2}$	(−∞,1]	1 (Higher is better)
Percent Bias (PBIAS)	$PBIAS = 100\frac{\sum_{i=1}^{n}(x_i-y_i)}{\sum_{i=1}^{n}(x_i)}$	(−∞,+∞)	0 (Lower absolute value is better)

* where n = total number of observations, x_i = observed data, y_i = simulated data, $\overline{x}$ = mean value of observed data, $\overline{y}$ = mean value of simulated data, j = 1 (proposed value in the literature), SDx = standard deviation of the observed data, SDy = standard deviation of the simulated data, r = correlation coefficient between observed and simulated data.

Moreover, KGE, PBIAS, RMSE, and RSR were employed for analyzing the performance of remotely sensed ET data in relation to SWAT-derived ET estimates. The rationale for applying this set of four (4) metrics is that, collectively, they encompass multiple dimensions of model performance, particularly in the context of ET comparisons, since (a) KGE combines three (3) key evaluation aspects into a single metric, that is, correlation, bias, and variability [37]; (b) PBIAS isolates a model's tendency to consistently overestimate or underestimate ET [33], and, when applied alongside KGE, aids in identifying whether strong overall performance (e.g., a high KGE value) may obscure systematic errors; (c) RMSE quantifies absolute errors; and, finally, (d) RSR normalizes these errors relative to data variability [33], which is particularly useful when ET values vary widely.

2.4. Hydrological Model Development

2.4.1. Description of the SWAT Model

The Soil and Water Assessment Tool (SWAT) is a hydrologic and water quality model developed by the United States Department of Agriculture, Agricultural Research Service (USDA-ARS) [38]. It is a physically based, semi-distributed model that operates on a continuous time scale with a daily or sub-hourly time step. The primary objective of the model is to forecast the impact of land management practices on water, sediment, and agricultural chemical yield at a river basin scale for various periods of time. In SWAT, the simulation of all hydrologic processes occurring in the basin is established upon the fundamental water balance equation (Equation (1)).

$$SW_t = SW_0 + \sum_{i=1}^{t}(R_{day} - Q_{surf} - ET_{day} - W_{day} - Q_{gw}) \quad (1)$$

where t is the simulation period, SW_t is the final soil water content (mm), SW_0 is the initial soil water content (mm), and R_{day}, Q_{surf}, ET_{day}, W_{day}, and Q_{gw} represent daily values (mm) for precipitation, runoff, evapotranspiration, infiltration, and groundwater flow, respectively.

In terms of spatial analysis, the watershed is partitioned into several subbasins, which are, in turn, subdivided into specific segments named Hydrologic Response Units (HRUs) that are defined by homogeneous soil properties, land use patterns, and slope attributes. Then, the water balance equation is utilized at each HRU and, thus, the water cycle components (i.e., runoff, infiltration, ET) are computed at the HRU level. Finally, results from all HRUs within a subbasin are aggregated to depict the overall water budget of each individual subbasin, before being routed through the entire basin from upstream subbasins to the watershed outlet [39].

SWAT model can be incorporated into a GIS environment, such as QGIS (QSWAT) and ArcGIS (ArcSWAT), thus rendering hydrologic simulations at the watershed scale far more intuitive. In this research, ArcSWAT 2012 (as an ArcGIS extension tool) was applied for delineating the water budget in the Upper Anthemountas Basin.

The model's requisite spatial and attribute data are classified into four (4) principal categories: (1) topography (DEM); (2) land cover; (3) soil; and (4) climate data. Among these, climate data constitute a vital input parameter, since their precision is paramount for the accurate simulation of the water budget [25,40,41]. The climate data incorporated into the model encompass measurements of precipitation, temperature, relative humidity, wind speed, and solar radiation. Moreover, depending on the accessibility of weather data, users may select from one of three evapotranspiration methods. Those are as follows: (a) the Hargreaves method, which incorporates mean min and max daily temperature; (b) the Priestly–Taylor method, which demands mean min and max daily temperature, mean daily humidity and solar radiation; and (c) the Penman–Monteith method, which requires mean min and max daily temperature, mean daily humidity and solar radiation, as well as mean daily wind speed [42].

With regard to land use, SWAT accommodates a default database that contains an extensive list of predefined land use and land cover categories, which facilitates users in classifying different land types for hydrological modeling [42,43]. This database includes land use codes, descriptions, and associated parameters that affect processes like surface runoff, infiltration, evapotranspiration, and erosion. When users introduce a land cover map, they match the land cover classes to SWAT's predefined codes. Then, the model assigns hydrological properties based on the default database. In case the default database does not fully represent the land cover observed in the study area, modelers have the option to modify the land use table by altering custom parameter values of existing land use types or even expand the table by adding new land cover types and new parameter values [42].

Finally, the soil data required by the SWAT model represent the most demanding category of input information, as they embody an extensive array of parameters that designate the physical properties of the various soil types that cover the study area. More specifically, the essential soil parameters include the soil texture (expressed as proportions of sand, silt, clay and rock fragments), organic carbon content (OC), bulk density (BD), available water capacity (AWC), saturated hydraulic conductivity (K_s), the soil hydrologic group (HYDGRP), albedo, and the soil erosion factor (USLE-K) [42].

2.4.2. Model Input Data Preparation

For the present study, an exceptionally comprehensive hydrological model was assembled by integrating and implementing high-quality measured data and high spatial resolution data (with grid cells 5 × 5 m) of topography, land cover, and soil features into SWAT. Details concerning the input data information applied to the SWAT model are presented in Table 3.

Table 3. Overview of input data sources acquired for applying the SWAT model in the Upper Anthemountas Basin.

Input Data in the SWAT Model	Data Source
Topography	Digitizing and editing twenty-three (23) Hellenic Military Geographical Service (HMGS) elevation map sheets (4 m interval, 1:5000 scale) by using GIS techniques
Land Cover	Applying various remote sensing techniques on satellite images (Landsat 5, 7, 8, and MODIS)
Soil	• Collecting soil samples (45) from the study area • Performing field and laboratory measurements of key soil properties • Acquiring additional soil information from existing soil surveys (e.g., Misopolinos et al. [32]) • Applying GIS techniques
Climate	Data acquired from three (3) stations: (1) National Agricultural Research Foundation (NAGREF) station (2) Galatista station (3) Aristotle University of Thessaloniki (AUTH) station

It should be noted that the generation of detailed soil and land use maps, which serve as inputs for the SWAT model, was a rather exhausting and rigorous process that extends beyond the scope of this research. Consequently, only a brief overview of their production is presented. Moreover, all GIS-based operations (such as digitization, interpolation) and satellite image processing were executed using the appropriate tools within the ArcGIS 10.5 platform.

Climate Data

Climate data presented a considerable challenge due to the limited number of operational meteorological stations within the study area. All available stations with continuous records were incorporated into the model to ensure optimal utilization of the existing climatic observations. In greater detail, two (2) primary weather stations provided the data required for SWAT. Within the study area, only a single station is located (Galatista station) (coordinates, X: 438,375, Y: 4,480,142), which, unfortunately, does not yield a satisfactory time series of recorded data. To be precise, the station effectively operated from 2005 to 2011. However, during this period, the cataloged time series were not continuous and exhibited significant month-long gaps. Notwithstanding, the available data from this station were utilized, as it constitutes the sole station within the basin under study.

Furthermore, another important weather post, the National Agricultural Research Foundation (NAGREF) station, is situated close to the study area, approximately 15 km from the center of the Upper Anthemountas Basin (coordinates, X: 421,897, Y: 4,484,442). The available recorded data from this station were primarily utilized in this study, since the meteorological unit: (a) represents the closest weather observatory to the study area; and (b) produces fairly comprehensive and reliable time series, displaying day-long to, rarely, week-long and month-long gaps. Additionally, a third weather unit maintained by the Aristotle University of Thessaloniki (AUTH station) also offers reliable data, although it is situated somewhat further from the study area (roughly 30 km) (coordinates, X: 410,690, Y: 4,498,110). Ultimately, taking into account all the available weather data, the daily time series employed in the SWAT model spans from 1 January 2002 to 31 December 2014. For this timeframe, data from both the Galatista station and the NAGREF station were combined. Both stations provide data pertaining to precipitation, temperature, relative humidity, wind speed, and solar radiation, parameters that are vital for building the hydrological model. Moreover, any gaps observed in the registered measurements of the aforementioned two (2) stations were supplemented by incorporating data from the AUTH station (primarily solar radiation). In Table 4, values of mean annual precipitation, temperature, relative humidity, wind speed, and solar radiation from 2002 to 2014, as estimated using data from all the aforementioned stations, are provided.

Table 4. Mean annual climate data input for the SWAT model, where Prec = precipitation, Temp = temperature, MIN = minimum, MAX = maximum, RH = relative humidity, WS = wind speed, and SR = solar radiation.

Mean Annual Weather Data							
Year	**Prec (mm)**	**Temp (°C)**	**MIN Temp (°C)**	**MAX Temp (°C)**	**RH (%)**	**WS (m/s)**	**SR (MJ/m^2)**
2002	730.4	15.8	10.3	21.2	66.12	1.11	11.55
2003	784.3	15.5	9.9	21.0	70.09	1.07	13.21
2004	544	15.2	9.4	21.0	68.29	1.12	15.64
2005	533.1	14.7	8.8	20.6	68.07	1.99	15.44
2006	630.2	14.2	8.2	20.2	71.91	2.35	15.11
2007	525	15.3	8.9	21.6	67.73	2.32	15.52
2008	383.4	15.7	9.5	21.9	69.50	1.65	14.29
2009	693.2	15.4	9.3	21.5	73.48	1.76	14.40
2010	623.6	15.7	9.8	21.5	73.70	1.06	14.87
2011	475.4	15.0	9.0	21.0	74.39	1.23	14.83
2012	504.4	16.2	10.4	22.0	68.97	1.64	15.46
2013	404.6	16.0	9.8	22.2	72.76	1.56	13.95
2014	931.6	15.9	10.5	21.3	76.60	1.30	14.65
Mean	**597.17**	**15.42**	**9.53**	**21.30**	**70.89**	**1.55**	**14.53**

Topography (DEM)

The initial stage of the ArcSWAT model setup involves the incorporation of a DEM, which serves as the basis for watershed delineation. During this process, the model analyzes the terrain to define the watershed boundary, extract the stream network, identify outlet points, and divide the area into subbasins based on flow direction and accumulation patterns. The resulting delineation establishes the spatial foundation upon which all subsequent components of the model are built, including the integration of land use, soil, and slope characteristics in later stages of the setup.

Therefore, in order to create a high-quality DEM as input for the model (Figure 3), the following two (2) steps were taken.

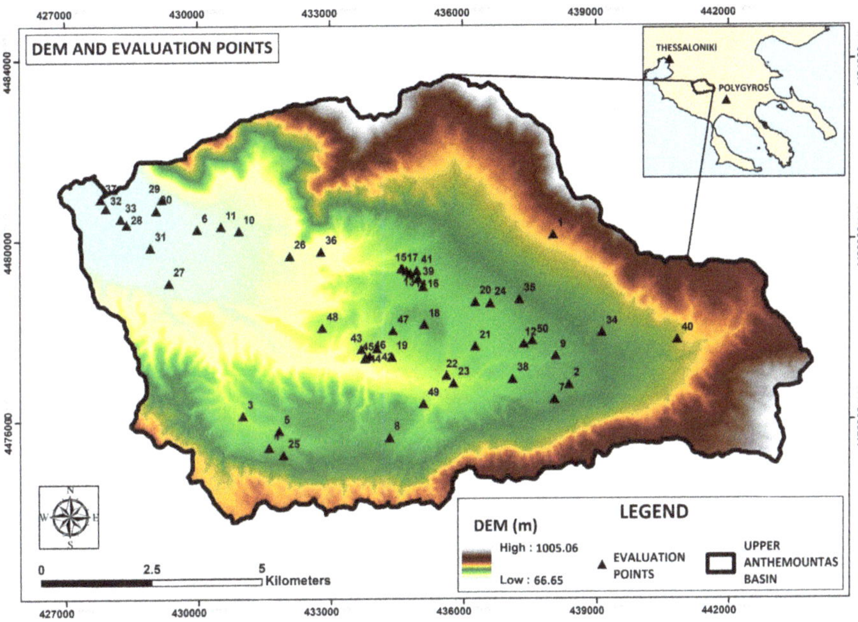

Figure 3. The Digital Elevation Model (DEM) of the study area (used as input in the SWAT model), along with the corresponding DEM evaluation points.

(1) First, a digital mosaic of twenty-three (23) Hellenic Military Geographical Service (HMGS) contour map sheets (scale 1:5000) that cover the entire basin was created. The mosaic was georeferenced in the Greek coordinate system GGRS87.

(2) Then, the contours of this mosaic (4 m interval) were digitized, and a high-quality DEM with a 5×5 m cell size grid was produced. For the verification and evaluation of the generated DEM, elevations from fifty (50) points scattered within the study area (Figure 3) were utilized, which were obtained using a high-precision GPS device (STONEX S900A–SE GNSS Receiver). The RMSE was employed as an evaluation statistic metric, since, as attested by Wise [44], "it is the most commonly applied measure of DEM quality and the one published by data suppliers". A value of RMSE = 0.892 m was obtained, indicating an overall excellent quality of the produced DEM.

Upon the introduction of the DEM into ArcSWAT, the delineation process was conducted, resulting in the subdivision of the watershed into seventeen (17) hydrologically distinct subbasins. Furthermore, the stream network previously illustrated in Figure 1 was generated, while various topographic features of the basin, previously presented in Table 1, were derived from the watershed delineation report.

Land Cover

The development of an accurate land cover map of the study area was a rigorous process and can be briefly summarized in the following three (3) steps:

(1) Initially, all land covers in the study area that are assumed to be constant over time (i.e., forests, residential areas, olive groves, grasslands, barren land, and limited extents of abandoned mines) were delineated.

(2) Then, the irrigated crops were distinguished from the non-irrigated crops within the study area.

(3) Finally, the different types of irrigated crops were discriminated.

Regarding the first step, a high-resolution satellite natural color image with a 0.6×0.6 m cell size was acquired. Urban settlements, barren land, and abandoned magnesite mines, along with olive orchards and scattered small-scale grassland sites, were delineated based on this image. These land classes were also identified through multiple on-site inspections. The forested areas were extracted based on the ratified forest maps provided by the Hellenic Cadastre (https://gis.ktimanet.gr/gis/forestfinal, accessed on 31 May 2020).

On the other hand, it is crucial to define the cultivated land in the study area, which includes non-irrigated or partially irrigated crops (mainly wheat) and irrigated crops, involving maize (corn), alfalfa, and cotton. For this purpose, satellite imagery from the year 2008 was utilized. The selection of this particular year was based on the following considerations: (a) satellite data needed to fall within the overall simulation period, which spans from 2002 to 2014; (b) satellite images were required to have less than 10% cloud cover and be sufficient in quantity, particularly during the summer months, when irrigated crops reach peak vegetative development (in contrast to non-irrigated cereals) and are easily distinguishable due to high chlorophyll activity; and (c) satellite images had to be acquired at a sufficient temporal distance from rainfall events to minimize the effects of short-term vegetation responses in rain-fed crops.

Considering all the above, the year 2008 satisfies adequately all the aforementioned conditions concerning the available satellite images. However, although it is relatively easy to distinguish irrigated from non-irrigated crops within the study area, differentiating between various irrigated crops is more challenging and requires the application of remote sensing techniques. Therefore, satellite images from the year 2019 were initially examined, as prior knowledge of specific irrigated crops for that period was available. The objective was to investigate the spectral behavior of these crops and subsequently apply the findings to the year 2008. In this regard, eight (8) Landsat 8 images (Collection 2, Level 2—C2L2) acquired between June and September 2019 were utilized. In comparison, six (6) Landsat 5 images (Collection 1, Level 1—C1L1, and Collection 2, Level 2—C2L2) acquired between July and September 2008, along with two (2) Landsat 7 images (C1L1 and C2L2) from June and July 2008, were also obtained. All images were retrieved through the USGS EarthExplorer platform (https://earthexplorer.usgs.gov/, accessed on 29 July 2020).

The remote sensing methods derived from the literature and applied to differentiate between irrigated and non-irrigated crops, as well as to identify all types of irrigated crops (for both years 2019 and 2008) and for all Landsat 5, 7, and 8 images, are as follows.

(1) Color Composites, which involve the combination of three (3) spectral bands from satellite imagery, assigned to the three (3) primary display colors, that is, Red, Green, and Blue (RGB), so as to generate composite images. Specifically, a false color composite was applied using the band combination RGB of 6,5,4 for Landsat 8 images, while RGB of 5,4,3 was applied for Landsat 5 and 7 images. This configuration is suitable for agricultural studies and vegetation analysis [45,46]. In this composite, healthy vegetation appears in shades of green, while bare soils (or plowed crops) are typically represented in magenta tones.

(2) The Tasseled Cap Transformation (TCT), which produces three (3) primary components [47]: the Brightness index, associated with soil reflectance, the Greenness index, related to vegetation vigor (chlorophyll content), and the Wetness index, indicative of soil

and vegetation moisture. In essence, the TCT is an orthogonal transformation that computes these three (3) components by applying specific linear equations to a combination of six (6) Landsat bands. The coefficients used in the TCT differ for each of the components as well as for each Landsat satellite and its product versions (Collections and Levels). Further details can be found in Crist and Kauth [48] for Landsat 5, Huang et al. [49] for Landsat 7, and Baig et al. [47] for Landsat 8.

(3) A composite image was also generated from TCT components, with RGB of Brightness, Greenness, Wetness, in order to distinguish the highest content of water and the lowest soil moisture [50].

Ultimately, the conclusions drawn from the visual interpretation of satellite imagery regarding crop behavior in the study area during the summer months (June through September) are summarized as follows: (a) non-irrigated crops (wheats) are easily identifiable, as their color tones in the satellite images do not exhibit significant variations throughout the summer period; (b) maize, which is intensively irrigated, becomes distinguishable from mid-June, with harvesting typically occurring between late August and mid-September; (c) cotton becomes identifiable from late July, with harvest taking place around late September; and, finally, (d) alfalfa is also easily recognized due to its recurring harvest cycle, which occurs approximately every 30 to 35 days. Figure 4 shows the detailed land cover map developed using the aforementioned procedure and incorporated into the SWAT model.

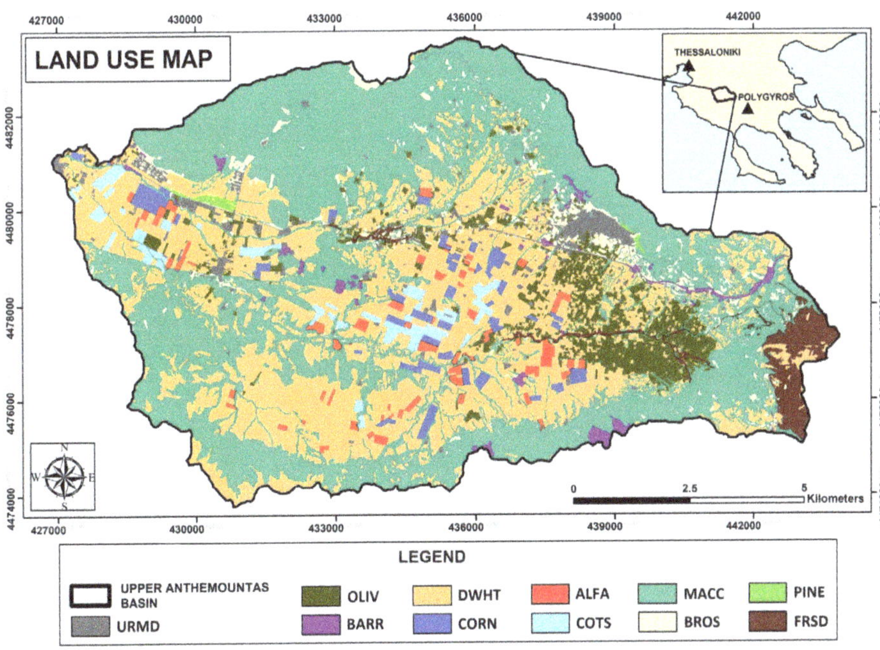

Figure 4. Land cover classification map of the study area (used as input in the SWAT model), where OLIV = olive trees, DWHT = durum wheat fields, CORN = corn fields, ALFA = alfalfa fields, COTS = cotton fields, MACC = Maquis forests, BROS = grasslands (brome grass), PINE = pine trees, FRSD = deciduous forests (Hungarian oak), BARR = barren land and abandoned magnesite mines, URMD = residential area.

Furthermore, as previously mentioned, the evergreen forest–shrub vegetation ("Maquis") accounts for the largest proportion of land use in the study area. However, this specific land use category was not directly available in the SWAT database. Instead, the database contained a very similar land use (labeled as "GRAR"), which refers to a comparable hard-leaved Mediterranean shrubland known as "Garrigue", primarily found in southern France and other Mediterranean regions. Essentially, this category pertains to formations of scattered, low (0.2 to 2 m) bushy shrubs and pseudo-maquis [51,52].

Based on the foregoing, a new land cover category was created in the SWAT database named "Macchia-Maquis shrubland" ("MACC"). The values for the physical parameters of this new land use were largely adopted from the corresponding and pre-existing "GRAR" classification. However, certain key physical parameters were refined, either based on the literature or through remote sensing methods, to better represent the natural environment of the maquis in the study area. These parameters include 'BLAI' (maximum leaf area index), 'ALAI_MIN' (minimum leaf area index), 'T_OPT' (optimal temperature for plant growth), and 'T_BASE' (minimum temperature for plant growth). It should be noted that improving the accuracy of LAI representation contributes to more reliable hydrological modeling [13,53]. Moreover, the assimilation of LAI data has proven effective in enhancing the estimation of key processes, including ET and biomass accumulation [53,54].

In this context, remote sensing products were employed to determine both the minimum leaf area index (ALAI_MIN) and the maximum leaf area index (BLAI) for the maquis scrubland biome. Specifically, LAI data were obtained from the MODIS MCD15A3H Version 6.1 product [55], which provides 4-day composite time series at 500-m resolution. The dataset (920 images) spans the period 2003–2012 and was accessed via the USGS EARTHDATA platform (https://lpdaac.usgs.gov/products/mcd15a3hv061/, accessed on 6 September 2020). Following Daneshvar et al. [56], by focusing on locations predominantly occupied by maquis, average minimum and maximum annual LAI cycles were generated (from 2003 to 2012), from which the ALAI_MIN and BLAI values were subsequently derived. The updated values for "MACC" classification, incorporated in the SWAT land use database, are presented in Table 5.

Table 5. Updated values for the parameters utilized in the SWAT database model with regard to maquis classification ("MACC") scrubland.

SWAT Parameters for "MACC" Classification	Description	Value Range	Selected Value	Source
BLAI	Maximum Leaf Area Index	2.4–3.1	2.8	Analyzing/Editing time series of MODIS LAI images
ALAI_MIN	Minimum Leaf Area Index	0.24–0.35	0.3	
T_OPT (°C)	Optimal Temperature for Plant Growth	-	20	[57]
T_BASE (°C)	Minimum Temperature for Plant Growth	-	4	

Soil

Developing an accurate digital soil map, alongside its corresponding attribute database, proved challenging and was executed in two (2) primary phases. In the first stage, all required soil physicochemical parameters for SWAT input were determined. These parameters were subsequently integrated into the model through a database that is associated with the spatial characteristics of the input soil map. To this task, forty-five (45) soil samples were systematically collected across the basin to ascertain the prevailing soil types and their physical properties. Samples were retrieved from depths of 0–30 cm and 30–60 cm (where feasible), while a limited number of cores were acquired from deeper soil layers (60–100 cm). In addition to these samples, another twenty (20) specimens and their physical characteristics were obtained from the research conducted by Misopolinos et al. [32].

Determination of particle size distribution (proportions of rock fragments > 2 mm, sand < 2000–50 μm, silt < 50–2 μm, and clay < 2 μm) was accomplished in two (2) stages. At first, Sieve Analysis was employed to separate coarse fragments exceeding 2 mm, followed by a hydrometer-based sedimentation process for the finer materials. Soil textural classes were subsequently categorized in accordance with the U.S. Department of Agriculture (USDA) soil textural triangle. Organic matter content was quantified using the Loss on Ignition (LOI) method as outlined by Nelson et al. [58], while the carbon content was derived from the conversion factor described by Pribyl [59] (Equation (2)).

$$\text{Organic Carbon (\%)} = \text{Organic Matter (\%)}/2 \tag{2}$$

In situ measurements of saturated hydraulic conductivity were performed using a single-ring infiltrometer technique suggested by Bagarello et al. [60]. Soil albedo values for each sample were assigned according to the surveys conducted by Penning De Vries et al. [61] and ten Berge [62]. Moreover, bulk density and available water capacity were determined using the Soil Water Characteristic–Hydraulic Properties Calculator (SWC-HPC) Model developed by Saxton and Willey [63]. Finally, the erosion factor (soil erodibility K factor) was calculated applying the equations of Williams [64]. Rock layers, representing fractured bedrock in the study area, were classified as "rock outcrop", a soil class that exists in the default soil database of the model.

Next, a GIS-based digital soil map was generated following a well-structured procedure. Initially, soil particle-size fractions (sand, silt, and clay fractions) were interpolated for each soil depth layer (0–30 cm, 30–60 cm, and 60–100 cm) through a three-step process:

(1) First, a centered log-ratio (CLR) transformation was applied to the fractions in each layer to eliminate the "closure effect" inherent in compositional data [65,66]. The CLR is expressed as follows [67,68]:

$$CLR_i = ln\left(\frac{Z_i}{\left(\sqrt[D]{\prod_{i=1}^{i=D} Z_i}\right)}\right) \tag{3}$$

where z = compositional data (particle-size fractions) and $i = 1, 2, \ldots, D$, $D = 3$ (the number of soil particle-size categories).

(2) Next, spline interpolation was conducted on the transformed sand, silt, and clay values to produce maps of their spatial distribution.

(3) Then, the interpolated values were back-transformed to recover the soil particle-size fractions (sand, silt, and clay). The inverse transformation equation is expressed as follows [65,66]:

$$Z_i = \frac{e^{CLR_i}}{\sum_{i=1}^{i=D} e^{CLR_i}} \tag{4}$$

where CLR_i are the transformed values produced from Equation (3).

(4) The back-transformed fractions were used in order to assign soil textural classes for each layer (0–30 cm, 30–60 cm, and 60–100 cm). In addition, the soil cover thickness of the study area was determined based on two (2) connected and digitized land capability map sheets [69].

Ultimately, all the above-produced maps, comprising the three (3) soil texture stratification maps and the soil cover thickness map, were subsequently overlaid. In total, twenty-five (25) soil profiles of 1 m depth were created and imported into SWAT along with their corresponding attribute database. Each soil profile represents a unique combination of three (3) soil units corresponding to the depths of 0–30 cm, 30–60 cm, and 60–100 cm, with attribute values derived from the averaged measurements of each soil textural class. The detailed soil map produced in SWAT is depicted in Figure 5.

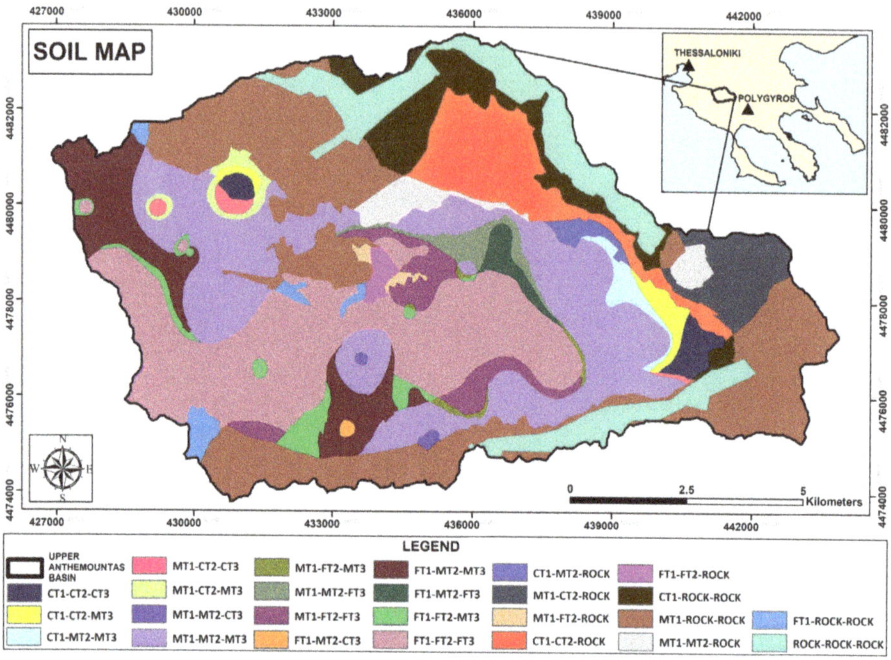

Figure 5. Soil map classification of the study area (used as input in the SWAT model), where CT = coarse texture soils, MT = medium texture soils, FT = fine texture soils, ROCK = rocks, while numbers 1, 2, 3 represent soil depth layers, with 1 = from 0 to 30 cm depth, 2 = from 30 to 60 cm depth and 3 = from 60 to 100 cm depth.

2.4.3. Set up and Initial Model Run

At this stage, the final configuration of the model is established. Essential spatial and attribute data, including slope, land use, and soil maps with accompanying feature tables, were imported into SWAT to ensure accurate representation of runoff and infiltration processes. In this analysis, three (3) slope thresholds (classes) were selected: (i) <10%; (ii) 10% to 35%; and (iii) >35% (Figure 6). This classification was derived from the literature, as it is considered that these threshold slopes may affect both the surface runoff and infiltration of water [70–72]. Furthermore, although practitioners may apply HRU thresholds (e.g., default land use 20%, soil 10%, slope 20%) to exclude minor areas, reduce model complexity, and focus on dominant classes [39], no such thresholds were applied in this study, resulting in a total of 1441 HRUs for the Upper Anthemountas Basin.

Climate inputs and simulation parameters were subsequently defined for the period spanning 1 January 2002 to 31 December 2014, with a monthly timestep, while the Penman–Monteith method was selected for evapotranspiration estimation. The simulation period was determined based on the following: (a) the availability of high-quality weather data series acquired from climate stations; and (b) the availability of measured streamflow data, which are crucial for the calibration of the model.

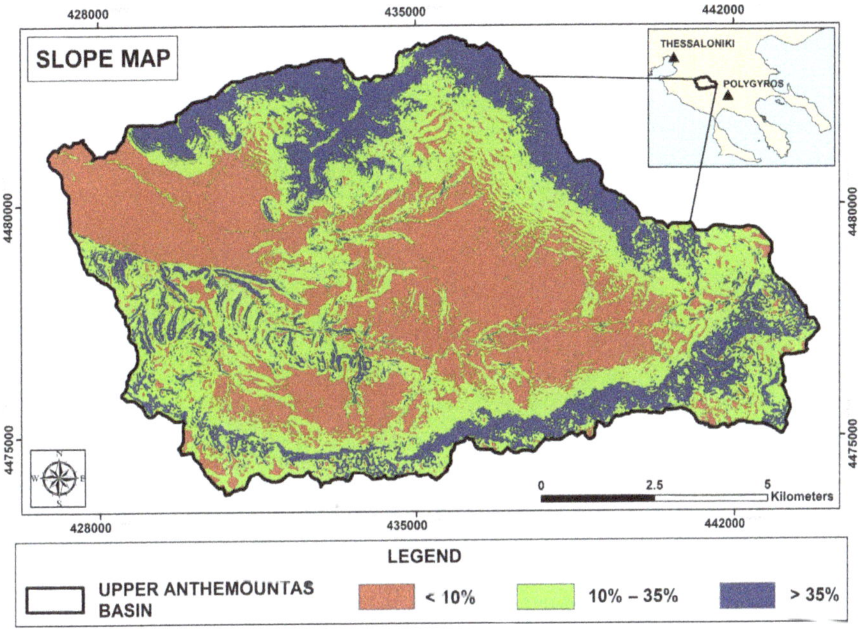

Figure 6. Slope classification map of the study area.

Finally, assigning a "warm-up" period (NYSKIP) in SWAT is vital, as it enables the model to stabilize key hydrological components, such as soil moisture and baseflow, before generating meaningful simulation outputs. In the present study, since the total simulation period is not considerably long (spanning 13 years), a two-year warm-up period was adopted, in accordance with practices recommended in the literature [73,74].

2.4.4. Model Calibration Procedure

The traditional approach to calibrating hydrological models such as SWAT relies on discharge measurements at hydrometric stations. To this task, model calibration was performed using forty-seven (47) available measured monthly streamflow values (from 2004 to 2013), observed on the main channel of the Upper Anthemountas Basin. The main strategy adopted for the calibration of the model is described in the following steps:

(1) The calibration of the SWAT model was performed using the semi-automated multisite and inverse modeling algorithm SUFI-2 in the SWAT-CUP software package [75]. The SUFI-2 algorithm was implemented, as it is recognized for its high efficiency and effectiveness [23,76], while simultaneously yielding more reasonable and well-balanced calibration results [77]. The Nash–Sutcliffe Efficiency (NSE) criterion was selected as the objective function, given its widespread use and reliability [77].

(2) The parameters pre-selected to be calibrated (about 30 parameters) were designated based on a thorough and exhausting literature review [78–98].

(3) Prior to their utilization in SWAT-CUP, many parameters were systematically grouped based on their impact on the water cycle, following the guidelines suggested by Abbaspour et al. [23]. Subsequently, the grouped parameters were analyzed and calibrated in a step-by-step procedure. The number of simulations (model runs) conducted at each iteration (step) was set to 500, in accordance with the recommendations of Abbaspour et al. [23]. Parameters that did not significantly affect the model's behavior were excluded from further consideration during the calibration process.

(4) During calibration, all adjusted parameter values retained consistency with the watershed hydrological behavior, thereby ensuring that the parameterization remained representative of actual field conditions.

(5) The quantitative performance of the ArcSWAT model was attained by employing seven (7) statistical criteria and score classification for a monthly time step, pursuant to the recommendations of Kouchi et al. [34] and Sao et al. [35].

2.5. Satellite-Derived ET Datasets

In this study, remotely sensed ET products were employed, and their performance was evaluated against ET estimates simulated by the ArcSWAT model. Selection of the ET data was governed by specific criteria: (a) data must be freely accessible (open access); (b) data must span ArcSWAT model's simulation period (from 2002 to 2014); (c) data must be purely remotely sensed; (d) spatial resolution of the data must be sufficiently high, given that the study area is a relatively small watershed; and, finally, (e) data must (spatially) engulf the study area.

Consequently, several widely recognized ET products, such as GLEAM, WaPOR (FAO), ALEXI, and TerraClimate, were excluded from this research, since they fail to meet one of the aforementioned criteria. In this study, three (3) globally recognized ET products (Table 6) that utilize satellite-derived inputs coupled with hybrid energy balance modeling approaches were used: (a) the "MOD16A2_MONTHLY.MERRA_GMAO_1kmALB" dataset (Collection 5); (b) the "MOD16A2GF" product (Collection 6.1); and (c) the "SSEBop" product (Version 5).

Table 6. Satellite-derived ET datasets used in the study.

Dataset Name	Dataset	Spatial Resolution	Temporal Resolution	Data Source
Moderate Resolution Imaging Spectroradiometer (MODIS) Global Terrestrial Evapotranspiration	MOD16A2 (Collection 5)	1 km	Monthly	http://files.ntsg.umt.edu/data/NTSG_Products/MOD16/MOD16A2_MONTHLY.MERRA_GMAO_1kmALB/ (accessed on 3 January 2023)
	MOD16A2GF (Collection 6.1)	0.5 km	8-day	https://appeears.earthdatacloud.nasa.gov (accessed on 7 January 2023)
Operational Simplified Surface Energy Balance	SSEBop (Version 5)	1 km	Monthly	https://app.climateengine.org/climateEngine (accessed on 18 January 2023)

The MOD16A2_MONTHLY.MERRA_GMAO_1kmALB (MOD16A2 Collection 5, named as MOD16A2-C5) dataset is a value-added, monthly global ET product developed by the Numerical Terradynamic Simulation Group (NTSG) at the University of Montana. It is derived from the MODIS MOD16A2 Collection 5 algorithm, which estimates terrestrial ET using the Penman–Monteith approach based on satellite-observed land surface variables [99]. In this enhanced version, the standard meteorological inputs have been replaced with data from the Modern-Era Retrospective Analysis for Research and Applications (MERRA), a reanalysis dataset produced by NASA's Global Modeling and Assimilation Office (GMAO). The product also incorporates a 1 km spatial albedo correction to refine surface energy balance calculations. These modifications improve the accuracy and consistency of evapotranspiration, latent heat flux, and gross primary productivity (GPP) estimates under varying surface and atmospheric conditions [100].

The MOD16A2GF Collection 6.1 (named as MOD16A2GF-C6.1) dataset is a gap-filled version of NASA's MODIS-based evapotranspiration product, providing 8-day global ET estimates at 500-m resolution. It is based on the Penman–Monteith energy balance algorithm, incorporating MODIS land surface variables such as LAI, FPAR, albedo, and surface temperature. The gap-filling procedure improves data continuity in cloudy regions [101].

The SSEBop (Simplified Surface Energy Balance Operational Product) version 5 dataset (named as SSEBop-V5) is a global, operational ET dataset developed by the USGS. It estimates actual ET by scaling reference ET with satellite-derived land surface temperature (LST), using a simplified energy balance framework. SSEBop is designed for large-scale monitoring applications, such as drought assessment, agricultural water use, and hydrologic modeling [102].

3. Results and Discussion

3.1. Hydrological Model Results

3.1.1. Model Calibration and Evaluation

The results of the SWAT model calibration are presented and analyzed in this section. Following calibration, the model exhibits performance levels ranging from satisfactory to very good, as depicted in Table 7. Specifically, based on the statistical indicators of R^2 (0.89), NSE (0.79), and RSR (0.46), the model is deemed very good, while evaluation using bR^2 (0.80), MNS (0.50), KGE (0.69), and PBIAS (−23) suggests a satisfactory performance.

Table 7. Model performance evaluation for monthly discharge simulation.

Index	Evaluation Performance Criteria				Model Performance Results	Model Performance Evaluation
	Very Good	Good	Satisfactory	Unsatisfactory		
R^2	$0.75 < R^2 \leq 1$	$0.65 < R^2 \leq 0.75$	$0.5 < R^2 \leq 0.65$	$R^2 \leq 0.5$	0.89	Very Good
bR^2	-	-	$bR^2 \geq 0.4$	$bR^2 < 0.4$	0.8	Satisfactory
NSE	$0.75 < NSE \leq 1$	$0.65 < NSE \leq 0.75$	$0.5 < NSE \leq 0.65$	$NSE \leq 0.5$	0.79	Very Good
MNS	-	-	$MNS \geq 0.4$	$MNS < 0.4$	0.50	Satisfactory
RSR	$0 \leq RSR \leq 0.5$	$0.5 < RSR \leq 0.6$	$0.6 < RSR \leq 0.7$	$RSR > 0.7$	0.46	Very Good
KGE	$0.9 \leq KGE \leq 1$	$0.75 \leq KGE < 0.9$	$0.5 \leq KGE < 0.75$	$KGE < 0.5$	0.69	Satisfactory
PBIAS	$PBIAS < \pm 10$	$\pm 10 \leq PBIAS < \pm 15$	$\pm 15 \leq PBIAS < \pm 25$	$PBIAS \geq \pm 25$	−23	Satisfactory

Figure 7 illustrates the comparison between observed and simulated river streamflow values after model calibration. Overall, there is a strong agreement between measured and simulated data. The model displays a tendency to slightly overestimate monthly baseflow, while its performance in capturing extreme discharge events is generally satisfactory. An exception is noted during the summer months of 2009, where a discrepancy between observed and simulated values is evident. A similar, though less prominent, deviation is also observed in the spring of 2010.

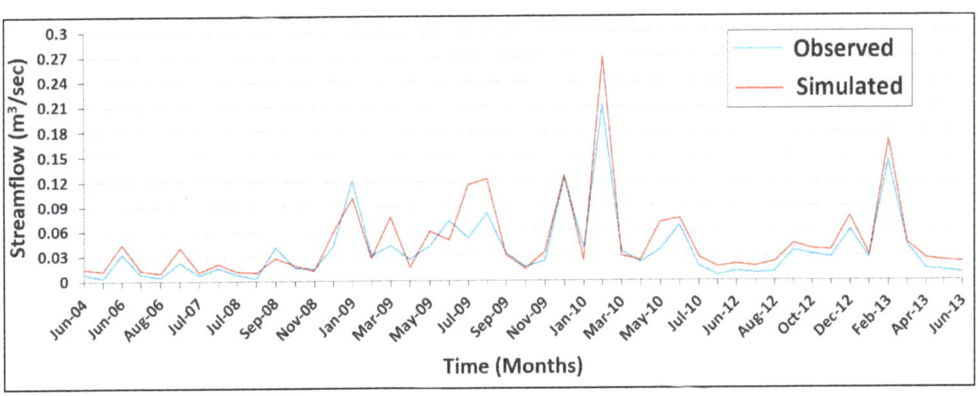

Figure 7. Observed vs. simulated monthly streamflow after model calibration.

As thoroughly discussed earlier, the Upper Anthemountas Basin is characterized by a semi-arid climate and a non-perennial river system, with flow occurring intermittently rather than consistently throughout the year. Additionally, the catchment is extensively covered by forest vegetation, which exerts a significant influence on hydrological processes. Based on the aforementioned considerations, various water losses are expected across all components of the hydrological cycle, providing a consistent rationale for the calibrated values. These water losses are parameterized through key variables listed in Table 8, and include: (a) subsurface flow from the river to the aquifer, as indicated with a high CH_K2 value; (b) deep-layer evapotranspiration (particularly intensified during the summer months), as reflected with lower ESCO rates; and (c) reduced surface runoff due to vegetation cover and land use patterns, as outlined with a higher OV_N value. Notably, higher values of GWQMN and GW_REVAP are associated with reduced baseflow and streamflow [89,103], a pattern observed both in the present analysis and consistently reported in other studies of Mediterranean semi-arid watersheds [104].

Table 8. Values of several SWAT model parameters as adjusted after calibration.

Parameter	Units	Description of the Parameter	Value Range		Fitted Value
			Min	Max	
ALPHA_BF	Days	Baseflow alpha factor	0	1	0.5
ALPHA_BNK	Days	Baseflow alpha factor for bank storage	0	1	0.000167
GWQMN	mm	Threshold depth of water in the shallow aquifer required for return flow to occur	0	5000	2383.33
REVAPMN	mm	Threshold depth of water in the shallow aquifer for "revap" to occur	0	500	433.67
RCHRG_DP	-	Deep aquifer percolation fraction	0	1	0.08
GW_REVAP	-	Groundwater "revap" coefficient	0.02	0.2	0.19
OV_N	-	Manning's "n" value for overland flow	0.01	1	0.342
CH_N2	-	Manning's "n" value for the main channel	0	0.3	0.213
CH_K2	mm/h	Effective hydraulic conductivity in main channel alluvium	0	500	357.5
GW_DELAY	Days	Groundwater delay time	0	500	49–192 *
ESCO	-	Soil evaporation compensation factor	0	1	0.7–0.91 *
EPCO	-	Plant uptake compensation factor	0	1	0.21–0.8 *
SOL_Z	mm	Depth from soil surface to bottom of layer	0	30,000	1500
BLAI (MACC)	(kg/ha)/(MJ/m^2)	Maximum potential leaf area index	0.5	10	2.55
T_BASE (MACC)	°C	Minimum (base) temperature for plant growth	0	18	3
T_OPT (MACC)	°C	Optimal temperature for plant growth	11	38	24
CN2	-	Runoff Curve Number	30	98	(~−5%) **

* Varies with slope; ** varies with land use/land cover classes and hydrologic soil groups.

With respect to Runoff Curve Number (CN2), calibration resulted in reductions ranging from approximately 3% to 6% (with an average decrease of 5%) across all land cover classes, while remaining within hydrologically and physically plausible bounds for the study area. In other words, the adjusted values (Table 8) continued to reflect the actual land use/cover and hydrological characteristics of the study area.

Notably, the maximum leaf area index (BLAI), as determined through model calibration, falls well within the range estimated from MODIS satellite imagery. Minor deviations were observed in the values of the T_BASE and T_OPT parameters for the maquis land cover class ("MACC"), compared to those reported in the literature [57].

3.1.2. Water Balance Components and Discharge Hydrograph

Following the calibration of the hydrological model, the average monthly water balance of the Upper Anthemountas Basin was established for the out-printed simulation period 2004–2014. The simulated hydrological behavior demonstrates strong alignment with the geomorphological, climatic, and hydrological attributes of the catchment, as these were outlined in prior sections. During the simulation period, mean precipitation was estimated at 568 mm, while mean actual evapotranspiration reached 367.8 mm. Percolation averaged 90.1 mm, surface runoff was calculated at 70.89 mm, and lateral flow contributed the smallest portion, averaging 40.31 mm. Overall, the runoff coefficient for the basin was determined to be 0.20, whereas the infiltration coefficient was 0.16. Notably, actual evapotranspiration emerged as the most significant component of the water balance, accounting for approximately 65% of total precipitation. The resulting water balance components are displayed in Figure 8.

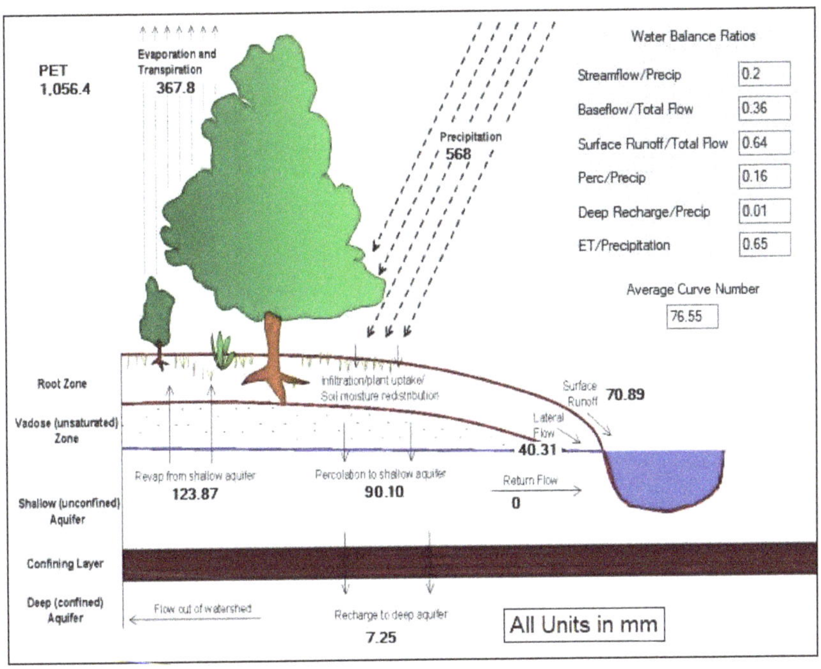

Figure 8. Water balance components and ratios for the study area as derived from the calibrated model across the out-printed simulation period (2004–2014).

Furthermore, Figure 9 shows the simulated hydrograph alongside precipitation data. The temporal correspondence between rainfall and discharge clearly indicates that the river follows an ephemeral flow regime, with runoff confined to short periods immediately following precipitation and an absence of flow for the majority of the hydrological year.

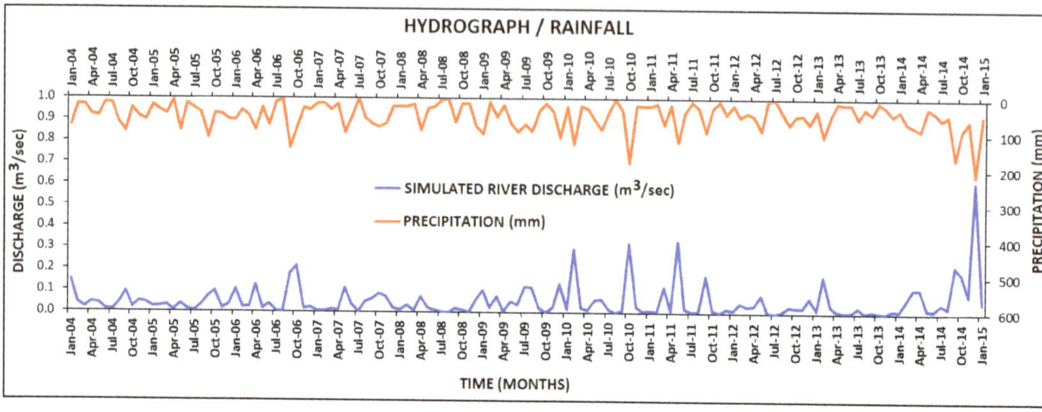

Figure 9. Hydrograph of the simulated streamflow vs. rainfall for the out-printed simulation period (2004–2014).

3.2. Performance Analysis Between ArcSWAT-Simulated and Remotely Sensed ET Products

3.2.1. Overall Monthly Evolution of ET

The overall temporal pattern of monthly ET simulated by the calibrated ArcSWAT model, alongside the corresponding estimates from three (3) remote sensing-derived products (MOD16A2-C5, MOD16A2GF-C6.1, and SSEBop-V5) over the period 2004–2014, is illustrated in Figure 10. In addition, Table 9 summarizes the comparative performance metrics for each satellite product against the ArcSWAT baseline.

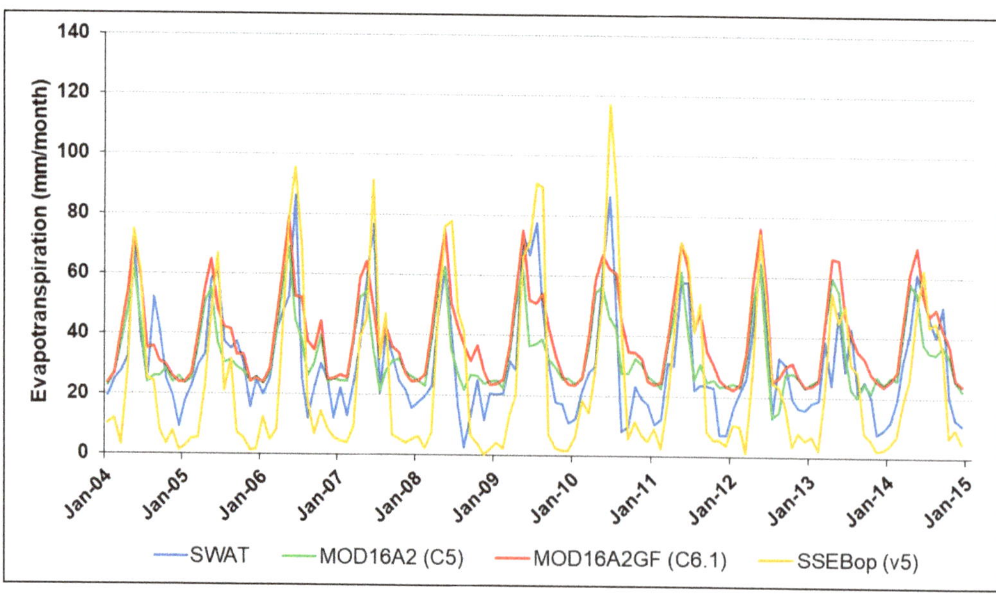

Figure 10. Comparison of ET between calibrated SWAT and remotely sensed products over the out-printed simulation period (2004–2014).

Table 9. Performance evaluation of remote sensing-derived ET products compared to simulated ET obtained by SWAT over the out-printed simulation period (2004–2014).

Index	Remote Sensing-Derived ET Products		
	MOD16A2-C5	MOD16A2GF-C6.1	SSEBop-V5
KGE	0.536	0.482	0.171
PBIAS	9.95	30.18	−16.81
RMSE	13.33	14.6	16.98
RSR	0.741	0.812	0.944

The results clearly indicate that the MOD16A2-C5 product shows the closest agreement with the calibrated SWAT estimates, achieving a KGE of 0.536, a PBIAS of 9.95%, a RMSE of 13.33 mm, and an RSR value of 0.741. The MOD16A2GF-C6.1 follows closely, with similar KGE, RMSE, and RSR values; however, its substantially higher PBIAS indicates a consistent overestimation of ET across the catchment. This result is somewhat unexpected, given the methodological enhancements introduced in the Collection 6.1 product. Conversely, the SSEBop Version 5 product delivers the poorest overall performance, demonstrating limited ability to reproduce observed ET dynamics.

3.2.2. Averaged Monthly ET Comparisons

The mean monthly ET of all dataset sources for the period 2004–2014 is provided in Table 10. In general, the calibrated SWAT model yielded a mean monthly ET of 30.65 mm. The corresponding averages derived from satellite products were 33.70 mm for MOD16A2-C5, 39.90 mm for MOD16A2GF-C6.1, and 25.50 mm for SSEBop-V5, indicating that MOD16A2-C5 outperforms the others in terms of mean monthly ET. Furthermore, a multi-product average of the satellite-derived ET estimates was calculated, which demonstrated an overall notable alignment with the ArcSWAT-simulated ET.

Table 10. Mean monthly ET estimations of all dataset sources over the out-printed simulation period (2004–2014).

Month	Mean Monthly ET (mm) of All Dataset Sources				Average ET (mm) of All Remote Sensing Products
	SWAT	MOD16A2-C5	MOD16A2GF-C6.1	SSEBop-V5	
Jan	16.92	24.48	24.18	7.03	18.56
Feb	20.02	24.85	26.31	6.50	19.22
Mar	30.32	38.00	40.82	10.88	29.90
Apr	35.67	53.17	58.57	33.97	48.57
May	60.83	59.72	70.99	61.61	64.11
Jun	58.54	38.92	53.96	73.72	55.53
Jul	34.05	28.88	42.94	46.82	39.55
Aug	29.13	27.34	41.19	40.00	36.18
Sep	28.98	28.86	34.98	10.68	24.84
Oct	23.53	29.49	33.94	6.18	23.20
Nov	15.82	25.78	26.70	5.10	19.19
Dec	14.01	24.94	24.26	3.52	17.58
Mean	**30.65**	**33.70**	**39.90**	**25.50**	**33.04**

More specifically, a detailed month-by-month comparison reveals that MOD16A2-C5 displays strong agreement with SWAT estimates during the months of May, August, and September. Satisfactory alignment is observed in February, July, and October, while November and December exhibit somewhat moderate consistency. The most pronounced discrepancy between the two (2) datasets occurs in June. The MOD16A2GF-C6.1 product follows a similar pattern, generally exhibiting partial to satisfactory agreement with the SWAT-derived ET, with the exception of April, where a significant deviation is observed. Notably, this product achieves near-exact alignment with the SWAT outputs in June. By contrast, SSEBop-V5 shows poor overall alignment with the calibrated model, offering values that differ substantially from those simulated by SWAT throughout most of the year. The sole exception occurs during mid to late spring (April and May), when a very close agreement is observed. On the other hand, the multi-product average of the satellite-derived ET estimates shows excellent agreement with the ArcSWAT-simulated ET, particularly during January, February, March, and October. For the remaining months, the consistency is predominantly satisfactory, with the exception of April, which displays a noticeable deviation.

3.2.3. Seasonal ET Comparisons

Seasonal ET values for the 2004–2014 period, derived from all datasets, are summarized in Table 11. The results demonstrate appreciable discrepancies between the calibrated SWAT model outputs and the satellite-derived ET estimates. Among the remote sensing products, MOD16A2-C5 presents the closest correspondence with SWAT-simulated ET, especially throughout the autumn months. The MOD16A2GF-C6.1 product attains moderate concordance with the model across most seasons, yet it surpasses all other products in replicating early summer (June) ET dynamics. In contrast, while the SSEBop-V5 product shows weak overall alignment with the SWAT simulations, it delivers relatively improved agreement during the spring season (April and May), where it outperforms all other satellite products in capturing the modeled ET. Strikingly, the averaged ET from all satellite products exhibits substantial correspondence with the SWAT-simulated ET, particularly during winter and fall, where the consistency between the datasets is especially apparent.

Table 11. Seasonal ET estimations of all dataset sources over the out-printed simulation period (2004–2014).

Season	Seasonal ET (mm) of All Dataset Sources				Average ET (mm) of All Remote Sensing Products
	SWAT	**MOD16A2-C5**	**MOD16A2GF-C6.1**	**SSEBop-V5**	
Winter	49.96	72.23	72.56	16.61	53.80
Spring	126.82	150.90	170.38	106.46	142.58
Summer	121.72	95.13	138.09	160.54	131.26
Fall	68.33	84.13	95.63	21.96	67.24

3.2.4. Annual ET Comparisons

Annual ET totals for the 2004–2014 interval, derived from each dataset, are presented in Table 12. Consistent with the preceding analyses, the MOD16A2-C5 product reveals strong concurrence with SWAT-derived annual ET estimates throughout most of the simulation period. However, this agreement weakened in the years 2008 and 2013, which are characterized by diminished precipitation (Table 4). Conversely, the SSEBop-V5 product for the most part fails to reproduce the SWAT-simulated annual ET, with the notable exceptions of 2008 and 2011, when its estimates align more closely with the model. A reasonable agreement is also evident in 2006 and 2010, though less distinct. On the other hand, the MOD16A2GF-C6.1 dataset demonstrates only moderate to poor correspondence with the SWAT outputs, tending predominantly to overestimate annual ET dynamics. By way of comparison, exceptional alignment between the averaged satellite-derived ET and the SWAT estimates is observed in 2007 and 2012, whereas the years 2004, 2009, and 2014 show substantial agreement. For the remaining years of the simulation period, the consistency is, as a rule, satisfactory, with the exception of 2008 and 2013, where pronounced discrepancies are noticed.

Table 12. Annual ET estimations of all dataset sources over the out-printed simulation period (2004–2014).

Year	Annual ET (mm) of All Dataset Sources				Average ET (mm) of All Remote Sensing Products
	SWAT	MOD16A2-C5	MOD16A2GF-C6.1	SSEBop-V5	
2004	385.88	395.30	460.10	258.71	371.37
2005	403.80	397.44	460.30	217.58	358.44
2006	399.12	444.92	510.64	361.54	439.03
2007	387.37	395.89	461.24	293.25	383.46
2008	298.90	391.13	476.54	317.13	394.94
2009	445.84	421.19	499.46	371.29	430.65
2010	366.71	432.50	511.20	396.10	446.60
2011	308.20	390.17	481.72	322.47	398.12
2012	344.75	362.77	434.38	258.22	351.79
2013	304.54	384.35	459.85	257.66	367.29
2014	401.02	433.21	511.91	312.11	419.07
Mean	**367.83**	**404.44**	**478.85**	**306.00**	**396.43**

3.2.5. Insights into Model–Satellite ET Agreement

The comparative analysis between SWAT-simulated ET and three (3) satellite-derived ET products revealed distinct performance patterns that can be largely attributed to technical differences in their underlying algorithms. Among the tested datasets, MOD16A2-C5 demonstrated the highest consistency with SWAT outputs, possibly due to its robust Penman–Monteith formulation and mature biome-specific parameterization [99]. Its reduced agreement during the driest years of the simulation period likely reflects challenges in accurately capturing water-stressed vegetation dynamics under moisture-limited conditions [105].

In contrast, the SSEBop-V5 product, while typically presenting weaker alignment with the modeled ET, displayed improved correspondence during these dry years. This is likely attributable to its thermal-based approach, which leverages land surface temperature to infer ET, especially under arid or semi-arid regimes [102]. The MOD16A2GF-C6.1 product, although more recent, showed moderate overall agreement, with a tendency to overestimate ET across most years. However, it surpassed all other products in capturing early summer ET dynamics, likely due to algorithmic refinements such as enhanced gap-filling and updated land cover inputs [1,105].

Significantly, averaging the values of all three (3) satellite-derived ET products results in a consistently strong alignment with the calibrated ArcSWAT-simulated ET, across all temporal scales. This agreement may be attributed to the compensatory effect of combining multiple products. As each satellite dataset is based on distinct algorithms and inputs, their errors may partially cancel out when averaged, producing a more balanced estimate. While this ensemble-like approach is thought to improve consistency, further research is needed to confirm its effectiveness in ET estimation.

Nevertheless, it is worth highlighting that the SWAT model employed in this comparison was calibrated focusing on streamflow (the hydrological component). As a result, the simulated ET was not directly calibrated against observed ET values. Although MODIS-derived LAI data were incorporated into the model's database, improving its representation of vegetation phenology and transpiration dynamics, they do not eliminate the potential for direct ET calibration. This may partially explain some of the discrepancies with satellite-based ET estimates, highlighting a common challenge in hydrological modeling studies where ET is assessed indirectly [106,107]. Nonetheless, the observed agreements and divergences provide valuable insights into the strengths and limitations of both modeling and remote sensing approaches, and suggest that bias-correction techniques or ensemble ET products could enhance consistency in future applications [1].

On the other hand, beyond serving as a benchmark for comparison, remote sensing-derived ET data play an active role nowadays in refining the SWAT model itself [14]. Incorporating satellite ET products into the calibration process has been demonstrated to improve the model's performance by enhancing not only streamflow simulations but also the overall accuracy of the water and energy balance components [5,6]. However, exclusive reliance on remotely sensed ET data for hydrological model calibration should be approached with considerable caution, as it carries the risk of producing inaccurate or misleading results.

3.2.6. Comparison of SWAT-Simulated and Satellite-Derived Evapotranspiration: Insights and Literature Context

The comparison between SWAT-simulated and satellite-derived ET revealed patterns of agreement and divergence, some of which align with findings from previous studies, while others reflect context-specific differences. Specifically, Herman et al. [108] compared ET results derived from the SWAT model with remotely sensed ET datasets, aiming to assess their differences and quantify the level of agreement between modeled and observed ET. The remote sensing datasets used for comparison included, among others, the MOD16A2 at 1 km resolution, the MOD16A2 at 500 m resolution, and the SSEBop dataset. The results indicated that, at the watershed scale, SWAT-derived ET exhibited stronger agreement with MOD16A2 (1 km) during the spring and summer seasons, while better alignment was observed with MOD16A2 (500 m) in the fall. In contrast, the SSEBop dataset showed higher agreement with SWAT ET during the winter months.

Chun et al. [109] evaluated ET estimates produced by the SWAT model in an arid region by comparing them with remotely sensed ET products, including SSEBop and MOD16. Their analysis demonstrated that the SWAT model had a stronger correlation with the MOD16 dataset compared to the SSEBop product. Abiodun et al. [110] compared ET estimates from the MOD16 dataset with those generated by the SWAT model across multiple spatial resolutions, focusing on a catchment characterized by complex topography in a semi-arid climate. The findings indicated a strong correspondence between the two ET sources at the catchment scale, with SWAT reproducing annual ET values within 6% of those reported by MOD16.

Furthermore, Parajuli et al. [14] conducted a comparative analysis of SWAT-simulated ET and MODIS-derived ET estimates. The study found that the SWAT model more accurately represented seasonal dynamics, particularly during the growing and dormant periods. However, MODIS-based ET was deemed sufficiently reliable when evaluated over an annual timescale. Consistently, Nguyen and Kappas [111] assessed ET outputs from the SWAT model against satellite-based estimates from the MODIS MOD16 product at 1 km resolution. Their results suggested that SWAT-derived ET may offer greater accuracy. Nonetheless, they acknowledged the value of MODIS ET data in supporting the validation of finer-scale ET assessments generated by both approaches. It is important to note that their findings were based on analyses conducted within a tropical watershed.

In contrast, Fathian et al. [112] conducted a performance comparison between the SWAT-derived ET estimates and those from the MOD16 product within a semi-arid basin. Their analysis indicated that the MODIS dataset tended to overstate ET values at both monthly and annual scales relative to the SWAT simulations. Additionally, Morsy et al. [113] applied the SWAT model to simulate ET and compared the results with the SSEBop and MOD16A2 products, in two different climatic zones (Mediterranean and hyper-arid climate conditions). Their findings revealed that SWAT-derived ET values frequently lay between those of the two satellite datasets, with the SSEBop product demonstrating marginally superior performance overall.

4. Conclusions

In this study, a high-resolution, well-developed SWAT model was implemented for the Upper Anthemountas Basin, with the primary objective of comparing simulated evapotranspiration outputs from the calibrated model to three (3) widely recognized remote sensing-based ET products. These products—the MODIS MOD16 Collection 5, the updated MODIS MOD16A2GF Collection 6.1, and the SSEBop Version 5 dataset—present several key advantages, such as open access availability, sufficient spatial resolution for basin-scale analysis, and use of the well-established Penman–Monteith method for evapotranspiration estimation. Both interval-based comparisons (monthly averages, seasonal, and annual) and continuous time series comparisons at a monthly time step were conducted between SWAT-simulated ET and satellite-derived products over the simulation period (2004–2014). Key features derived from the present study can be summarized as follows:

- The hydrological model was calibrated using observed monthly streamflow records spanning 2004 to 2013, and the resulting performance was classified as satisfactory to very good, based on multiple established statistical evaluation metrics.
- Over the simulation period, mean actual evapotranspiration was estimated, on average, at 367.8 mm and emerged as the most significant component of the water balance, accounting for approximately 65% of total precipitation.
- Of all remote sensing-based ET datasets, MOD16A2-C5 exhibited the greatest agreement with the SWAT model outputs.
- MOD16A2GF-C6.1 demonstrated moderate overall agreement and tended to overestimate ET in most years, yet it outperformed all other products in representing early summer ET dynamics.
- The SSEBop-V5 product, while generally indicating weaker alignment with the modeled ET, displayed improved correspondence during the driest years of the simulation period.
- Across all temporal ET comparisons, averaging the values of all satellite-derived products yields a consistently strong agreement with the calibrated SWAT-simulated ET.

These findings emphasize the importance of critical evaluation when using remotely sensed ET products as calibration data for hydrological models. While such datasets offer valuable insights, their variable performance underscores the importance of careful selection and context-specific interpretation. This consideration is particularly critical in model development and water resource planning for studies conducted over relatively small spatial domains, such as the one considered in the study.

Author Contributions: Conceptualization, S.S.; methodology, S.S. and I.S.; software, S.S.; validation, S.S.; formal analysis, S.S. and I.S.; investigation, S.S.; resources, S.S.; writing—original draft preparation, S.S.; writing—review and editing, S.S., I.S. and Z.M.; visualization, S.S. and I.S.; supervision, Z.M. All authors have read and agreed to the published version of the manuscript.

Funding: This research received no external funding.

Data Availability Statement: The data used in this study are contained within the article. Additional data are available upon request from the corresponding author.

Acknowledgments: Part of the research was conducted by Stefanos Sevastas at the School of Civil Engineering, Aristotle University of Thessaloniki, Thessaloniki, Greece. The authors would like to thank Grigorios Sevastas for providing a high-resolution satellite natural color image covering the entire study area.

Conflicts of Interest: The authors declare no conflicts of interest.

References

1. Pan, S.; Pan, N.; Tian, H.; Friedlingstein, P.; Sitch, S.; Shi, H.; Arora, V.K.; Haverd, V.; Jain, A.K.; Kato, E.; et al. Evaluation of global terrestrial evapotranspiration using state-of-the-art approaches in remote sensing, machine learning and land surface modeling. *Hydrol. Earth Syst. Sci.* **2020**, *24*, 1485–1509. [CrossRef]
2. Ghiat, I.; Mackey, H.R.; Al-Ansari, T. A review of evapotranspiration measurement models, techniques and methods for open and closed agricultural field applications. *Water* **2021**, *13*, 2523. [CrossRef]
3. Rientjes, T.H.M.; Muthuwatta, L.P.; Bos, M.G.; Booij, M.J.; Bhatti, H.A. Multi-variable calibration of a semi-distributed hydrological model using streamflow data and satellite-based evapotranspiration. *J. Hydrol.* **2013**, *505*, 276–290. [CrossRef]
4. Tobin, K.J.; Bennett, M.E. Constraining SWAT calibration with remotely sensed evapotranspiration data. *JAWRA J. Am. Water Resour. Assoc.* **2017**, *53*, 593–604. [CrossRef]
5. Rajib, A.; Evenson, G.R.; Golden, H.E.; Lane, C.R. Hydrologic model predictability improves with spatially explicit calibration using remotely sensed evapotranspiration and biophysical parameters. *J. Hydrol.* **2018**, *567*, 668–683. [CrossRef]
6. Parajuli, P.B.; Jayakody, P.; Ouyang, Y. Evaluation of using remote sensing evapotranspiration data in SWAT. *Water Resour. Manag.* **2018**, *32*, 985–996. [CrossRef]
7. Herman, M.R.; Nejadhashemi, A.P.; Abouali, M.; Hernandez-Suarez, J.S.; Daneshvar, F.; Zhang, Z.; Anderson, M.C.; Sadeghi, A.M.; Hain, C.R.; Sharifi, A. Evaluating the role of evapotranspiration remote sensing data in improving hydrological modeling predictability. *J. Hydrol.* **2018**, *556*, 39–49. [CrossRef]
8. Sirisena, T.J.G.; Maskey, S.; Ranasinghe, R. Hydrological model calibration with streamflow and remote sensing based evapotranspiration data in a data poor basin. *Remote Sens.* **2020**, *12*, 3768. [CrossRef]
9. Dile, Y.T.; Ayana, E.K.; Worqlul, A.W.; Xie, H.; Srinivasan, R.; Lefore, N.; Dile, Y.T.; Ayana, E.K.; Worqlul, A.W.; Xie, H.; et al. Evaluating satellite-based evapotranspiration estimates for hydrological applications in data-scarce regions: A case in Ethiopia. *Sci. Total Environ.* **2020**, *743s*, 140702. [CrossRef]
10. Shah, S.; Duan, Z.; Song, X.; Li, R.; Mao, H.; Liu, J.; Ma, T.; Wang, M. Evaluating the added value of multi-variable calibration of SWAT with remotely sensed evapotranspiration data for improving hydrological modeling. *J. Hydrol.* **2021**, *603*, 127046. [CrossRef]
11. Sun, Z.; Lotz, T.; Huang, Q. An ET-based two-phase method for the calibration and application of distributed hydrological models. *Water Resour. Manag.* **2021**, *35*, 1065–1077. [CrossRef]
12. Bennour, A.; Jia, L.; Menenti, M.; Zheng, C.; Zeng, Y.; Asenso Barnieh, B.; Jiang, M. Calibration and validation of SWAT model by using hydrological remote sensing observables in the Lake Chad Basin. *Remote Sens.* **2022**, *14*, 1511. [CrossRef]
13. Alemayehu, T.; van Griensven, A.; Woldegiorgis, B.T.; Bauwens, W. An improved SWAT vegetation growth module and its evaluation for four tropical ecosystems. *Hydrol. Earth Syst. Sci.* **2017**, *21*, 4449. [CrossRef]
14. Parajuli, P.B.; Risal, A.; Ouyang, Y.; Thompson, A. Comparison of SWAT and MODIS evapotranspiration data for multiple timescales. *Hydrology* **2022**, *9*, 103. [CrossRef]
15. Koltsida, E.; Kallioras, A. Multi-variable SWAT model calibration using satellite-based evapotranspiration data and streamflow. *Hydrology* **2022**, *9*, 112. [CrossRef]
16. Dangol, S.; Zhang, X.; Liang, X.Z.; Anderson, M.; Crow, W.; Lee, S.; Moglen, G.E.; McCarty, G.W. Multivariate calibration of the SWAT model using remotely sensed datasets. *Remote Sens.* **2023**, *15*, 2417. [CrossRef]
17. Immerzeel, W.A.; Droogers, P. Calibration of a distributed hydrological model based on satellite evapotranspiration. *J. Hydrol.* **2008**, *349*, 411–424. [CrossRef]
18. Verma, N. Simulation of hydrologic processes through SWAT and Modis ET for Sirsa River Basin in Western Himalaya. *Int. J. Water Resour. Arid Environ.* **2018**, *7*, 21–35.
19. Odusanya, A.E.; Mehdi, B.; Schürz, C.; Oke, A.O.; Awokola, O.S.; Awomeso, J.A.; Adejuwon, J.O.; Schulz, K. Multi-site calibration and validation of SWAT with satellite-based evapotranspiration in a data-sparse catchment in southwestern Nigeria. *Hydrol. Earth Syst. Sci.* **2019**, *23*, 1113–1144. [CrossRef]
20. Bárdossy, A.; Huang, Y.; Wagener, T. Simultaneous calibration of hydrological models in geographical space. *Hydrol. Earth Syst. Sci.* **2016**, *20*, 2913–2928. [CrossRef]
21. Oliveira, A.M.; Fleischmann, A.S.; Paiva, R.C.D. On the contribution of remote sensing-based calibration to model hydrological and hydraulic processes in tropical regions. *J. Hydrol.* **2021**, *597*, 126184. [CrossRef]
22. Michailovsky, C.I.; Coerver, B.; Mul, M.; Jewitt, G. Investigating sources of variability in closing the terrestrial water balance with remote sensing. *Hydrol. Earth Syst. Sci.* **2023**, *27*, 4335–4354. [CrossRef]
23. Abbaspour, K.C.; Rouholahnejad, E.; Vaghefi, S.; Srinivasan, R.; Yang, H.; Klove, B. A continental scale hydrology and water quality model for Europe: Calibration and uncertainty of a high-resolution large-scale SWAT model. *J. Hydrol.* **2015**, *524*, 733–752. [CrossRef]
24. Singh, G.; Kumar, E. Input data scale impacts on modeling output results: A review. *J. Spat. Hydrol.* **2017**, *13*, Article 4.

25. Sevastas, S.; Siarkos, I.; Theodossiou, N.; Ifadis, I. Evaluating the use of measured and/or open access data in watershed modeling through an integrated modeling procedure. *Desalination Water Treat.* **2018**, *133*, 268–282. [CrossRef]
26. Kumar, S.; Merwade, V. Impact of watershed subdivision and soil data resolution on SWAT model calibration and parameter uncertainty. *JAWRA J. Am. Water Resour. Assoc.* **2009**, *45*, 1179–1196. [CrossRef]
27. Gregar, J.; Petrů, J.; Novotná, J. Evaluation of the SWAT model as an integrated management tool in the Švihov drinking water supply catchment. *Soil Water Res.* **2019**, *14*, 76–83. [CrossRef]
28. Sevastas, S.; Siarkos, I.; Theodossiou, N.; Ifadis, I. Establishing wellhead protection areas and managing point and non-point pollution sources to support groundwater protection in the aquifer of Upper Anthemountas, Greece. *Water Util. J.* **2017**, *16*, 81–95.
29. Sevastas, S.; Siarkos, I.; Theodossiou, N.; Ifadis, I.; Kaffas, K. Comparing hydrological models built upon open access and/or measured data in a GIS environment. In Proceedings of the 6th International Conference on Environmental Management, Engineering, Planning and Economics (CEMEPE) and SECOTOX Conference, Thessaloniki, Greece, 25–30 June 2017; pp. 377–386.
30. Sevastas, S. Applying Geo-Information Systems, Numerical Models and Genetic Algorithms in Optimizing Groundwater Risk Assessment: A Case Study from the Upper Anthemountas River Basin. Ph.D. Thesis, Aristotle University of Thessaloniki, Thessaloniki, Greece, 2023. (In Greek)
31. Sevastas, S.; Gasparatos, D.; Botsis, D.; Siarkos, I.; Diamantaras, K.I.; Bilas, G. Predicting bulk density using pedotransfer functions for soils in the Upper Anthemountas basin, Greece. *Geoderma Reg.* **2018**, *14*, e00169. [CrossRef]
32. Misopolinos, N.; Silleos, N.; Kosmas, K.; Bilas, G.; Karapetsas, N.; Dionisiou, N. *Development of an Integrated System for Soil Geographic Data and Delineation of Agricultural Zones in Greece*; Payment and Control Agency for Guidance and Guarantee Community Aid (OPEKEPE) of the Greek Ministry of Agricultural Development and Food: Athens, Greece, 2015.
33. Moriasi, D.N.; Arnold, J.G.; Van Liew, M.W.; Bingner, R.L.; Harmel, R.D.; Veith, T.L. Model evaluation guidelines for systematic quantification of accuracy in watershed simulations. *Trans. ASABE* **2007**, *50*, 885–900. [CrossRef]
34. Kouchi, D.H.; Esmaili, K.; Faridhosseini, A.; Sanaeinejad, S.H.; Khalili, D.; Abbaspour, K.C. Sensitivity of calibrated parameters and water resource estimates on different objective functions and optimization algorithms. *Water* **2017**, *9*, 384. [CrossRef]
35. Sao, D.; Kato, T.; Tu, L.H.; Thouk, P.; Fitriyah, A.; Oeurng, C. Evaluation of different objective functions used in the SUFI-2 calibration process of SWAT-CUP on water balance analysis: A case study of the Pursat river basin, Cambodia. *Water* **2020**, *12*, 2901. [CrossRef]
36. Verma, S.; Verma, M.K.; Prasad, A.D.; Mehta, D.J.; Islam, M.N. Modeling of uncertainty in the estimation of hydrograph components in conjunction with the SUFI-2 optimization algorithm by using multiple objective functions. *Model. Earth Syst. Environ.* **2024**, *10*, 61–79. [CrossRef]
37. Gupta, H.V.; Kling, H.; Yilmaz, K.K.; Martinez, G.F. Decomposition of the Mean Squared Error and NSE performance criteria: Implications for improving hydrological modelling. *J. Hydrol.* **2009**, *377*, 80–91. [CrossRef]
38. Arnold, J.G.; Srinivasan, R.; Muttiah, R.S.; Williams, J.R. Large area hydrologic modeling and assessment part I: Model development. *JAWRA J. Am. Water Resour. Assoc.* **1998**, *34*, 73–89. [CrossRef]
39. Winchell, M.; Srinivasan, R.; Di Luzio, M.; Arnold, J.G. *Arcswat Interface for SWAT2012: User's Guide*; Blackland Research Center, Texas AgriLife Research: College Station, TX, USA, 2013; pp. 1–464.
40. Monteiro, J.A.; Strauch, M.; Srinivasan, R.; Abbaspour, K.; Gücker, B. Accuracy of grid precipitation data for Brazil: Application in river discharge modelling of the Tocantins catchment. *Hydrol. Process.* **2016**, *30*, 1419–1430. [CrossRef]
41. Roth, V.; Lemann, T. Comparing CFSR and conventional weather data for discharge and soil loss modelling with SWAT in small catchments in the Ethiopian Highlands. *Hydrol. Earth Syst. Sci.* **2016**, *20*, 921–934. [CrossRef]
42. Arnold, J.G.; Kiniry, J.R.; Srinivasan, R.; Williams, J.R.; Haney, E.B.; Neitsch, S.L. *SWAT Input/Output Documentation Version 2012*; Texas Water Resources Institute: College Station, TX, USA, 2012; p. 654.
43. Abbaspour, K.C.; Vaghefi, S.A.; Yang, H.; Srinivasan, R. Global soil, landuse, evapotranspiration, historical and future weather databases for SWAT Applications. *Sci. Data* **2019**, *6*, 263. [CrossRef]
44. Wise, S. Cross-validation as a means of investigating DEM interpolation error. *Comput. Geosci.* **2011**, *37*, 978–991. [CrossRef]
45. Acharya, T.D.; Yang, I. Exploring Landsat 8. *Int. J. IT Eng. Appl. Sci. Res. (IJIEASR)* **2015**, *4*, 4–10.
46. Wang, D.; Zhan, Y.; Yu, T.; Liu, Y.; Jin, X.; Ren, X.; Chen, X.; Liu, Q. Improving meteorological input for surface energy balance system utilizing mesoscale weather research and forecasting model for estimating daily actual evapotranspiration. *Water* **2019**, *12*, 9. [CrossRef]
47. Baig, M.H.A.; Zhang, L.; Shuai, T.; Tong, Q. Derivation of a tasselled cap transformation based on Landsat 8 at-satellite reflectance. *Remote Sens. Lett.* **2014**, *5*, 423–431. [CrossRef]
48. Crist, E.P.; Kauth, R.J. The tasseled cap de-mystified. *Photogramm. Eng. Remote Sens.* **1986**, *52*, 81–86.
49. Huang, C.; Wylie, B.; Yang, L.; Homer, C.; Zylstra, G. Derivation of a tasselled cap transformation based on Landsat 7 at-satellite reflectance. *Int. J. Remote Sens.* **2002**, *23*, 1741–1748. [CrossRef]
50. Ghazali, M.F.; Susantoro, T.M.; Wikantika, K.; Harto, A.B.; Nurtyawan, R. Soil moisture mapping at paddy field in Indramayu Residence using Landsat 8 OLI/TIRS. *HAYATI J. Biosci.* **2020**, *27*, 71–81. [CrossRef]

51. Aliaume, C.; Bodini, A.; Bondavalli, C.; Carreño, F.; Chapelle, A.; Duarte, P.; Esteve-Selma, M.A.; Falco, M.; Fiandrino, A.; Giordani, G.; et al. *Model Benchmarks for Coastal Lagoons*; EUR 22216EN.; European Commission, Joint Research Centre: Ispra, Italy, 2006; p. 82.
52. Plus, M.; La Jeunesse, I.; Bouraoui, F.; Zaldívar, J.M.; Chapelle, A.; Lazure, P. Modelling water discharges and nitrogen inputs into a Mediterranean lagoon: Impact on the primary production. *Ecol. Model.* **2006**, *193*, 69–89. [CrossRef]
53. Igder, O.M.; Alizadeh, H.; Mojaradi, B.; Bayat, M. Multivariate assimilation of satellite-based leaf area index and ground-based river streamflow for hydrological modelling of irrigated watersheds using SWAT+. *J. Hydrol.* **2022**, *610*, 128012. [CrossRef]
54. Ma, T.; Duan, Z.; Li, R.; Song, X. Enhancing SWAT with remotely sensed LAI for improved modelling of ecohydrological process in subtropics. *J. Hydrol.* **2019**, *570*, 802–815. [CrossRef]
55. Myneni, R.; Knyazikhin, Y.; Park, T. *MODIS/Terra+ Aqua Leaf Area Index/FPAR 4-Day L4 Global 500 m SIN Grid V061*; The Land Processes Distributed Active Archive Center (LP DAAC): Sioux Falls, SD, USA, 2021.
56. Daneshvar, F.; Frankenberger, J.R.; Bowling, L.C.; Cherkauer, K.A.; de Lima Moraes, A.G. Development of strategy for SWAT hydrologic modeling in data-scarce regions of Peru. *J. Hydrol. Eng.* **2021**, *26*, 05021016. [CrossRef]
57. Emberson, L.D. *Development and Application of Methods for Modeling and Mapping Ozone Deposition and Stomatal Flux*; Final Report; Department of Environment, Food and Regional Affairs: London, UK, 2003.
58. Nelson, D.W.; Sommers, L.E. Total carbon, organic carbon, and organic matter. In *Methods of Soil Analysis–Part 3 Chemical Methods*; Sparks, D.L., Ed.; Soil Science Society of America: Fitchburg, MA, USA, 1996; pp. 961–1010.
59. Pribyl, D.W. A critical review of the conventional SOC to SOM conversion factor. *Geoderma* **2010**, *156*, 75–83. [CrossRef]
60. Bagarello, V.; Di Prima, S.; Iovino, M.; Provenzano, G. Estimating field-saturated soil hydraulic conductivity by a simplified Beerkan infiltration experiment. *Hydrol. Process.* **2014**, *28*, 1095–1103. [CrossRef]
61. Penning de Vries, F.W.T.; Jansen, D.M.; ten Berge, H.F.M.; Bakema, A. *Simulation of Ecophysiological Processes of GROWTH in Several Annual Crops*; Centre for Agricultural Publishing and Documentation (Pudoc): Wageningen, The Netherlands, 1989.
62. ten Berge, H.F.M. *Heat and Water Transfer in Bare Topsoil and the Lower Atmosphere*; Centre for Agricultural Publishing and Documentation (Pudoc): Wageningen, The Netherlands, 1996.
63. Saxton, K.E.; Willey, P.H. The SPAW model for agricultural field and pond hydrologic simulation. In *Watershed Models*; Singh, V.P., Frevert, D.K., Eds.; CRC Press: Boca Raton, FL, USA, 2009; pp. 401–435.
64. Williams, J.R. Chapter 25: The EPIC model. In *Computer Models of Watershed Hydrology*; Singh, V.P., Ed.; Resources Publications: Highlands Ranch, CO, USA, 1995; pp. 909–1000.
65. Sheng, J.; Ma, L.; Jiang, P.A.; Li, B.; Huang, F.; Wu, H. Digital soil mapping to enable classification of the salt-affected soils in desert agro-ecological zones. *Agric. Water Manag.* **2010**, *97*, 1944–1951. [CrossRef]
66. Akpa, S.I.; Odeh, I.O.; Bishop, T.F.; Hartemink, A.E. Digital mapping of soil particle-size fractions for Nigeria. *Soil Sci. Soc. Am. J.* **2014**, *78*, 1953–1966. [CrossRef]
67. Boente, C.; Albuquerque, M.T.D.; Fernández-Braña, A.; Gerassis, S.; Sierra, C.; Gallego, J.R. Combining raw and compositional data to determine the spatial patterns of potentially toxic elements in soils. *Sci. Total Environ.* **2018**, *631*, 1117–1126. [CrossRef]
68. Li, J.; Wan, H.; Shang, S. Comparison of interpolation methods for mapping layered soil particle-size fractions and texture in an arid oasis. *Catena* **2020**, *190*, 104514. [CrossRef]
69. Lakaphossi, A.; Ziangas, E.; Nakos, G. *Soil Map of Greece, Scale 1:50000*; Forest Service Publications, Ministry of Agriculture: Athens, Greece, 1992.
70. Danacova, M.; Vyleta, R.; Valent, P.; Hlavcova, K. The impact of slope gradients on the generation of surface runoff in laboratory conditions. In Proceedings of the 17th International Multidisciplinary Scientific GeoConference SGEM2017, Water Resources, Forest, Marine and Ocean Ecosystems, Vienna, Austria, 27–29 November 2017; pp. 677–684.
71. Komatsu, Y.; Kato, H.; Zhu, B.; Wang, T.; Yang, F.; Rakwal, R.; Onda, Y. Effects of slope gradient on runoff from bare-fallow purple soil in China under natural rainfall conditions. *J. Mt. Sci.* **2018**, *15*, 738–751. [CrossRef]
72. Zhang, X.; Zhu, Q.; Sang, J.; Ding, X. Effects of rainfall and the slope gradient on the soil and water loss in a purple soil area. *Soil Water Res.* **2022**, *17*, 232–242. [CrossRef]
73. Yacoub, C.; Foguet, A.P. Slope effects on SWAT modeling in a mountainous basin. *J. Hydrol. Eng.* **2013**, *18*, 1663–1673. [CrossRef]
74. Daggupati, P.; Pai, N.; Ale, S.; Douglas-Mankin, K.R.; Zeckoski, R.W.; Jeong, J.; Parajuli, P.B.; Saraswat, D.; Youssef, M.A. A recommended calibration and validation strategy for hydrologic and water quality models. *Trans. ASABE* **2015**, *58*, 1705–1719.
75. Abbaspour, K.C.; Yang, J.; Maximov, I.; Siber, R.; Bogner, K.; Mieleitner, J.; Zobrist, J.; Srinivasan, R. Modelling hydrology and water quality in the pre-alpine/alpine Thur watershed using SWAT. *J. Hydrol.* **2007**, *333*, 413–430. [CrossRef]
76. Yang, J.; Reichert, P.; Abbaspour, K.C.; Xia, J.; Yang, H. Comparing uncertainty analysis techniques for a SWAT application to the Chaohe Basin in China. *J. Hydrol.* **2008**, *358*, 1–23. [CrossRef]
77. Wu, H.; Chen, B. Evaluating uncertainty estimates in distributed hydrological modeling for the Wenjing River watershed in China by GLUE, SUFI-2, and ParaSol methods. *Ecol. Eng.* **2015**, *76*, 110–121. [CrossRef]

78. Niraula, R.; Norman, L.M.; Meixner, T.; Callegary, J.B. Multi-gauge calibration for modeling the semi-arid Santa Cruz Watershed in Arizona-Mexico border area using SWAT. *Air Soil Water Res.* **2012**, *5*, ASWR-S9410. [CrossRef]
79. Xie, H.; Longuevergne, L.; Ringler, C.; Scanlon, B.R. Calibration and evaluation of a semi-distributed watershed model of Sub-Saharan Africa using GRACE data. *Hydrol. Earth Syst. Sci.* **2012**, *16*, 3083–3099. [CrossRef]
80. Me, W.; Abell, J.M.; Hamilton, D.P. Effects of hydrologic conditions on SWAT model performance and parameter sensitivity for a small, mixed land use catchment in New Zealand. *Hydrol. Earth Syst. Sci.* **2015**, *19*, 4127–4147. [CrossRef]
81. Ramos, M.C.; Martínez-Casasnovas, J.A. Climate change influence on runoff and soil losses in a rainfed basin with Mediterranean climate. *Nat. Hazards* **2015**, *78*, 1065–1089. [CrossRef]
82. Zuo, D.; Xu, Z.; Peng, D.; Song, J.; Cheng, L.; Wei, S.; Abbaspour, K.C.; Yang, H. Simulating spatiotemporal variability of blue and green water resources availability with uncertainty analysis. *Hydrol. Process.* **2015**, *29*, 1942–1955. [CrossRef]
83. Gholami, A.; Roshan, M.H.; Shahedi, K.; Vafakhah, M.; Solaymani, K. Hydrological stream flow modeling in the Talar catchment (central section of the Alborz Mountains, north of Iran): Parameterization and uncertainty analysis using SWAT-CUP. *J. Water Land Dev.* **2016**, *30*, 57. [CrossRef]
84. Jung, C.G.; Lee, D.R.; Moon, J.W. Comparison of the Penman-Monteith method and regional calibration of the Hargreaves equation for actual evapotranspiration using SWAT-simulated results in the Seolma-cheon basin, South Korea. *Hydrol. Sci. J.* **2016**, *61*, 793–800. [CrossRef]
85. Leta, O.T.; Van Griensven, A.; Bauwens, W. Effect of single and multisite calibration techniques on the parameter estimation, performance, and output of a SWAT model of a spatially heterogeneous catchment. *J. Hydrol. Eng.* **2017**, *22*, 05016036. [CrossRef]
86. Romagnoli, M.; Portapila, M.; Rigalli, A.; Maydana, G.; Burgués, M.; García, C.M. Assessment of the SWAT model to simulate a watershed with limited available data in the Pampas region, Argentina. *Sci. Total Environ.* **2017**, *596*, 437–450. [CrossRef] [PubMed]
87. Shannak, S. Calibration and validation of SWAT for sub-hourly time steps using SWAT-CUP. *Int. J. Sustain. Water Environ. Syst.* **2017**, *9*, 21–27.
88. Kumar, N.; Singh, S.K.; Srivastava, P.K.; Narsimlu, B. SWAT Model calibration and uncertainty analysis for streamflow prediction of the Tons River Basin, India, using Sequential Uncertainty Fitting (SUFI-2) algorithm. *Model. Earth Syst. Environ.* **2017**, *3*, 1–13. [CrossRef]
89. Chilkoti, V.; Bolisetti, T.; Balachandar, R. Multi-objective autocalibration of SWAT model for improved low flow performance for a small snowfed catchment. *Hydrol. Sci. J.* **2018**, *63*, 1482–1501. [CrossRef]
90. Dash, P. Simulation of hydrologic processes through calibration of SWAT model with MODIS evapotranspiration data for an ungauged basin in Western Himalaya, India. In *Geospatial Applications for Natural Resources Management*; Singh, C.K., Ed.; CRC Press: Boca Raton, FL, USA, 2018; pp. 223–241.
91. Nkiaka, E.; Nawaz, N.R.; Lovett, J.C. Effect of single and multi-site calibration techniques on hydrological model performance, parameter estimation and predictive uncertainty: A case study in the Logone catchment, Lake Chad basin. *Stoch. Environ. Res. Risk Assess.* **2018**, *32*, 1665–1682. [CrossRef]
92. Kannan, N.; Santhi, C.; White, M.J.; Mehan, S.; Arnold, J.G.; Gassman, P.W. Some challenges in hydrologic model calibration for large-scale studies: A case study of SWAT model application to Mississippi-Atchafalaya River Basin. *Hydrology* **2019**, *6*, 17. [CrossRef]
93. Dawit, M.; Olika, B.D.; Muluneh, F.B.; Leta, O.T.; Dinka, M.O. Assessment of surface irrigation potential of the Dhidhessa River Basin, Ethiopia. *Hydrology* **2020**, *7*, 68. [CrossRef]
94. Pérez-Sánchez, J.; Senent-Aparicio, J.; Martínez Santa-María, C.; López-Ballesteros, A. Assessment of ecological and hydro-geomorphological alterations under climate change using SWAT and IAHRIS in the Eo River in Northern Spain. *Water* **2020**, *12*, 1745. [CrossRef]
95. Duan, Y.; Luo, M.; Guo, X.; Cai, P.; Li, F. Study on the relationship between snowmelt runoff for different latitudes and vegetation growth based on an improved SWAT model in Xinjiang, China. *Sustainability* **2021**, *13*, 1189. [CrossRef]
96. Musyoka, F.K.; Strauss, P.; Zhao, G.; Srinivasan, R.; Klik, A. Multi-step calibration approach for SWAT model using soil moisture and crop yields in a small agricultural catchment. *Water* **2021**, *13*, 2238. [CrossRef]
97. Saade, J.; Atieh, M.; Ghanimeh, S.; Golmohammadi, G. Modeling impact of climate change on surface water availability using SWAT model in a semi-arid basin: Case of El Kalb River, Lebanon. *Hydrology* **2021**, *8*, 134. [CrossRef]
98. Soulis, K.X.; Generali, K.A.; Papadaki, C.; Theodoropoulos, C.; Psomiadis, E. Hydrological response of natural mediterranean watersheds to forest fires. *Hydrology* **2021**, *8*, 15. [CrossRef]
99. Mu, Q.; Zhao, M.; Running, S.W. Improvements to a MODIS global terrestrial evapotranspiration algorithm. *Remote Sens. Environ.* **2011**, *115*, 1781–1800. [CrossRef]
100. Mu, Q.; Zhao, M.; Running, S.W. *MODIS Global Terrestrial Evapotranspiration (ET) Product (NASA MOD16A2/A3). Algorithm Theoretical Basis Document, Collection, 5*; Numerical Terradynamic Simulation Group Publications: Missoula, MT, USA, 2013; pp. 381–394.

101. Running, S.W.; Mu, Q.; Zhao, M.; Moreno, A. *MODIS/Terra Net Evapotranspiration Gap-Filled 8-Day L4 Global 500m SIN Grid V061. 2021, by NASA EOSDIS Land Processes DAAC*; NASA Land Processes Distributed Active Archive Center: Sioux Falls, SD, USA, 2021. [CrossRef]
102. Senay, G.B.; Parrish, G.E.; Schauer, M.; Friedrichs, M.; Khand, K.; Boiko, O.; Kagone, S.; Dittmeier, R.; Arab, S.; Ji, L. Improving the operational simplified surface energy balance evapotranspiration model using the forcing and normalizing operation. *Remote Sens.* **2023**, *15*, 260. [CrossRef]
103. Kannan, N.; White, S.M.; Worrall, F.; Whelan, M.J. Sensitivity analysis and identification of the best evapotranspiration and runoff options for hydrological modelling in SWAT-2000. *J. Hydrol.* **2007**, *332*, 456–466. [CrossRef]
104. Jimeno-Sáez, P.; Senent-Aparicio, J.; Pérez-Sánchez, J.; Pulido-Velazquez, D. A comparison of SWAT and ANN models for daily runoff simulation in different climatic zones of peninsular Spain. *Water* **2018**, *10*, 192. [CrossRef]
105. Zhang, K.; Zhu, G.; Ma, J.; Yang, Y.; Shang, S.; Gu, C. Parameter analysis and estimates for the MODIS evapotranspiration algorithm and multiscale verification. *Water Resour. Res.* **2019**, *55*, 2211–2231. [CrossRef]
106. Fisher, J.B.; Tu, K.; Baldocchi, D.D. Global Estimates of the Land–Atmosphere Water Flux Based on Monthly AVHRR and ISLSCP-II Data, Validated at 16 FLUXNET Sites. *Remote Sens. Environ.* **2008**, *112*, 901–919. [CrossRef]
107. Piani, C.; Weedon, G.P.; Best, M.; Gomes, S.M.; Viterbo, P.; Hagemann, S.; Haerter, J.O. Statistical bias correction of global simulated daily precipitation and temperature for the application of hydrological models. *J. Hydrol.* **2010**, *395*, 199–215. [CrossRef]
108. Herman, M.R.; Nejadhashemi, A.P.; Hernandez-Suarez, J.S.; Sadeghi, A.M. Analyzing the variability of remote sensing and hydrologic model evapotranspiration products in a watershed in Michigan. *JAWRA J. Am. Water Resour. Assoc.* **2020**, *56*, 738–755. [CrossRef]
109. Chun, J.A.; Baik, J.; Kim, D.; Choi, M. A comparative assessment of SWAT-model-based evapotranspiration against regional-scale estimates. *Ecol. Eng.* **2018**, *122*, 1–9. [CrossRef]
110. Abiodun, O.O.; Guan, H.; Post, V.E.; Batelaan, O. Comparison of MODIS and SWAT evapotranspiration over a complex terrain at different spatial scales. *Hydrol. Earth Syst. Sci.* **2018**, *22*, 2775–2794. [CrossRef]
111. Nguyen, H.Q.; Kappas, M. Modeling surface runoff and evapotranspiration using SWAT and beach for a tropical watershed in North Vietnam, compared to MODIS products. *Int. J. Adv. Remote Sens. GIS* **2015**, *4*, 1367–1384. [CrossRef]
112. Fathian, F.; Ahmadzadeh, H.; Bagheri Harooni, M.H. Comparison of evapotranspiration estimation methods based on remote sensing technique and SWAT model. *Irrig. Water Eng.* **2024**, *15*, 163–182.
113. Morsy, M.; Sayad, T.; Abdou, M.I.; Aboelkhair, H. Comparative study of evapotranspiration from the SWAT model and MODIS-derived remote-sensing data in two climatic zones in Egypt. *J. Water Clim. Change* **2024**, *15*, 5219–5241. [CrossRef]

MDPI AG
Grosspeteranlage 5
4052 Basel
Switzerland
Tel.: +41 61 683 77 34

Hydrology Editorial Office
E-mail: hydrology@mdpi.com
www.mdpi.com/journal/hydrology

www.ingramcontent.com/pod-product-compliance
Lightning Source LLC
Chambersburg PA
CBHW040818250926
51850CB00007BA/1206
* 9 7 8 3 7 2 5 8 5 3 6 9 4 *